Technology

Shaping Our World

by

John Gradwell
McGill University

Malcolm Welch
Queen's University

Eugene Martin
Southwest Texas State University

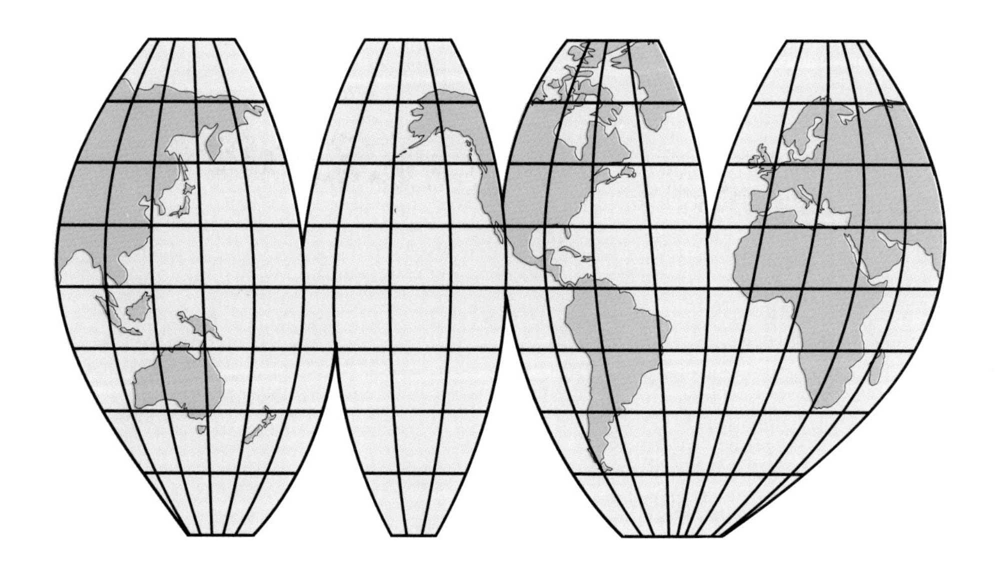

Publisher
The Goodheart-Willcox Company, Inc.
Tinley Park, Illinois

Library of Congress Catalog Card Number 95-12533
International Standard Book Number 1-56637-217-8

3 4 5 6 7 8 9 10   96   99 98 97

Illustrations by Nadia Graphics

Library of Congress Cataloging in Publication Data

Gradwell, John B.
Technology—shaping our world / by John Gradwell, Malcolm Welch, Eugene Martin.—[Ref. ed.]
p. cm.
Includes index.
ISBN 1-56637-217-8
1. Technology. I. Welch, Malcolm. II. Martin Eugene. III. Title.
T45.G67 1996
600--dc20 95-12533
CIP

Preface

TECHNOLOGY: SHAPING OUR WORLD is written to help you understand the technological world around you. This book introduces you to the various technologies and shows how they have used basic scientific principles.

First you will study the solving of problems using the design process. Various chapters will introduce the major technologies: Communications, Manufacturing, Construction, and Transportation. The importance of materials, energy, people, and information to *all* technological activity becomes clear as you study. You will also learn about the job opportunities in technology.

The technology of our earliest ancestors was simple; still, it helped them to live better lives. You will see how the invention of simple machines marked the advance of humans struggling to control their environment. Often, new developments affected only a small group of people. Travel being limited, a new plow might be developed and used in only one village. Even so, life was improved for the village through greater crop yields.

Technology today is usually more complicated. A technological decision might affect an entire city, a whole continent, or even the entire planet. Millions of people have experienced great changes in their lives as a result of new technology. Consider the new technologies affecting your life: computers, robots, surgical techniques, medicines, composites, microelectronics, biotechnology, and telecommunications. Among these technologies there are sure to be new inventions unheard of today.

It is very important that you begin to understand technology so that you can make intelligent decisions about its use. While a new technology may improve your life, you will need to consider its side effects. Your understanding of technology will help you care for our Earth and its people. In this way you will help shape our world. Your decisions will help provide a healthier environment.

Technology is an exciting subject. It affects what you wear, how you travel, what you eat, how you communicate with others, the comfort of your home, and much more. Nearly every day you will be faced with new technology. It is important to understand how and why technological changes come about. This book is written to help you become more involved.

John Gradwell
Malcolm Welch
Eugene Martin

Contents

Technology and You—An Introduction 7
Understanding Technology, Technology: Both Old and New, How Science and Technology Are Related, Key Terms, Test Your Knowlege, Apply Your Knowledge.

1 **Solving Problems in Technology** 13
Solving A Problem, The Design Process, Elements of Design, Principles of Design, Summary, Key Terms, Test Your Knowledge, Apply Your Knowledge.

2 **Communicating Ideas** 35
Communication System, Forms of Communication, Drawings and Their Types, Drawing Techniques, Computer-Aided Design, Summary, Key Terms, Test Your Knowledge, Apply Your Knowledge.

3 **Materials** 51
Properties of Materials, Types of Materials, Summary, Key Terms, Test Your Knowledge, Apply Your Knowledge.

4 **Processing Materials** 71
Safety, Shaping Materials, Joining Materials, Finishing Materials, Summary, Key Terms, Test Your Knowledge, Apply Your Knowledge.

5 **Structures** 95
What Structures Have in Common, Types of Structures, Static and Dynamic Loads, Designing Structures to Withstand Loads, Forces Acting on Structures, Summary, Key Terms, Test Your Knowledge, Apply Your Knowledge.

6 **Construction** 109
The Structure of a House, Planning for a Home, Systems in Structures, Summary, Key Terms, Test Your Knowledge, Apply Your Knowledge.

7 **Machines** 127
Simple Machines, Gears, Pressure, Friction, Summary, Key Terms, Test Your Knowledge, Apply Your Knowledge.

8 **Transportation** 153
Modes of Transportation, Engines and Motors, The Impact of Transportation, Summary, Key Terms, Test Your Knowlege, Apply Your Knowledge.

9 **Energy** **165**
Forms of Energy, Energy Sources, Nonrenewable Sources of Energy, Renewable Sources of Energy, Summary, Key Terms, Test Your Knowledge, Apply Your Knowledge.

10 **Electricity and Magnetism** **183**
Generating Electricity, Transmission and Distribution of Electricity, What is Electricity?, Magnets and Magnetism, The Generation of Electricity with Magnetism, Electric Motors, Cells and Batteries, Summary, Key Terms, Test Your Knowledge, Apply Your Knowledge.

11 **Using Electricity and Electronics** **203**
Electric Circuits, Conductors and Insulators, Measuring Electrical Energy, Electronics, Summary, Key Terms, Test Your Knowledge, Apply Your Knowledge.

12 **The World of Work** **221**
The Primary Sector: Processing of Raw Material, The Secondary Sector: Manufacturing Products, The Tertiary Sector: Providing Services, Summary, Key Terms, Test Your Knowledge, Apply Your Knowledge.

13 **Communication in an Information Society** **245**
Old and New Technology, Microelectronics, Telecommunications, Summary, Key Terms, Test Your Knowledge, Apply Your Knowledge.

14 **Past, Present, and Future** **253**
The Beginning of Technology, The Present: How Technology Affects Us, Attacking Pollution Problems, The Future: Clues from the Leading Edge of Technology, A Last Word, Summary, Test Your Knowledge, Apply Your Knowledge.

Dictionary of Terms **269**

Index **277**

NASA

Technology shapes and expands our world.

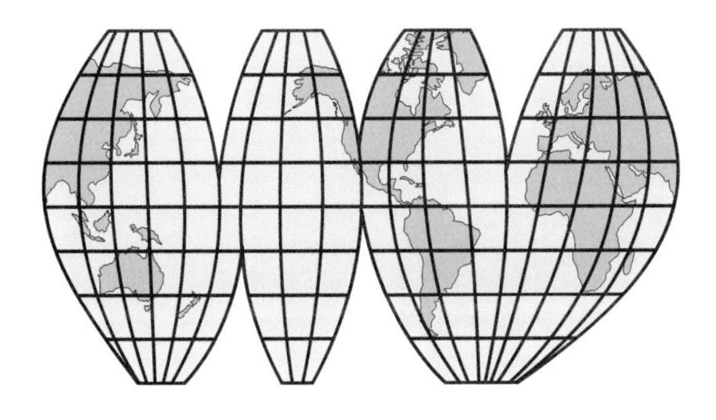

Technology and You– An Introduction

OBJECTIVES

After reading this introduction you will be able to:

- O Define "technology."
- O Describe benefits of technology and list some of the products it has developed.
- O Appreciate that technological change is occurring at an ever-increasing rate.
- O Explain the relationship that exists between technology and science.

Space stations and robots . . . synthetic skin and artificial organs . . . supersonic aircraft . . . composite materials . . . "high-tech" sports equipment. These are all part of our lives. We hear about them every day. We use some of these items ourselves. We take them for granted. What do they have in common? Just this: they are all products of technology. See figure 1 through figure 4.

NASA

Figure 1. Space station. Construction in space is a reality.

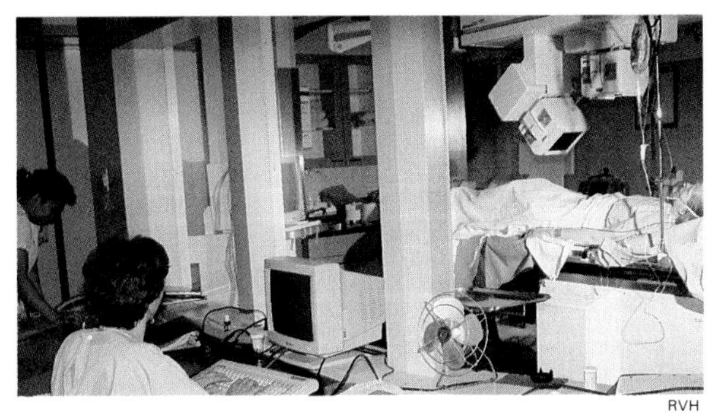

RVH

Figure 2. Electronic equipment produced by high technology is necessary in the modern hospital operating room. We are healthier because of computers and modern electronics.

BRITISH AIRWAYS

Figure 3. Modern jetliners like the Concorde whisk us through the air at speeds above 1500 miles an hour (2400 Km/h).

LASER

Figure 4. Our leisure time is more enjoyable with sports equipment made from modern synthetics.

UNDERSTANDING TECHNOLOGY

Technology is a process we use to solve problems by designing and making products or structures. The things technologists make satisfy real needs. All around you there are objects made to meet your needs.

It is exciting when rockets hurl satellites into space. It is just as amazing when powerful telescopes discover unknown stars. Most of us are also interested in learning about the development of an artificial heart. However, our daily lives are more directly affected by less spectacular products of technology. For example, we all benefit from the use of pocket calculators, microcomputers, microwave ovens, video cassette recorders, and digital watches.

How technology affects us

Imagine that you had lived thousands of years ago. You would probably wake up when the sun rose. There would have been no alarm clocks to ring. You would crawl out of a pile of animal skins spread over branches cut from trees. Your animal skin cloth would hang loosely around your body. There were no zippers, Velcro™, or even strings to tie them.

Leaving your cave, you might find that a pile of rocks had partially blocked the entrance. You would need to move the rocks by hand. There would be no carts. The wheel had not yet been invented.

What about breakfast? Want to make waffles or pour a bowl of cereal? No chance! There might be a leftover bone from yesterday's kill, but no microwave oven to reheat it.

Feeling the urge to "go," you would head for the bushes. If your mouth still tasted of breakfast, too bad. Toothbrushes and toothpaste were still unknown. You could always reach for a stick of gum to freshen your breath, couldn't you? More likely you would scrape some of the congealed sap off the bark of a tree and chew that. The day being warm and sunny, you might feel like taking a holiday. Not if you wanted to eat! The entire family used most of its energy and time each day collecting enough food to survive.

Are you starting to appreciate more the comforts and conveniences provided by modern technology? Technology plays a crucial role in our lives. Almost everything we do depends in some way on the products and services that, together, form our technological society. These products and services are frequently the result of very sophisticated technologies. Often we take them for granted, without really thinking about them. It is only when we stop to consider what we do each day that we realize how important technology is to us.

One day with technology. Do you understand the importance of technology and your dependence on it? Think carefully about some of the items found in a teen's room, figure 5.

ECRITEK

Figure 5. How many examples of modern conveniences do you see on these shelves?

Your day might begin with your being awakened by an electronic digital alarm clock/radio. The music you hear is transmitted (sent) from a station a distance away. You stumble out of bed into a hot shower. The flow and temperature of the water can be exactly controlled. You dry and style your hair and brush your teeth with objects made from plastics. The clothes you wear are made from a mixture of natural and synthetic (artificial) fibers.

In the kitchen, figure 6, a variety of automatic appliances help prepare your breakfast. Toast is ejected from the toaster. The electric kettle shuts off when the water boils. A microwave oven cooks your bacon for a preselected period of time.

ECRITEK

Figure 6. What would a cave dweller think of all the appliances in today's kitchens?

Before leaving the house, you dress in waterproof synthetic fibers. As you leave, the timer built into the thermostat automatically lowers your home's temperature. If you live in an apartment building, you press a button to summon an elevator. Pressing a second button will bring you to the ground floor. Once there, the doors will open automatically. You leave the building through a front door controlled by an electromagnetic lock.

Out on the street, vehicles use energy to move people and goods. You may use a bus or subway train traveling to school, figure 7.

ECRITEK

Figure 7. How has technology changed the way students travel to school?

Your classroom uses electric lighting and may be heated by natural gas. The tables, desks, chairs, and cupboards have been made in a factory and transported to the school. The room is probably equipped with an overhead projector or a filmstrip projector, and a public address system.

Normally, we give little thought to these technological products and services on which we rely. One thing is certain. Without them your morning would have been far less comfortable and less safe.

Are you starting to recognize your dependency on the products of technology? Today most people lead very "technological" lives. You have seen that, even during the first hours of your day, technology was basic to your comfort and way of life. Were there any activities that did not depend on technology?

As the day progresses you will continue, of course, to make use of many other products and services. For example, your school, figure 8, is an artificial environment created by technology. Schools, as well as homes, are heated in the winter and cooled in the summer. They may be insulated to conserve energy.

Technology's effect on health. Your health depends on medical technology. Dentists use a wide variety of miniaturized tools and equipment to repair, replace, straighten, and keep

ECRITEK

Figure 8. How does technology affect your school day?

your teeth in the best possible condition, figure 9. Doctors are able to replace a damaged heart valve with one made of metal and plastic. Diabetics can wear a tiny, computer-controlled infusion pump, figure 10. It automatically delivers a supply of insulin to the wearer. Such advances in medical care help us to lead healthier and happier lives. At the same time, they reduce the number of diseases.

Technology for leisure. Even in your leisure time you are surrounded by the products of technology. Road racing bicycle frames, figure 11, are now made of epoxy-glued aluminum or carbon fiber, and have a mass (weight) of less than half the average steel frame. Computer-designed tennis racquets use fiberglass, graphite, or ceramic to replace the conventional laminated wood. Golf clubs use a super-hard aircraft alloy to replace hardwood. Jogging and ski suits are made of fabrics that keep out wind, rain, and snow, yet allow perspiration vapor to escape.

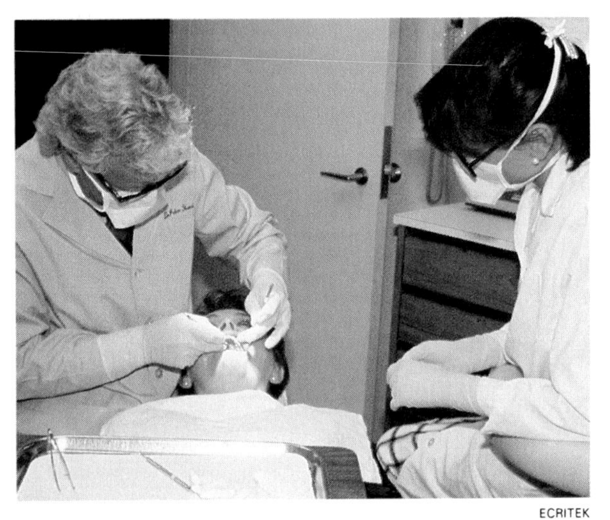

ECRITEK

Figure 9. Technology has done much to improve our health and care for our teeth.

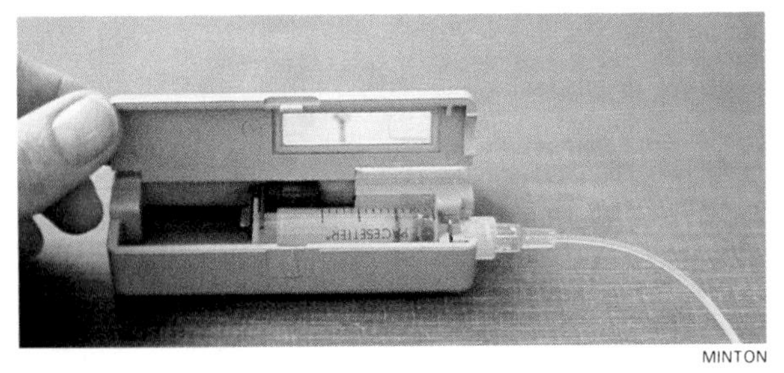

MINTON

Figure 10. This small infusion pump delivers medication slowly over a long period of time.

ECRITEK

Figure 10. This modern racing bicycle uses high-tech materials. Technology has made the bicycle an affordable alternative to walking.

Technology for work. A personal computer can be programmed to help you with mathematics problems. It may also help you word process an assignment, keep records, or act as a flight simulator, figure 12. The same computer may be linked through the telephone to computers in other homes, other cities, or even other countries.

TECHNOLOGY: BOTH OLD AND NEW

Technology is not really new but new technology is discovered daily. When early humans developed the bow and arrow they were using technology. They solved a problem by designing and making something. The objects technologists make are often good solutions to the problems. However, sometimes they are barely adequate. Sometimes they are miserable failures.

HOW SCIENCE AND TECHNOLOGY ARE RELATED

Some people have suggested that technology and science are very similar. It is thought by some that technology is applied science.

Figure 12. Do you know people who use computers to make their work easier?

TEC

While it is true that technology and science are complementary (help one another) they are very different. Science is concerned with the laws of nature. **Scientists** (biologists, chemists, and physicists) seek to discover and understand these laws. Technology designs and manufactures a human-made world. Technologists (designers, inventors, engineers, and craftspeople) use nature's laws to adapt nature to human needs. They use their knowledge and skill to create objects that fulfill our needs.

Often, people have thought that scientific knowledge came first—before technology. This is not true. Many technological advances preceded (came before) the understanding of the principles behind them. For example, the wheel and axle was used long before humans understood the physical principles governing levers.

Some examples will help you to understand the relationship between technology and science:

- Technologists invented and built the early telescopes. Scientists used these telescopes to observe and calculate the distance from earth to the planets. In turn, these scientific observations were used by technologists in the design of space vehicles.
- Scientists study the flow and formation of rivers. Technologists design and build dams across rivers.
- Technologists built the first steam engines. Scientists studied these engines to develop the laws of thermodynamics.
- Scientists study the causes and control of diabetes. Technologists design and build portable computer-controlled insulin pumps for diabetics.
- Technologists were able to shape glass into tubes, bottles, and flasks. Scientists used these objects in experiments to analyze the chemical composition of substances.
- Scientists study the atomic theory. Technologists use the theory to build nuclear power stations.

This book explains and explores some of the knowledge and skills you will need:

- To understand technology.
- To use technology wisely.
- To create simple technological products for yourself and others.

KEY TERMS

Science
Scientist
Technologist
Technology

TEST YOUR KNOWLEDGE

Write your answers to these review questions on a separate sheet of paper.

1. Identify three situations in which technological advances are affecting our daily lives.
2. Review the description in the text of a typical day in today's world. Then:
   a. List those objects that you sometimes use during the first hour of the day.
   b. List five other objects not included in the description that you, or a member of your family, may use.

3. Technology is best defined as ________.
   a. Discovering and understanding laws of nature.
   b. Designing machines and tools.
   c. The skilled use of hand tools.
   d. Solving problems by designing and making objects.
4. Describe, in your own words, the differences between technology and science.
5. Technology is an important subject to study because ________.
   a. Most students will become engineers.
   b. We live in a technological society.
   c. It teaches you how to repair appliances.
   d. It teaches you how to sketch.
6. Give one technology related word for each of the following letters of the alphabet.
   C ______________________________
   E ______________________________
   M ______________________________
   S ______________________________

APPLY YOUR KNOWLEDGE

1. Explain the difference between science and technology.
2. Describe how technology affected your life during lunch hour today.
3. Collect pictures to illustrate different aspects of technology. Paste your pictures onto a sheet of paper to form a collage.

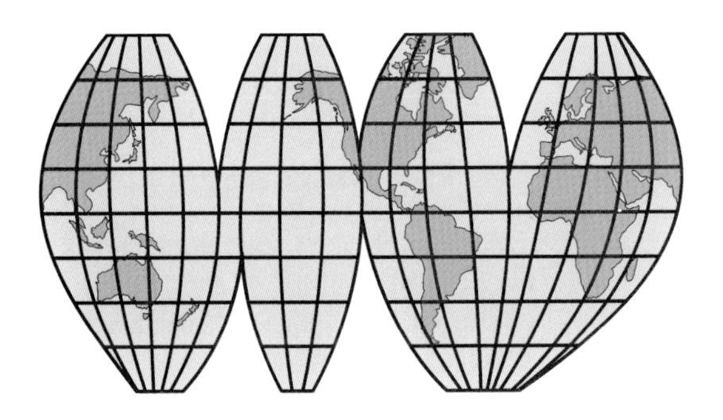

Chapter 1
Solving Problems in Technology

OBJECTIVES

After reading this chapter you will be able to:
- Distinguish between objects found in nature and those made by humans.
- List the steps in the design process.
- Identify and define a design problem.
- Apply the steps of the design process to develop a solution to a problem.
- Use the elements and principles of design.
- List the order of steps to build a prototype.
- Evaluate a prototype.

The first person to shape a rock making it a sharper tool was solving a problem in technology. He or she was designing a product to meet a need.

Today every product we use has to be designed by someone, somewhere. These products may be as simple as a paper clip or as complex as a high-speed train. In most cases, the product was designed because either the designer thought the product was needed for survival, or that it would improve the quality of life.

In today's technological world, a **designer** is one who creates and carries out plans for new products and structures. She or he will use special techniques. These help assure a successful product or structure.

Most people, at some time in their lives, have solved a problem in technology. By so doing, they acted as designers, figure 1.1. Building a sand castle, a tree house, constructing shelves for books or trophies, and decorating a birthday cake all involve designing. In each case there was a need. In each case the designer worked through a series of steps to arrive at a solution. This chapter will describe all the steps technologists use to solve problems. Figure 1.2 describes a typical design problem and how the designer solved it.

ECRITEK

Figure 1.1. Before this tree house could be built, someone had to solve design problems. What problems could you foresee in supporting the tree house?

■ SOLVING A PROBLEM

Imagine you have been given a stack of small pieces of different colored paper. You have decided to use these to take telephone messages. Some were placed at the side of

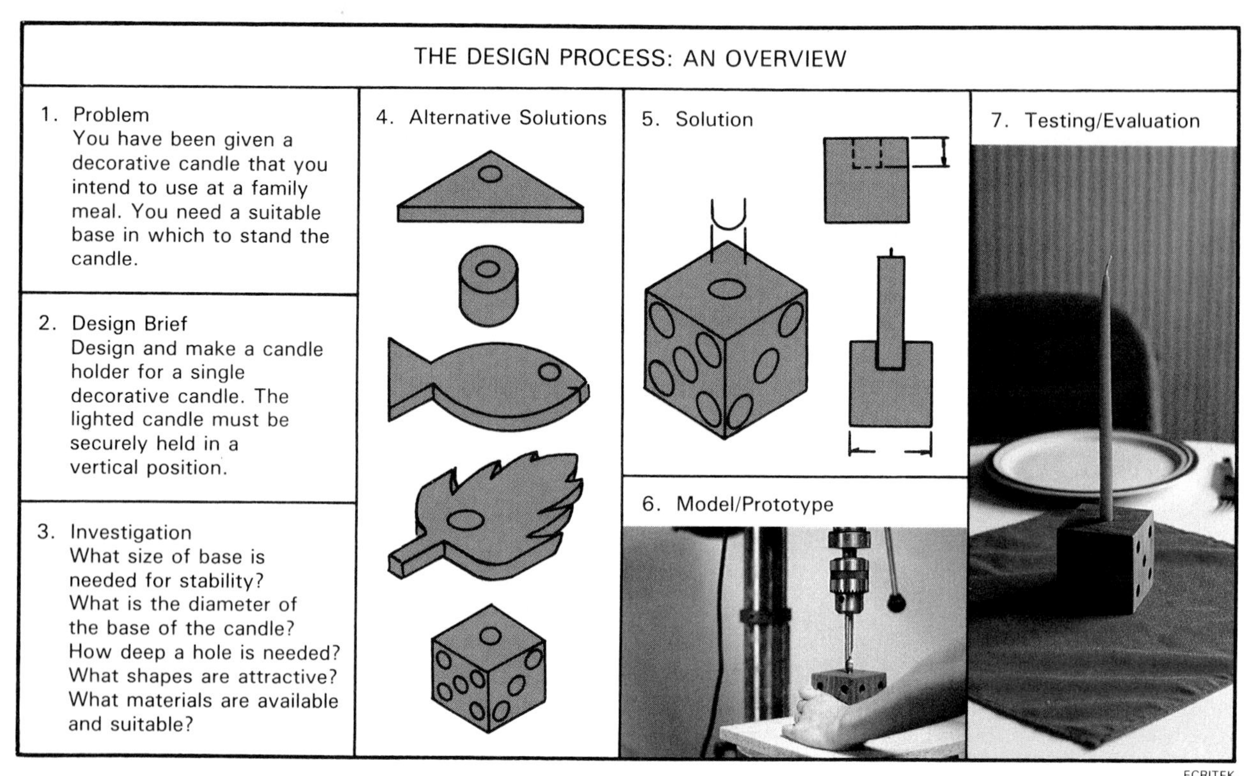

THE DESIGN PROCESS: AN OVERVIEW
1. Problem
You have been given a decorative candle that you intend to use at a family meal. You need a suitable base in which to stand the candle.
2. Design Brief
Design and make a candle holder for a single decorative candle. The lighted candle must be securely held in a vertical position.
3. Investigation
What size of base is needed for stability?
What is the diameter of the base of the candle?
How deep a hole is needed?
What shapes are attractive?
What materials are available and suitable?
4. Alternative Solutions
5. Solution
6. Model/Prototype
7. Testing/Evaluation
ECRITEK

Figure 1.2. This is a typical design problem. Perhaps you would have solved it another way.

the telephone. When you came back an hour later they were scattered over the floor. What could have happened? Perhaps, as in figure 1.3, the wind blew them onto the floor when the door was opened. Someone may have bumped into the table knocking them to the floor. Obviously there is a **problem.**

Figure 1.3. Every design project begins as a problem.

Defining need

Let's think about exactly what is needed. You need a method of holding a stack of paper neatly and securely next to a telephone so that a message can be written on the top sheet. This statement describes clearly what is needed. Such a statement is called a **design brief.** It describes the problem.

Asking questions

The first step in solving the problem is fun. You become a bit of a detective! You investigate. The kinds of questions you might ask include:

- What is the paper size?
- How many sheets are to be held?
- How much space is there beside the telephone?
- Should a pen or pencil be attached?
- Will it be easy to write on the top sheet?
- Can the sheets be easily removed one at a time?
- What materials are available to make the product?
- How much material is available?
- Which material would be best for appearance and strength?

Finding solutions

As you think about these questions, a number of **alternative solutions** may come to mind. Pehaps you would consider:

- Gluing the sheets together to make a pad.
- Using some kind of spring clip.
- Punching holes in the sheets and hanging them from a peg.
- Making a small box or container.

To remember all of these ideas, it is useful to sketch each one. You could also add some notes. Then, each sketch is better understood. As a general rule, record your ideas on paper as they occur to you, figure 1.4.

Suppose that you have sketched a number of alternative solutions. You can choose the one you think will work best. Let us assume that you prefer a container. The next step is to develop the idea of a container further. Figure 1.5 shows a number of shapes and details.

Once again it is time to make a decision. What you choose will combine the best elements of these shapes and details. The container shown in figure 1.5C is the preferred solution.

Model making

At this point, you should make a full-size cardboard **model,** figure 1.6. This model allows you to test and evaluate (see if it works) your solution. The best way to do this is to put it by the telephone and try it out, figure 1.7. Does it satisfy the need stated in the design brief? Is the paper held neatly and securely? Can you easily write a message on the top sheet?

If the model works, the next step is to make and test a prototype (working model). You would use an appropriate material, such as plastic.

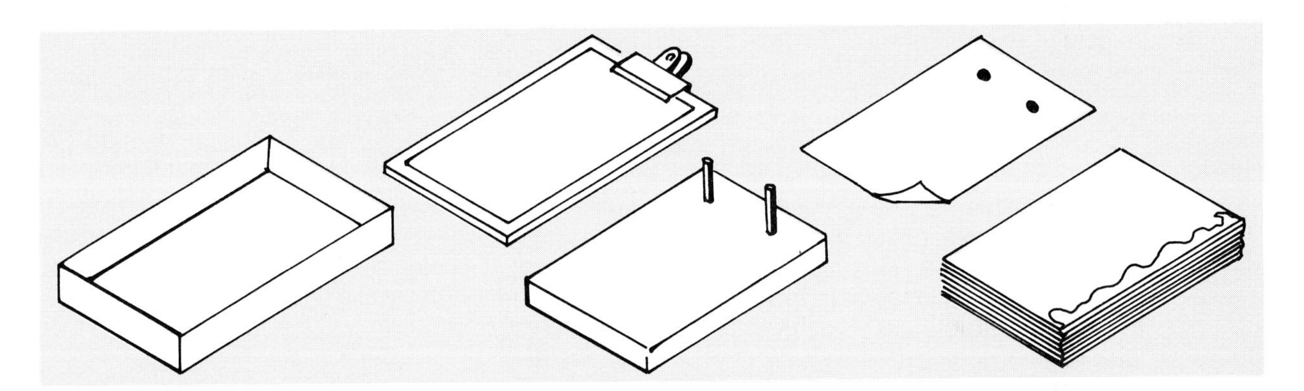

Figure 1.4. These sketches show possible solutions to the paper holder problem.

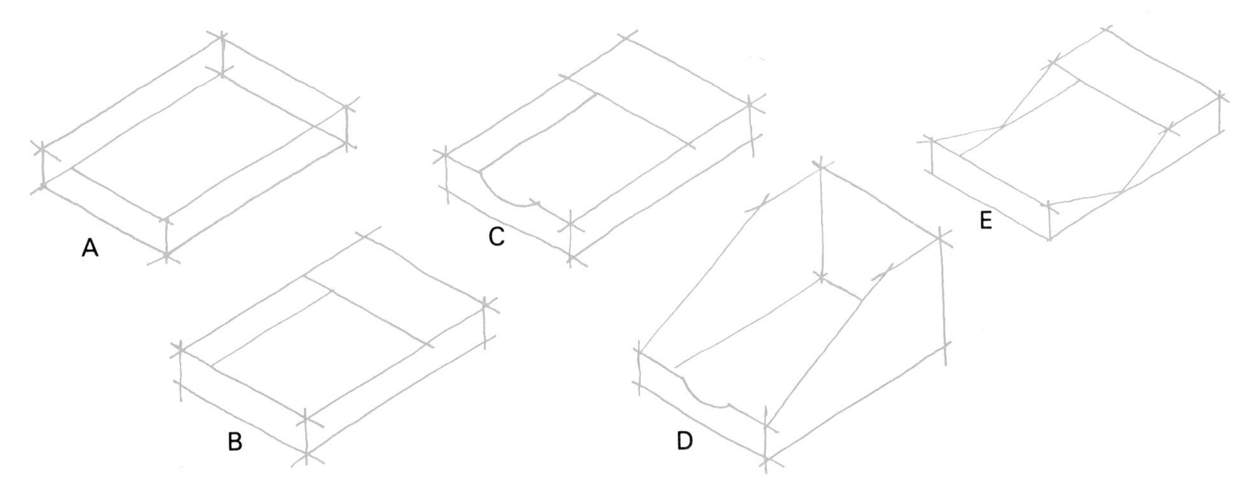

Figure 1.5 These sketches further refine the box idea in figure 1.4. They show more details and more shapes.

Figure 1.6. Making a model. This is a likeness of the actual box.

Figure 1.7. Testing a prototype means trying out a working model.

If you were a manufacturer you would make a pre-production series. You'd make a small number of samples to be tested by typical consumers. These users would tell you how the product works. They would also tell if further modifications (changes) are required. Further, they would tell how much they would pay for it. Such information is called **feedback.**

The final step is to manufacture the paper holder. (Manufacturing is making products in a workshop or factory.) Often manufacturing is done by mass production.

■ THE DESIGN PROCESS

Do you see that solving a problem involves working through a number of steps? It does not just happen. It is a careful and thoughtful procedure called the **design process,** figure

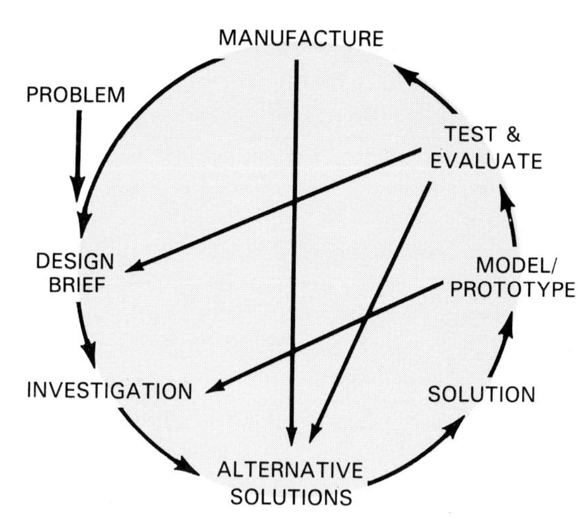

Figure 1.8. Every time we design we follow these steps. We call the steps the design process.

1.8. Now, let us look at each of these steps in greater detail.

The problem

As you've just learned, the process of designing begins when there is a need. Wherever there are people there are problems needing solutions. In some cases the designer may have to invent a product. An example might be a game for blind persons.

At other times the designer may change an existing design. (If the handle of a pot becomes too hot to touch, it must be redesigned.)

Designers also improve existing products. They make the product work even better. Could the chair in the waiting room of a bus or train station be altered so that waiting seems shorter?

The design brief

A design brief should describe simply and clearly what is to be designed. The design brief cannot be vague. Some examples of problems and design briefs follow:

Problem: Blind people cannot play many of the indoor games available to sighted people.
Design Brief: Design a game of dominoes that can be played by blind people.

Problem: The handle of a pot becomes too hot to hold when the pot is heated.
Design Brief: Design a handle that remains cool when the pot is heated.

Problem: Waiting time in a bus or train station seems too long. There is nothing to do.
Design Brief: Modify the seats so that a small television can be attached.

Investigation

Writing a clearly stated design brief is just one step. Now you must write down all the information you think you may need. Just write down your thoughts as they occur. Some things to consider are the following: function, appearance, materials, construction, safety. Let us consider what to ask about these topics:

Function. No matter how beautiful, an object that does not **function** (work) well should never have been made. A functional object must solve the problem described in the design brief. The basic question to ask is: "What, exactly, is the use of the article?"

Human beings like to be comfortable in the things they do. The products they use should be both easy and efficient to use. This is sometimes difficult to achieve. Humans vary in many ways. What suits one person often is not right for someone else. The study of how a person, the products used, and the environment (our surroundings) can be best fitted together is called **ergonomics.** Ergonomics includes these considerations:

- Body sizes—can people fit the object?
- Body movement—can everything be reached easily?
- Sight—can everything be seen easily?
- Sound—can important sounds be heard and are annoying ones eliminated (removed)?
- Touch—are parts that a person touches comfortable?
- Smell—are there any unpleasant smells?
- Taste—are any materials toxic (harmful)?
- Temperature—is the environment too hot, too cold, or comfortable?

Not all of these apply to every product. Look at figure 1.9. In the design of a computer console, the seat, keyboard, and screen adjust in various directions. Different sizes of people can use the same console.

Figure 1.10 shows two pairs of scissors. Those on the right have been designed to fit most peoples' hands. Their color makes them easily seen. The plastic is warm to the touch.

Appearance. How will the object look? The shape, color, and texture should make the object attractive.

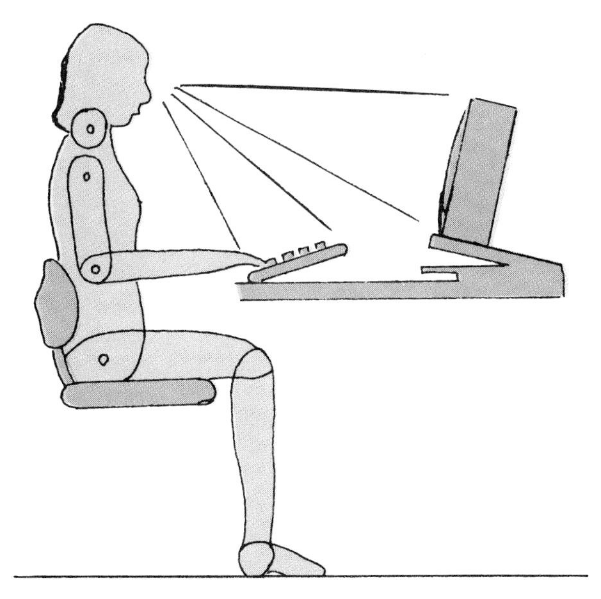

Figure 1.9. Can you think of ways to make this computer station fit people of different heights?

TEC

Figure 1.10. Two scissors designs. Which would you rather use?

Materials. What materials are available to you? You should think about the cost of these materials. Are they affordable? Do they have the right physical properties, such as strength, rigidity, color, and durability?

Construction. Will it be hard to make? Consider what methods you will need to cut, shape, form, join, and finish the material.

Safety. The object you design must be safe to use. It should not cause accidents.

Now that you know the questions to ask, you can begin looking for the answers. Where do you begin? Consider these sources:

- Existing solutions. Look around you for similar articles. Examine them. Collect pictures showing examples of other people's solutions.

- Libraries. Search in your school or local library for magazines, books, and catalogs with relevant (fits your needs) information and pictures.
- Experts. Seek out people in industries, schools, and colleges who have this type of problem in their daily work.

Developing alternative solutions

Information in hand, you are ready to develop your own designs. You should produce a number of solutions. In fact, as you were gathering your information, some ideas may have come to mind. Perhaps some were completely new ideas. Others may have been variations of existing ideas.

It is very important that you write or draw every idea on paper as it occurs to you. This will help you remember and describe them more clearly. It is also easier to discuss them with other people if you have a drawing.

These first sketches do not have to be very detailed or accurate. They should be made quickly. The important thing is to record **all**

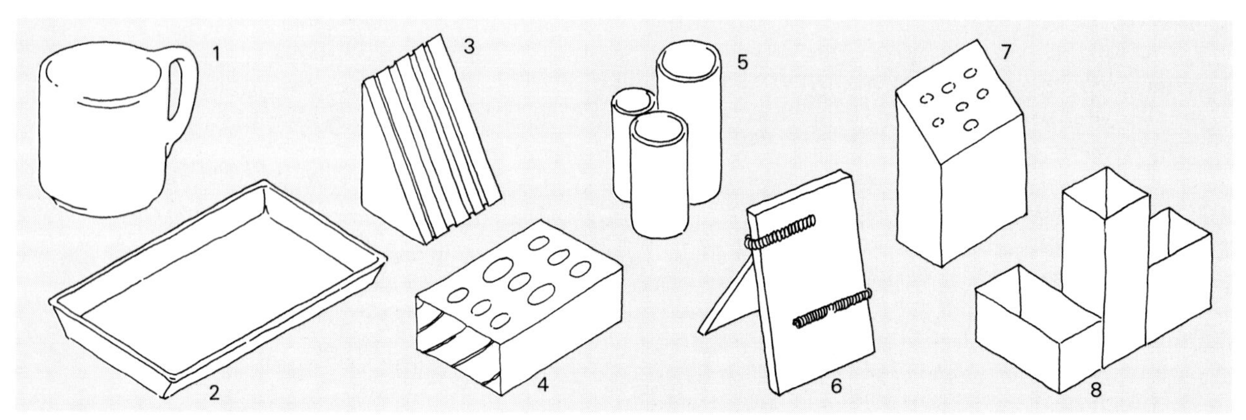

Figure 1.11. Eight possible solutions for a pencil holder design problem. Can you think of others?

Design requirements	ALTERNATIVE SOLUTIONS 1	2	3	4	5	6	7	8
Holds 4 pens?	✓	✓	✓	✓	✓	✓	✓	✓
Holds 3 pencils?	✓	✓	✓	✓	✓	✓	✓	✓
Pens and pencils separated?			✓	✓	✓		✓	✓
Are pens and pencils easily removed and replaced?	✓	✓	✓	✓	✓		✓	✓
Is container stable?	✓	✓	✓	✓	✓		✓	✓
Attractive?					✓		✓	✓
Possible to make?		✓	✓	✓	✓	✓	✓	✓
Uses appropriate materials?	✓	✓	✓	✓	✓	✓	✓	✓
Tools are available?		✓	✓	✓	✓	✓	✓	✓
Materials are available?	✓	✓	✓	✓	✓	✓	✓	✓

Figure 1.12. Evaluating solutions. This chart lets you compare them at a glance.

your ideas. Do not be critical. Try to think of lots of ideas, even some wild ones. The more ideas you have, the more likely you are to end up with a good solution.

Figure 1.11 shows a page from a designer's notebook. The design brief reads: ''Design a container to hold at least four pens and three pencils. Items must be easily identified and removed.'' The designer thought of eight different solutions.

Choosing a solution

You may find that you like several of the solutions. Eventually, you must choose one. Usually, careful comparison with the original design brief will help you to select the best.

You must also consider:

- Your own skills.
- The materials available.
- Time needed to build each solution.
- Cost of each solution.

Deciding among the several possible solutions is not always easy. Then it helps to summarize the design requirements and solutions. Put the summary in a chart, figure 1.12.

Three solutions, Nos. 5, 7, and 8, satisfy all of the design requirements. Which would you choose? In cases like this, let it be the one you like best. The designer chose No. 7.

In the next step, make a detailed drawing of the chosen solution. This drawing must include all of the information needed to make the pencil holder, figure 1.13. It should include:

- The overall dimensions.
- Detail dimensions.
- The material to be used.
- How it will be made.
- What finish will be required.

Now you can choose what to do next. You can make a model and later a prototype, or, you can go directly to making a prototype.

Models and prototypes

A model is a full-size or small-scale simulation (likeness) of an object, figure 1.14. Architects, engineers, and most designers use models.

Models are one more step in communicating an idea. It is far easier to understand an idea when seen in three-dimensional form. A scale model is used when designing objects that are

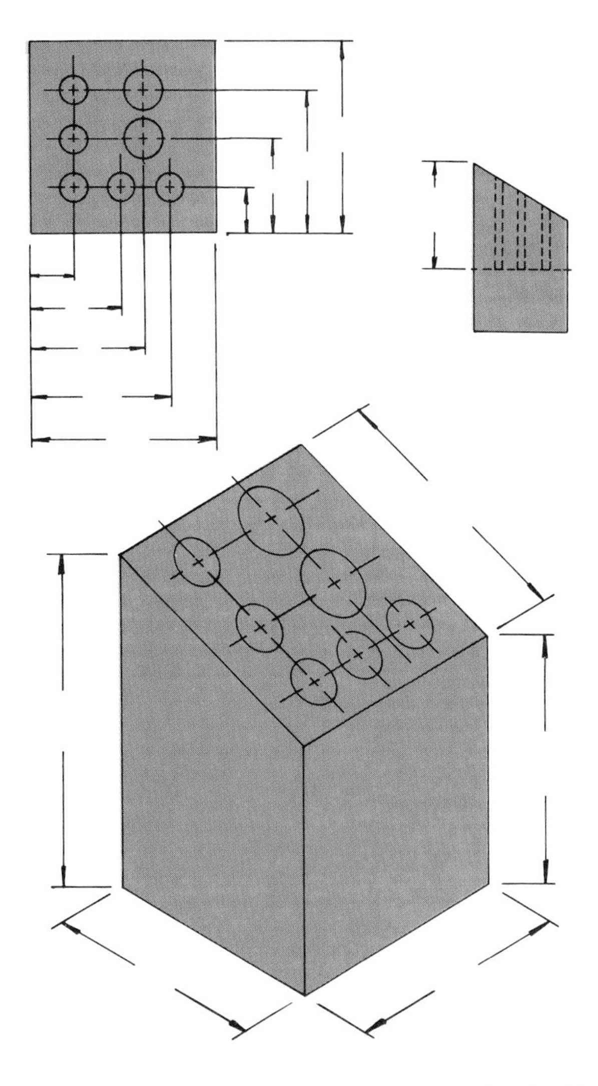

Figure 1.13. When dimensions are added, this detailed drawing will tell what size to make the holder and where to place the holes.

BUICK

Figure 1.14. Clay model of a new car. Models can be viewed from various angles. Then you can correct errors you see.

very large. Buildings, ships and planes are a few examples. In a scale model the size is reduced. For example, a scale model could be built at one-tenth of full size.

A prototype is the first working version of the designer's solution. It is generally full-size and often handmade. For a simple object such as the pencil holder, the designer probably would not make a model. He or she may go directly to a prototype, figure 1.15. However, the designer would plan the steps for making the object. For example, she or he would:

- Select the materials.
- Plan the steps for cutting and shaping the material.
- Choose the correct tools.
- Cut and shape material.
- Apply finish.

The steps will vary depending on the object you are making. Some products have many parts. They must be assembled. The important thing is that you plan ahead.

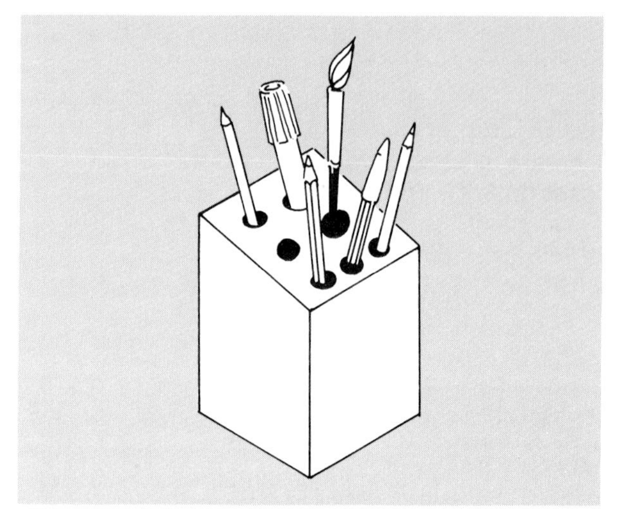

Figure 1.15. A prototype can be tested in a real-life situation.

Testing and evaluating

Testing and evaluating answers three basic questions:

- Does it work?
- Does it meet the design brief?
- Will modifications improve the solution?

The question, "does it work?" is basic to good design. It has to be answered. This same question would be asked by an engineer designing a bridge, by the designer of a subway car, or by an architect planning a new school. If you were to make a mistake in the final design of the pencil holder what would happen? The result might simply be unattractive. At worst, the holder would not work well. Not so if a designer makes mistakes in a car's seat belt design. Someone's life may be in danger!

More questions. Testing and evaluating the pencil holder will provide answers to other questions:

- Will it hold at least four pens and four pencils?
- Do the pens and pencils fit the holes? They must be neither too tight nor too loose.
- Is the container stable (not easy to tip)?
- Is it attractive?

Still more questions. Other products may raise special questions. These must be answered. For example, if the product has several parts we may need to ask:

- How efficiently (well) does it work?
- Will it last?
- Does it need maintenance?
- Will it need spare parts?
- Is it attractive?

All of these questions should be answered "yes." If not, the designer changes the design.

With corrections made to the prototype, it is time to make up a small number of samples. This is called a **pre-production series.** The samples are given to typical consumers. The consumers report their experiences to the manufacturer. Did it work well? How could it be improved? Is it attractive? Is it priced right? Designers use this feedback to make final changes. Those making design changes must also remember that the product must be sold at a reasonable profit.

Manufacturing

The company is satisfied with the design. It knows that it is marketable (will sell). It must decide how many to make. Products may be mass-produced in low volume or high volume, figure 1.16. Specialized medical equipment is produced in the hundreds. Other products, for example nuts and bolts, are produced in large volume. Millions may be made.

The task of making the product is divided into jobs. Each worker trains to do one job.

As workers complete their special jobs, the product takes shape. Mass production saves time. Since workers train to do a particular job, each becomes skilled in that job. Also, automatic equipment does such things as:
- Cut and shape materials.
- Weld parts together.
- Spray on final finishes.

Figure 1.16. Are all the products mass-produced in the same volume? Compare airplane manufacture to making toys!

□ ELEMENTS OF DESIGN

People who design use the term **elements of design.** The term means the things you see when you look at an object. There are five elements. They are line, shape and form, texture, and color. You will find them combined in every object. The following pages will describe each of the elements.

Products are designed to satisfy needs. As we saw in the previous section, it is important that products function well. An easy chair must be comfortable. A pen must write. An airplane must fly. At the same time they must be attractive.

All objects appeal to our senses. We buy clothing that *looks* good. We enjoy the smooth *feel* of the polished wooden arm of a chair. A meal on a plate must not only look attractive but should *taste* and *smell* good. The *sound* of musical chimes is preferable to the harsh sound of a door buzzer.

When you see something you like, ask yourself what is it you like about the product. Is it the color? Is it the shape or form? Is it the texture? Think about how each of these elements affects the appearance of the product.

Line

Lines describe the edges or contours (outlines) of shapes. They show how an object will look when it has been made. Lines can also be used to create some special effects. For example, straight lines suggest strength, direction, and stability, figure 1.17. What

A ECRITEK

B

C CHRISTOPHARO

Figure 1.17. How we use straight lines in our designs. A—Vertical lines show strength. B—Diagonal lines give a sense of movement. C—Horizontal lines give a feeling of stability (firmness).

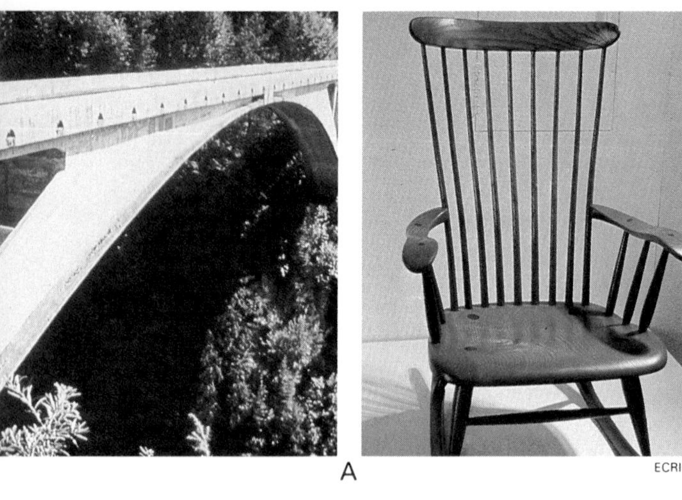

A ECRITEK

ECRITEK B

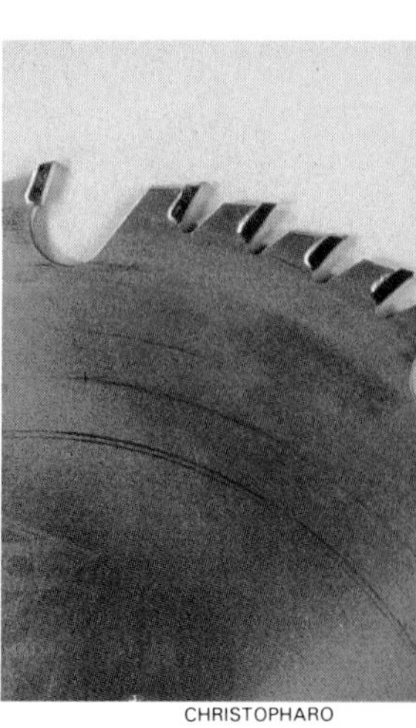

CHRISTOPHARO

Figure 1.18. Curved and jagged lines. A—Look at the lines in the bridge and in the rocking chair. Curved lines give a sense of grace and softness. B—A mountain ridge and a saw blade have jagged lines. Do they seem harsh and unfriendly to you?

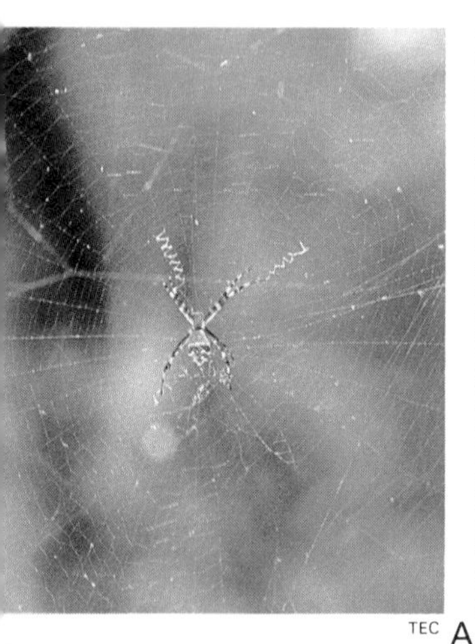

TEC A

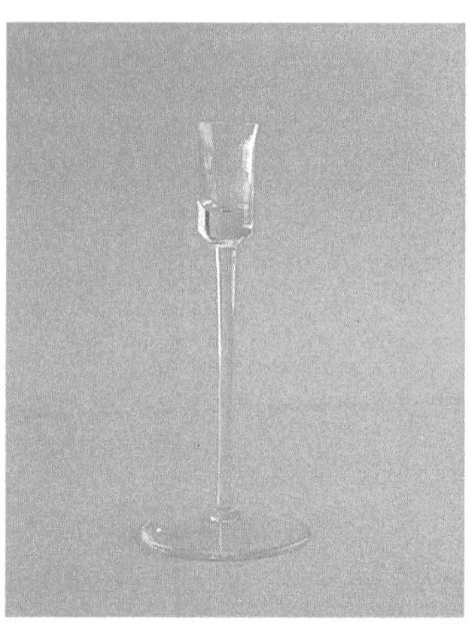

TEC

ECRITEK B

CHRISTOPHARO

Figure 1.19. Thin lines and heavy lines. A—We find thin lines in nature and in manufactured products. We think of them as weak. B—Heavy lines show extra strength.

feeling do you get from curved and jagged lines, figure 1.18? What do heavy or thin lines suggest, figure 1.19?

Shape and form

All objects occupy space or possess volume. We say that **shape** is two-dimensional, figure 1.20, and **form** is three-dimensional, figure 1.21.

Shapes and forms may be geometric, organic, or stylized. Geometric shapes (made up of straight lines and circles) can be drawn

Figure 1.20. In design, shape means two dimensions—width and height.

using rulers, compasses, or other instruments, figure 1.22. Organic and stylized shapes and forms are usually drawn freehand. Figures 1.23 and 1.24 show both types.

Texture

Texture refers to the way a surface feels or looks. We can describe a surface as rough, smooth, hard, slippery, fuzzy, coarse, or repulsive. Sandpaper feels rough. Glass feels smooth. Rock feels and looks hard. Ice looks cold and slippery. A fur coat feels warm and

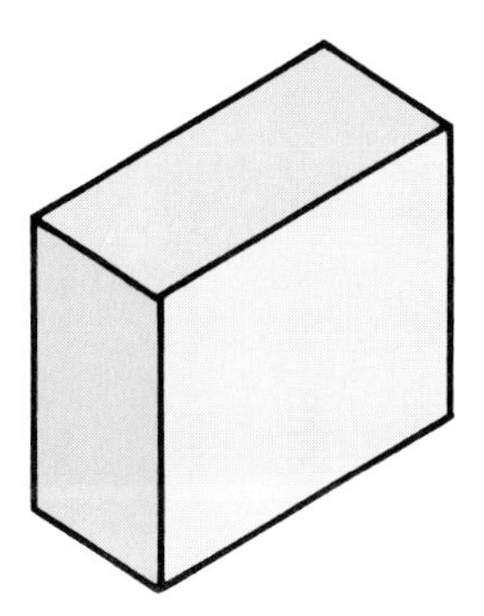

Figure 1.21. Form has three dimensions—width, height, and depth.

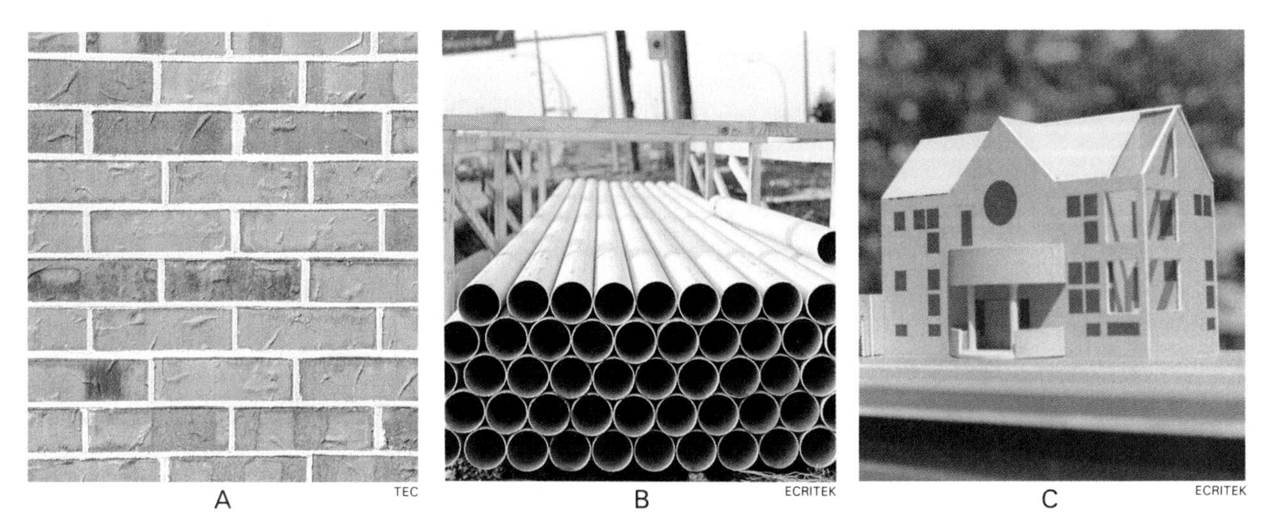

Figure 1.22. A few examples of geometric shapes. A–Bricks are rectangles. B–Pipes are circles. C–A house may combine several geometric shapes. Do you see rectangles and triangles?

Figure 1.23. Organic shapes and forms. A—A blossom is a natural organic shape. B—In the past, architecture has copied natural organic shapes found in leaves and flowers. C—The curved, free-form shape of this chair makes it a good example of organic shape.

A ECRITEK

B CHRISTOPHARO

Figure 1.24. Stylized shapes and forms. A—Anyone will recognize these stylized signs without the need for language. B—Stylized art form.

fuzzy. Sand feels coarse. Mud looks repulsive to adults but attractive to young children! Figures 1.25 through 1.27 show texture in a variety of materials.

A designer can choose materials according to their natural texture. She or he might also

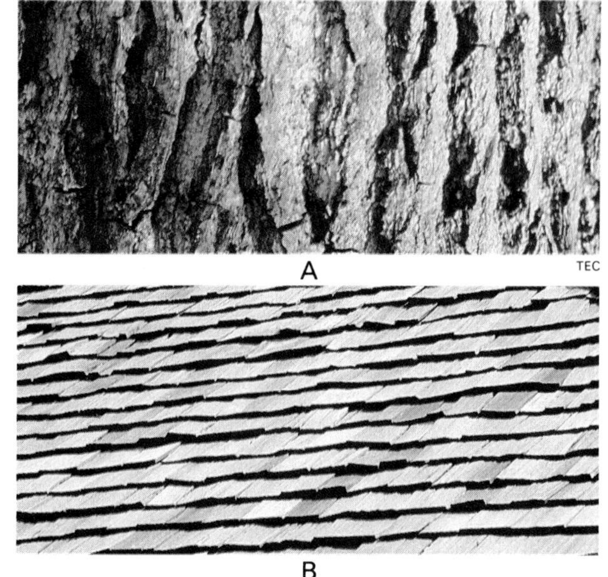

A TEC

B

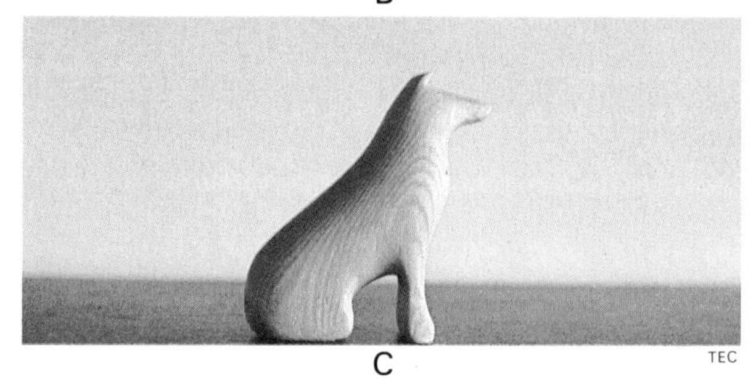

C TEC

Figure 1.25. The texture of wood is pleasing. A—In its natural state. B—Sawed and nailed to a roof, wood shingles have a different texture than a tree. C—Wood, sanded, stained, and polished has still another texture.

HANRAHAN

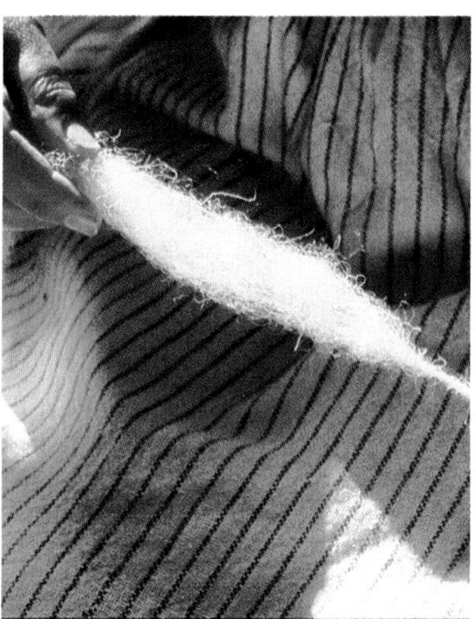

NFB

Figure 1.26. The texture of wool is also valued as a design element.

Figure 1.27. The texture of stone is attractive in scenery, in a wall, or used as art or jewelry.

choose materials because of the way the texture can be changed.

Color

We only see color when light shines on objects. Sunlight appears to be white. In fact, it is a mixture of seven different colors. When a beam of light shines through a glass prism the path of light is bent. Each color bends at a different angle. They can then be seen individually. These seven colors form a spectrum, figure 1.28.

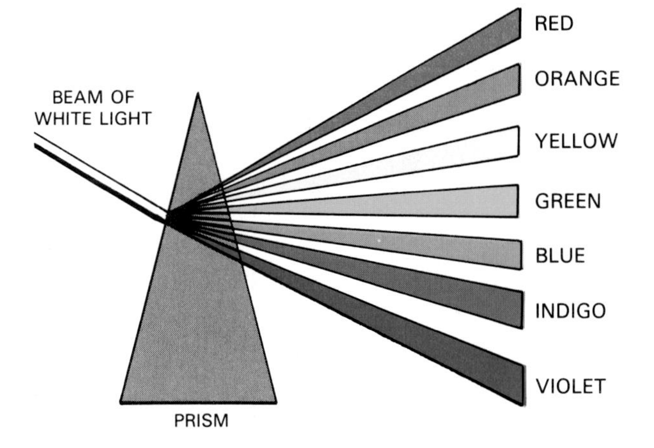

Figure 1.28. The seven colors in white light separate when shone through a glass prism. Where do you see this happen in nature?

The three most important colors are red, yellow, and blue. These are called **primary colors.** If you mix equal parts of two primary colors you obtain a **secondary color.** Red plus yellow gives orange. Yellow plus blue gives green. Blue plus red gives violet. Mixing equal parts of a primary and a secondary color creates a tertiary (third) color, figure 1.29.

When you are selecting colors you may want them to harmonize or to contrast. **Harmony** means the colors naturally go together. You will find harmonizing colors next to one another on the color wheel, figure 1.30. They are similar.

For example, if you first choose orange then the harmonizing colors are red-orange and yellow-orange.

On the other hand you may want your colors to contrast with each other. In that case you would select colors that are at the opposite side of the color wheel. For example,

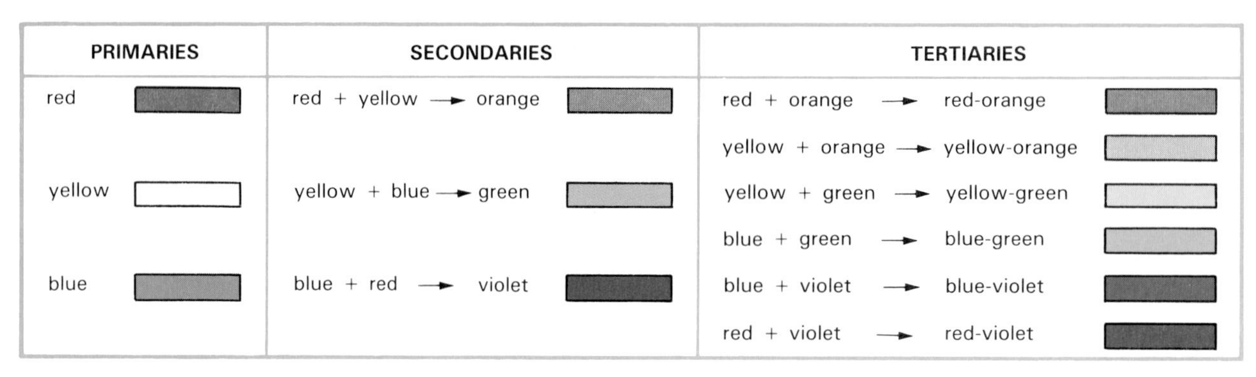

PRIMARIES	SECONDARIES	TERTIARIES
red	red + yellow → orange	red + orange → red-orange
		yellow + orange → yellow-orange
yellow	yellow + blue → green	yellow + green → yellow-green
		blue + green → blue-green
blue	blue + red → violet	blue + violet → blue-violet
		red + violet → red-violet

Figure 1.29. Mixing of colors. Which do you like best?

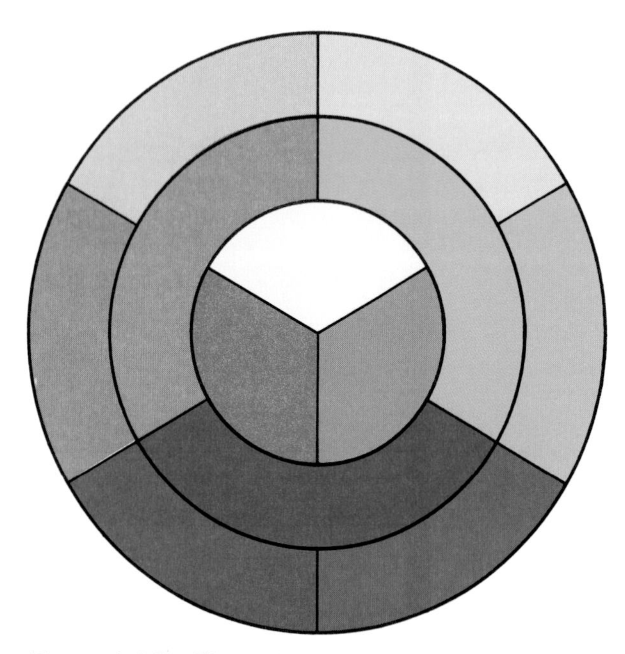

Figure 1.30. The color wheel can be used to pick colors that harmonize or which are a good contrast to each other.

blue contrasts with orange. Contrasting colors are also called complementary colors.

Designers use colors to produce certain reactions or effects, figure 1.31. Traffic signs use red to indicate danger. Yellow serves as a traffic warning. We also associate colors with objects. An apple is red. Grass is green. A hearse is black and a nurse's uniform is white.

Colors also produce emotions. Some are bright, others dull. Some are exciting, others boring. Colors are very personal. What is eventually right or wrong may depend on your own choice.

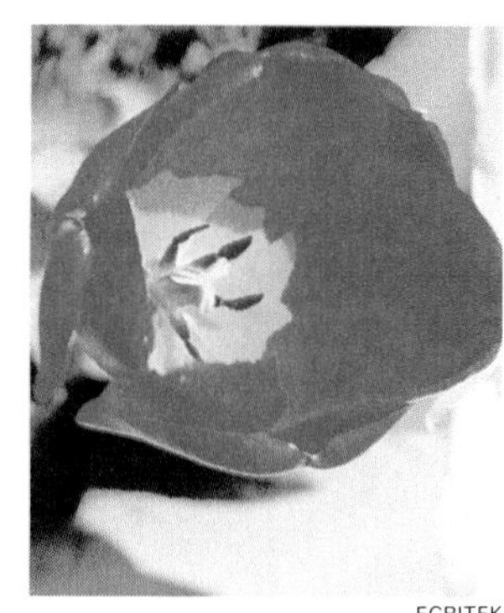

TO ATTRACT ECRITEK

TO CAMOUFLAGE CANADAIR

TO IDENTIFY CHRISTOPHARO

TO BRIGHTEN ECRITEK

TO EXCITE TEC

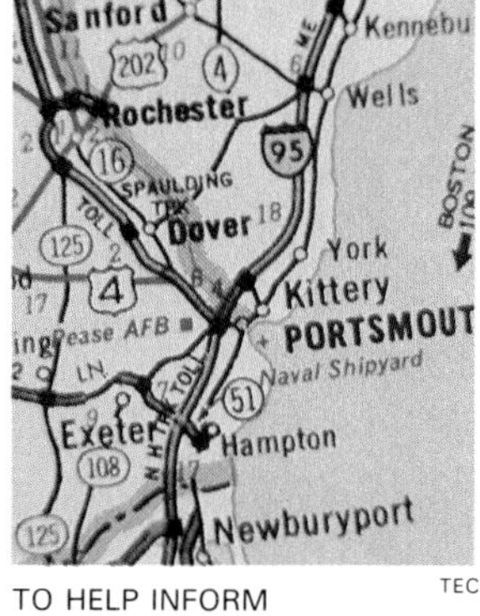

TO HELP INFORM TEC

Figure 1.31. Color can be used to create special effects.

□ PRINCIPLES OF DESIGN

You learned earlier that line, shape and form, texture, and color are the elements of design. You can think of these as building blocks that can be put together in many different ways. There are also guidelines for combining these elements. These guides are called the principles of design. They include balance, proportion, harmony and contrast, pattern, movement and rhythm, and unity and style.

Balance

You can think of balance as a tightrope-walker moving along a cable. She or he keeps balance using arms and a balancing pole. It is important to match or balance the mass (weight) of the body on both sides, figure 1.32. **Balance** is also very important in design. It means that the mass is evenly spread over the space used. There are three types of balance: symmetrical, asymmetrical, and radial, figures 1.33 to 1.35.

TEC

Figure 1.32. Think of balance in design like balance for a tightrope walker. Mass must be distributed evenly on each side of a centerline.

ECRITEK

Figure 1.33. An object, such as this church building, is symmetrical if one half is a mirror image of the other half.

Figure 1.34. In an asymmetrical design the two sides are in balance visually but are not mirror images.

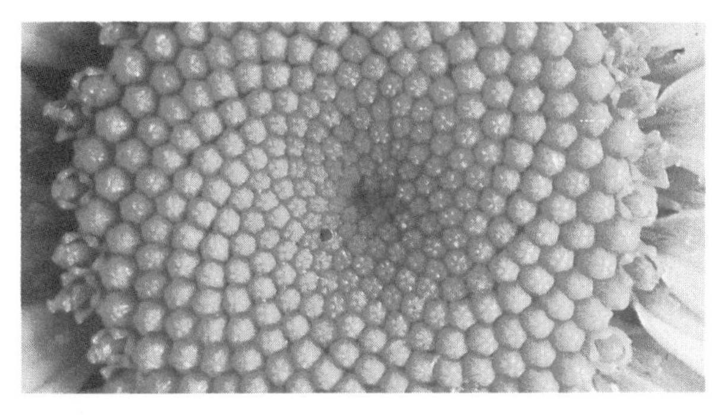

Figure 1.35. In radial balance the mass moves outward in all directions from a point at or near the middle. Two examples are shown.

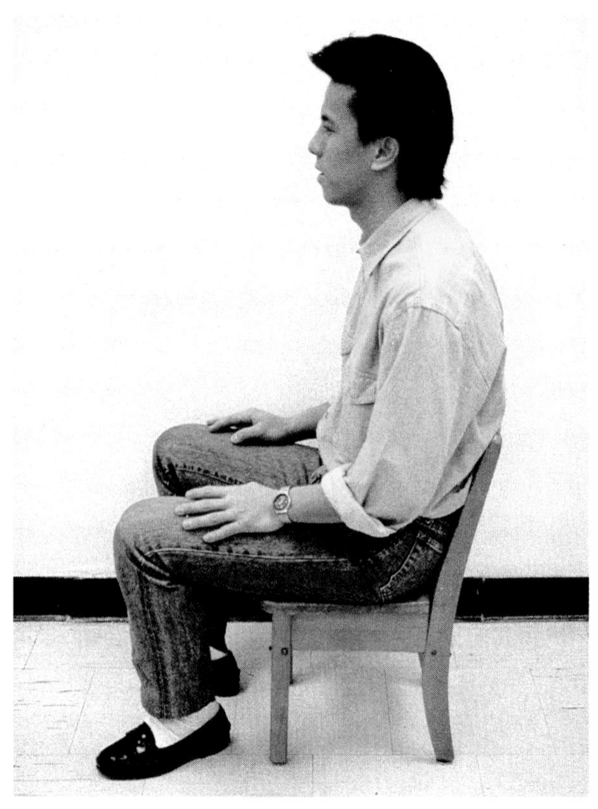

Figure 1.36. When one object is too large or too small for another object, they are out of proportion.

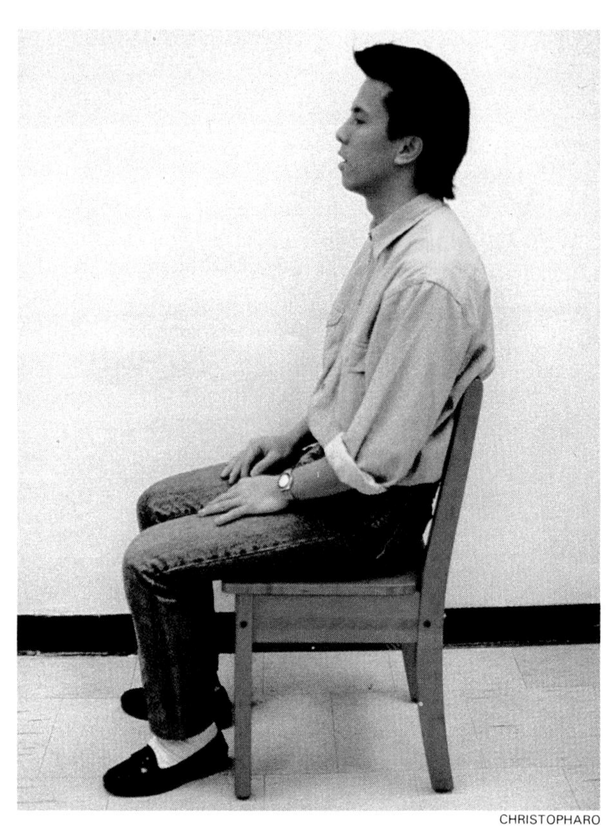

CHRISTOPHARO

Figure 1.37. When two related objects in a design are right for each other they are in proportion.

Proportion

Look at figure 1.36. Something seems to be wrong. Obviously the person is too big for the chair. Now look at figure 1.37. The person appears to be very comfortable. The relationship between the person and the chair seems to be right. The relationship between the sizes of two things is called **proportion.** In figure 1.36 the chair and the person are not in proportion. In figure 1.37 they are in proportion.

Proportion may apply to the relationship between objects. It can also apply to the parts of an object. Look at the doors and drawers of the cabin in an executive jet plane, figure 1.38. Their size is related to the overall size of the cabin. They are in proportion.

For thousands of years people have admired the proportions found in nature. The Greeks worked out a mathematical formula. It describes proportions found in nature. They called this formula the "golden mean." The

CANADAIR

Figure 1.38. Cabinetry in a jet airplane. Doors and drawers must be scaled down in size so they are in proportion to the space.

golden mean has a ratio of 1:1.618. (In the case of a rectangle, the long side is a little more than 1 1/2 times longer than the shorter side.) A golden rectangle may be drawn using the following procedure:

1. Draw a base line.
2. Draw a square. The length of one side of the square is the length of the short side of the rectangle.
3. Measure halfway along the base of the square as shown in figure 1.39. Put the point of your compass here. Now draw an arc from the top corner of the square to the base line.
4. The point where the arc touches the base line is the right hand corner of the rectangle. Draw a vertical line upwards from it. Then extend the top line of the square to complete the rectangle.

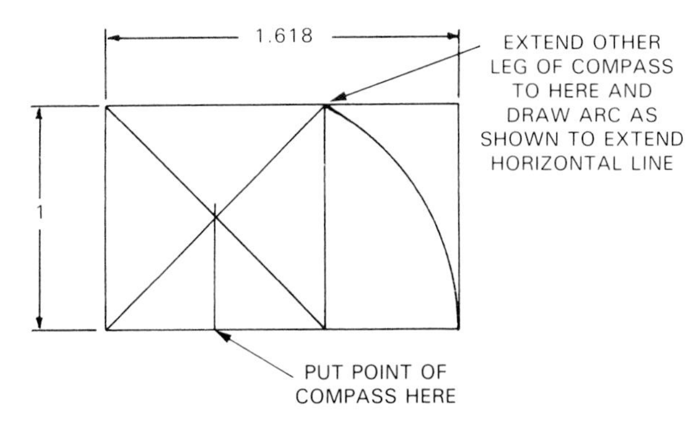

Figure 1.39. To draw a golden rectangle, start with a square.

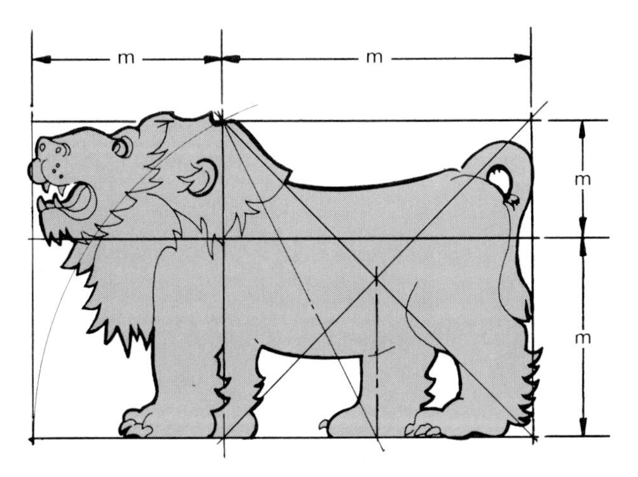

Figure 1.40. The lion's body fits the golden mean so often found in nature.

The golden mean also appears in the human body and many living things. In figure 1.40, the lion's proportions fit the golden mean.

Mathematics is important to designers. Still, they do not rely on mathematics alone to decide the proportions of an object. They must adjust the proportions until they look right. Look at the chest of drawers, figure 1.41. Notice that the drawers at the bottom are deeper than those at the top. If the drawers were of the same depth the chest of drawers would seem to be top heavy.

Harmony and contrast

Observe the best figure skaters and you will notice that their movements seem to flow with the music. We say they are in harmony with the music.

Designers use the idea of harmony in the objects they create. Buildings and their environment should be in harmony. The dishes in figure 1.42 go together well. Their shapes and colors are in harmony.

ECRITEK

Figure 1.41. Why should the bottom drawers be deeper than the upper ones?

Sometimes designers want to surprise you. They may want to make you feel excited about what you see. They may simply want to catch your attention. To do this a designer creates an obvious difference between things. This difference is called contrast.

Figure 1.42. Objects in harmony. Both color and shape go well together.

A DRURY

B

Figure 1.43. Looking at two types of contrast. A—Contrast through use of color. B—Contrast through use of lines and shapes.

You may wear bright clothes with contrasting colors, figure 1.43A. The red cross on an ambulance contrasts with its white background. The jagged mountain contrasts with the calm waters of the lake. Both harmony and contrast are used to make a designer's work attractive, figure 1.43B. Harmony makes you feel comfortable. Contrast adds excitement.

Pattern

What does the word pattern mean to you? To most people it means a design in which a shape is repeated many times. Look around you. Where can you see patterns? They are in unexpected places. As shown in figures 1.44 through 1.46, patterns are found in nature and in objects people have designed and made. Sometimes they are used to make an uninteresting surface appear more attractive. At other times the pattern may serve a particular function. Why does a tiger have stripes?

TEC

Figure 1.44. Do you see a pattern in this orange? Can you think of other patterns in nature?

Figure 1.45. Sometimes a pattern has a function. Here the squares are needed to play a game of chess.

TEC

Figure 1.46. Some patterns are created by arrangement of parts, others by the material. Some patterns are applied. Which type is this?

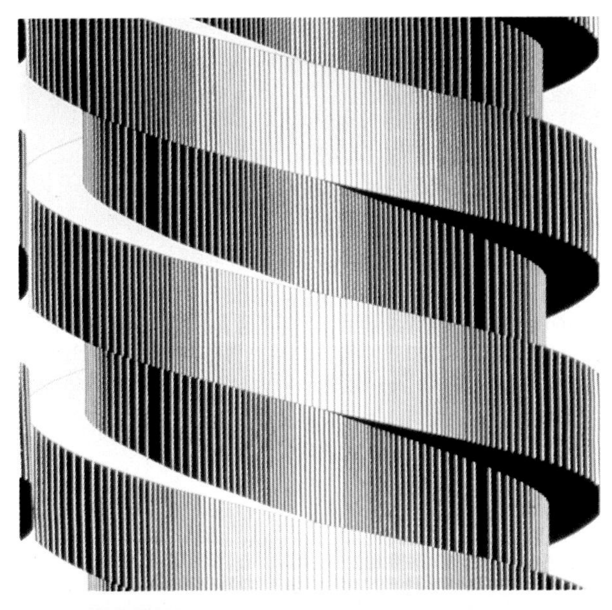

Movement and rhythm

Ocean waves create a pattern. The lines on an oscilloscope have a similar pattern, figure 1.47. Both suggest movement. Because they are repeating patterns, they are also said to have **rhythm.**

In figure 1.48, a feeling of movement is suggested by the spiral of the printed pattern. The tulip bowl creates a sense of movement through its use of shape and line. Both the pattern and the bowl also have rhythm.

ECRITEK

Figure 1.48. Shapes and lines give a sense of movement in objects created by humans.

LASER

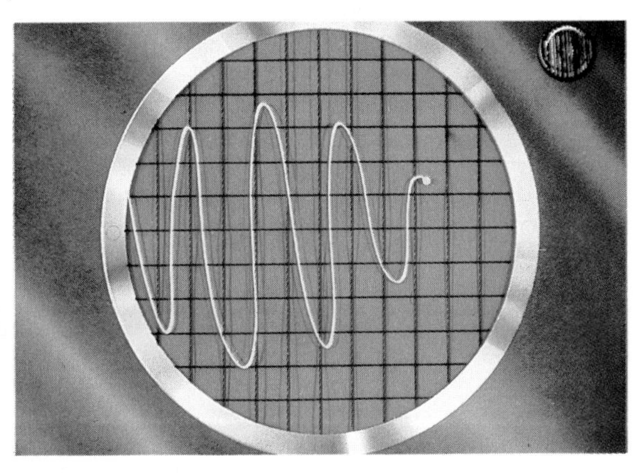

Figure 1.47. Patterns suggest movement. Both the ocean waves and the wavy pattern on the oscilloscope have rhythm because the pattern is repeated many times.

Unity and style

To sum up what you have learned:

- The elements of visual design are line, shape and form, texture, and color.
- The principles of visual design are balance, proportion, harmony and contrast, pattern, and movement and rhythm.

All of these are used when designing an object. However do not think of them as separate. Remember that they are all related. A well-designed product must have a sense of unity. Within the design there must be a sense of belonging or similarity among the parts, figure 1.49. Remember, also, that the principles do not provide hard and fast rules. They are only a guide. You must be the judge as to what is right or wrong. You must make the design decisions.

You can design an object any way you like. You can design according to your own style. There is no need to copy what other people have done. Each designer has his or her unique style. For example, Alvar Aalto developed a line of bent plywood furniture. Notice how different his chair is made from the one you might ordinarily see in a home, figure 1.49.

Style depends on many things:

- The availability and cost of materials.
- The tools and techniques available to shape the materials.
- Cultural preferences.
- A knowledge of the elements and principles of design, figure 1.50.

ECRITEK

ECRITEK

Figure 1.49. Each of these structures has a sense of unity.

Figure 1.50. What factors do you think have influenced the design of these telephones?

SUMMARY

Most people have solved problems and acted as designers. This chapter has described how problems in technology are solved by working through a series of steps. These steps first involve identifying a problem and writing a design brief. Investigating the problem further to find out all the information needed is followed by producing a number of alternative solutions. From these alternatives one solution is chosen. Detailed drawings are then made. A model or prototype is built and tested. After modification the product is ready for mass-production.

Products must function well. They must also look and feel good. Their attractiveness is a result of their line, shape, form, texture, and color. These elements of design are combined using the principles of design. These principles are balance, proportion, harmony, contrast, pattern, movement, rhythm and style.

KEY TERMS

Alternative solutions	Model
Balance	Pattern
Design brief	Pre-production series
Designer	Primary color
Design process	Problem
Elements of design	Proportion
Ergonomics	Prototype
Feedback	Rhythm
Form	Secondary color
Function	Shape
Harmony	Style
Investigation	Texture
Lines	

TEST YOUR KNOWLEDGE

Write your answers to these review questions on a separate sheet of paper.

1. Why are most new products invented?
2. List the eight steps in the design process.
3. The first step in the design of a storage container is to ______.
   a. Buy the wood, glue, nails, and hinges.
   b. Prepare the tools you will need.
   c. List and measure all the items to be stored.
   d. Decide its color and shape.
4. The most important information for a designer planning a new seat for a bus is the ______.
   a. Type of metal or plastic to be used.
   b. Color of the seat material.
   c. Time taken to manufacture each one.
   d. Average size of people using it.
5. Given a design problem, an engineer would sketch several possible solutions because ______.
   a. There is a good range of ideas from which to choose.
   b. It is difficult to decide which is the best solution.
   c. Similar objects can be made using different materials.
   d. Many people want to see the sketches.
6. What is a prototype and why would one be built?
7. The study of how a person, the products used, and the environment can be best fitted together is called ______.
8. What specific questions would you ask if you were testing and evaluating a new wheelchair?
9. Think of one object that you have seen and that you find attractive. Describe in your own words, (a) the object and (b) how the elements of design make it attractive to you.
10. Construct a golden rectangle with a short side of 2 in. (50 mm).

APPLY YOUR KNOWLEDGE

1. Collect or draw pictures illustrating three natural objects. Collect three more pictures to show the equivalent technical objects. For example, if you collected a picture of a bird's nest (a natural object) the equivalent technical object would be a house.
2. Collect four pictures to illustrate two elements of design and two principles of design.
3. Carefully observe some activities in your home. Make a list of technological problems that need to be resolved. For example, storing spices in the kitchen or making sure that your baby sister doesn't fall downstairs.
   a. From the list of problems above, iden-

tify one that you will try to resolve. Write a design brief for the problem.

b. Make a list of the questions you will have to answer to solve the design problem.

c. Generate a number of solutions to the problem and select the most appropriate solution.

d. Make a list of the steps involved to build a prototype of your solution.

e. Build, test, and evaluate your prototype. Keep a good record, in the form of notes and sketches, of all changes and modifications. Make recommendations for further improvements.

4. Design and build a device that will make one task easier in the life of a person with a special need, that is, a handicapped, sick, or elderly person.

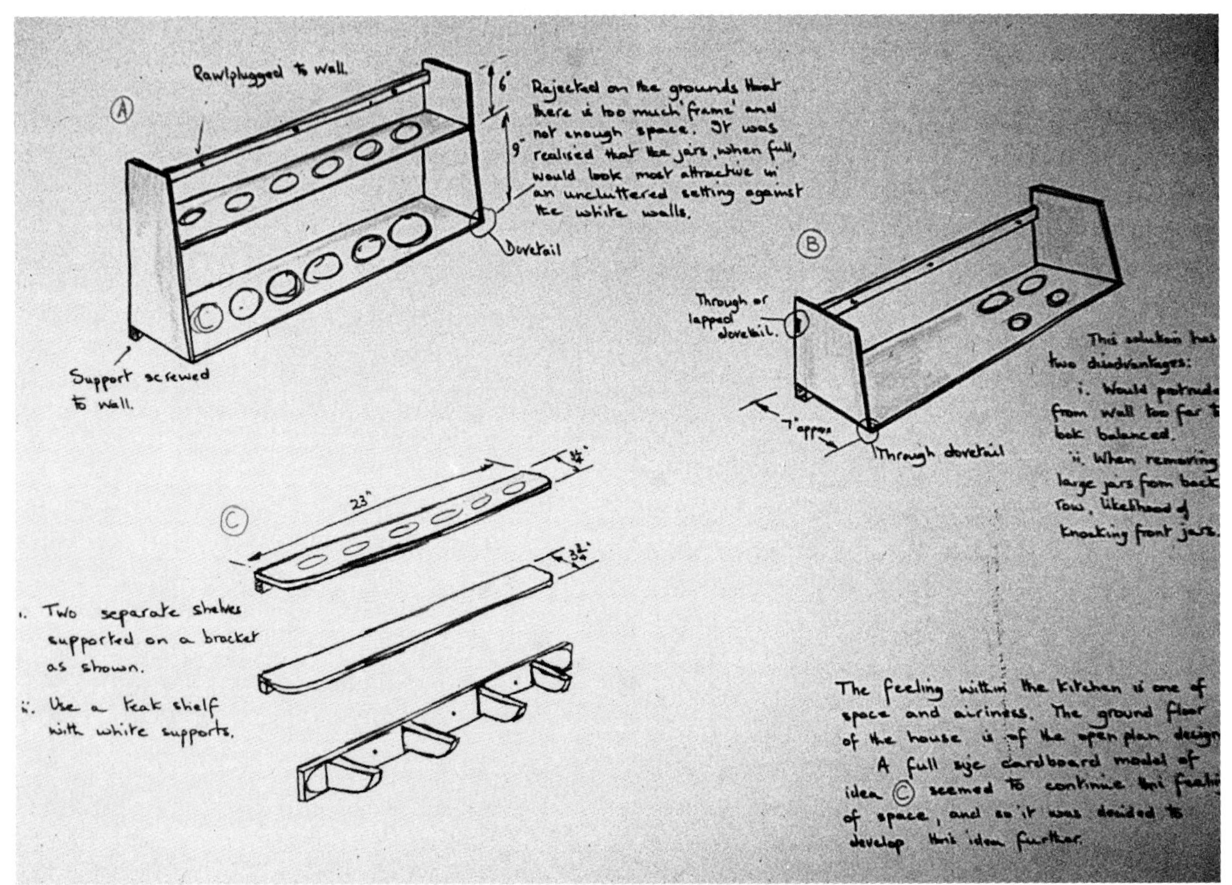

This page from a designer's notebook shows alternative solutions for a spice rack. TEC

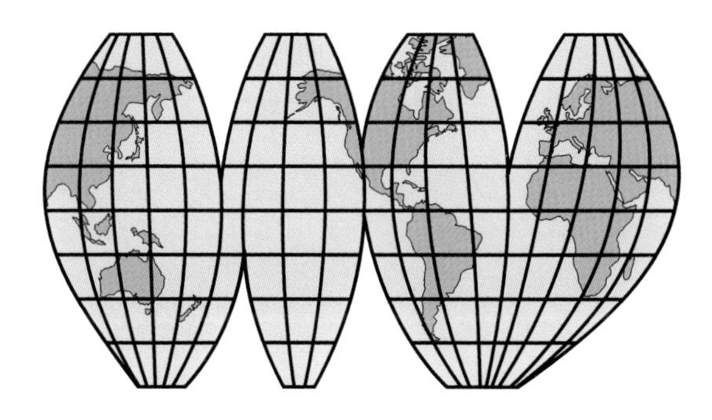

Chapter 2

Communicating Ideas

OBJECTIVES

After reading this chapter you will be able to:

- O Define communication technology.
- O Explain the three basic types of communication technology.
- O List various codes used to communicate information.
- O Explain how drawings "get across" ideas more efficiently than do words.
- O Communicate ideas by means of isometric or perspective sketches.
- O Draw simple objects using the principles of orthographic projection.
- O Determine which form of drawing is best for a given situation.
- O Describe how the computer is changing drafting practices.

Telling other people about our own ideas is known as **communication.** This is a big word for a simple act. It came from the Latin, ''communo.'' It means to ''pass along.''

Communication is more than sending a message. The message must be received and understood. If this does not happen, there is no communication. Communication technology is the transmitting and receiving of information using technical means. Tools and equipment are used to assist in the delivery of the message. There are many kinds of communication. Their elements form a system.

□ COMMUNICATION SYSTEM

A complete communication system has several parts, figure 2.1:

1. A source. This is the starting point for the message.
2. A means of transmission (moving the message from one place to another).
3. A destination or receiver. This is something or someone who gets the message.

There are other parts that we'll talk about later. For now, let's go to an example. Suppose that you have an idea to take in a movie. You say to a friend, ''Let's take in a movie tonight.'' Do you see the system? Your idea is the source. Your telling the idea to a friend is transmission. Your friend hears you talk. She or he is the destination or receiver of the message.

Communication systems also have two other parts.

4. Feedback.
5. Storage.

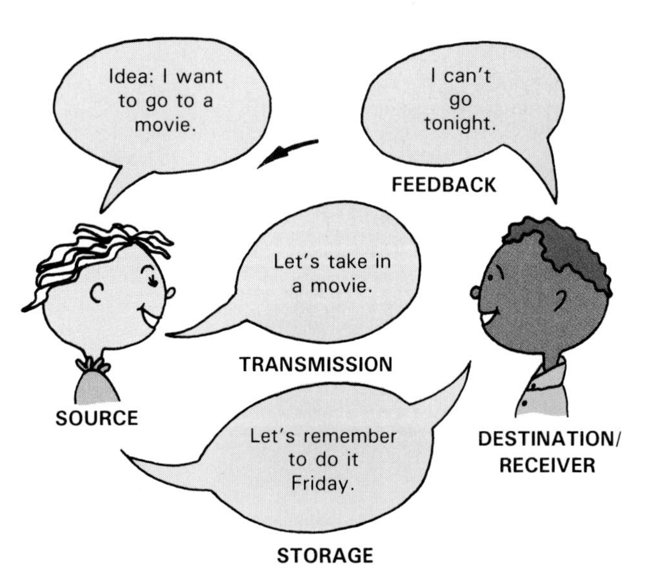

Figure 2.1. Talking to a friend is an example of a simple communication system.

What are they? We will use the same example. Suppose your friend cannot go to the movie. ''Sorry, I can't. I have to go with my father tonight.'' That is feedback. It indicates that your message was heard and understood. Then your friend says, ''But let's remember to do it Friday night.'' Then both of you will store the information and remember to go to the movie on Friday.

■ FORMS OF COMMUNICATION

All forms of communication use a code or symbols. For example, D-O-G and 犬 are letter symbols that communicate the idea of a certain animal. However, the same message can be given in a quite different ''language.'' See figure 2.2.

Hand signals and sounds

Simple movements or sounds can replace spoken and written messages. Look at figure 2.3. These are signals that most people anywhere in the world would understand. Can you think of other signals that you might use? What about the signal to be quiet?

What about sound signals? If you know any Morse code, you will recognize the dots and dashes being sent out by the ship in figure 2.3 as the international distress signal. It is known by people of all languages.

Humans, however, are not the only earth dwellers that exchange messages with sounds and ''body'' language. Sea animals such as dolphins and whales have systems of sound to exchange messages among their own kind. Deer and beaver use their tails to signal danger.

Figure 2.3. Sounds and hand signals have meaning. What message is being sent in each of these pictures?

Figure 2.2. What language is being used here?

Humans developed nonverbal (no words) ways of communicating long before formal language was developed. Cave dwellers drew on walls to tell their experiences. We still use such methods.

Symbols and signs

Simple pictures and shapes are one of the most effective methods of communication. **Symbols** can warn, instruct, and direct without using words. They ''speak'' in a hundred languages all at the same time, figure 2.4.

Figure 2.4. Who do you think would use these symbols? What do they mean?

Ways of communication

There are three basic ways or types of communication. All are based on our sense of hearing and sight.

1. Visual communication presents ideas in a form we can see. Thoughts are changed into words, symbols, and pictures. Do you understand that a stoplight, a street sign, a photograph, a computer, and a book are giving out visual messages?
2. Audio communication has messages that can be heard. However, they cannot be seen. Examples? How about the buzzer that tells you class is over? How about telephones, radios, and cassette recorders? Do you have a doorbell at your home that tells you someone is at the door?
3. Some communication can be both seen and heard. This is known as "audiovisual." You are receiving audiovisual messages when you watch and listen to television, videos, and movies.

Drawings and their types

Objects and ideas can also be represented (pictured) using lines and shapes such as the ones shown in figure 2.5. These are known as "line drawings." Designers, drafters, technicians, engineers, and architects must be able to make such line drawings. How do you suppose the designer would ever have explained how to build the parts in figure 2.6 without drawings? When discussing technical details, "a picture is worth a thousand words."

Drawings of the ideas to be communicated are called drafting. Drafting has always been known as the "language of industry." It prevents confusion about the size and shape

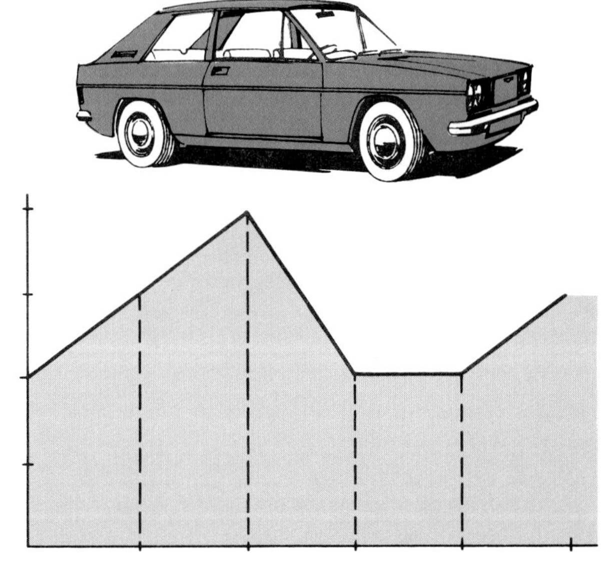

Figure 2.5. Designers use line drawings such as these to communicate designs to others.

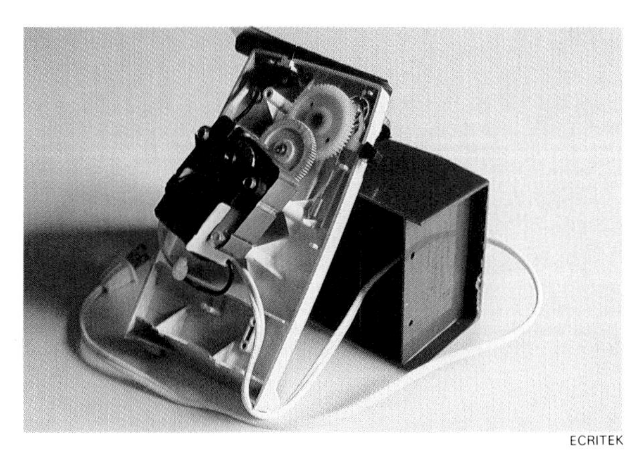

ECRITEK

Figure 2.6. A new product requires one or more drawings to describe it well enough so others can build it.

of an object or structure. Some types of drawings look a great deal like photographs. Others show only one surface or an object or structure.

There are three types of drawings: isometric, perspective, and orthographic projection. These three types are summarized in figure 2.7. All three may be sketched freehand, or drawn using manual drafting equipment or computer-aided drafting systems.

Isometric sketching. Sketching is the simplest type of drawing and one of the quickest ways to share an idea. While ideas are developing in your imagination, sketches will record them for further discussion, figure 2.8.

ISOMETRIC

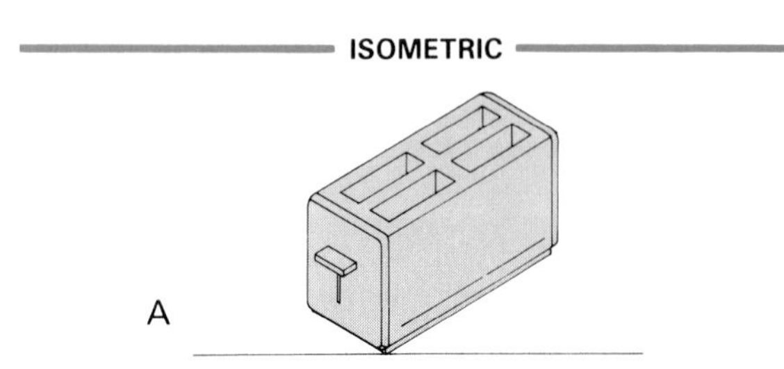

An isometric drawing gives a picture of three sides of an object. However, the picture does not look real.

PERSPECTIVE

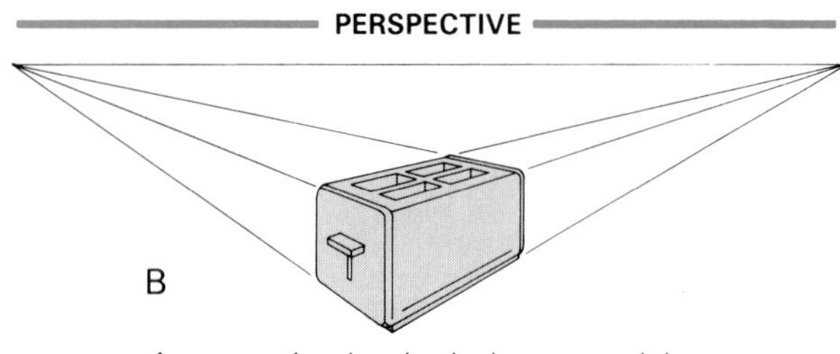

A perspective drawing looks more real than an isometric. Like railroad tracks, parallel lines appear to converge.

ORTHOGRAPHIC PROJECTION

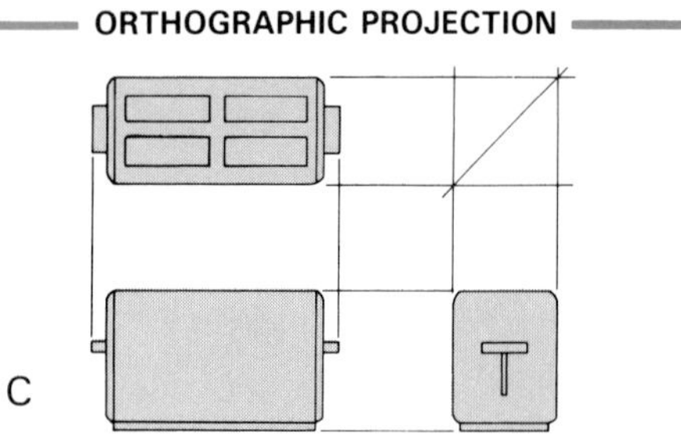

An orthographic projection shows the drawing from three directions as if it were flat. Front, top, and right side views are drawn.

Figure 2.7. Which drawing looks most realistic?

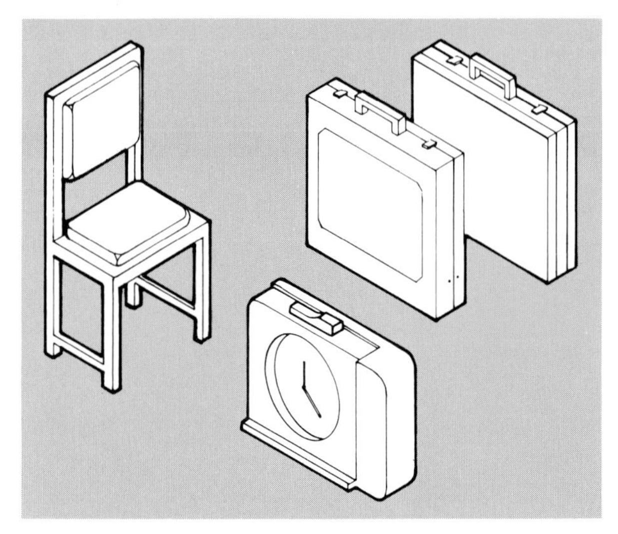

Figure 2.8. Designers often communicate their ideas using isometric sketches.

Isometric sketches can be easily drawn on **isometric paper.** This paper has a grid of vertical lines. Other lines are drawn at 30° to the horizontal. To sketch an isometric box that is six squares long, three squares wide, and four squares high:

1. First draw the front edge of the block (line 1). Draw lines 2 and 3 to show the bottom edges of the box. These three lines represent isometric axes. See figure 2.9.
2. Add dimensions to the three axes, as shown in figure 2.10.
3. Draw the vertical edges of the box as shown in figure 2.11.
4. As in figure 2.12, draw the two top edges of the box.
5. Complete the sketch, figure 2.13. Remove unnecessary construction lines. Darken the outline of the object.

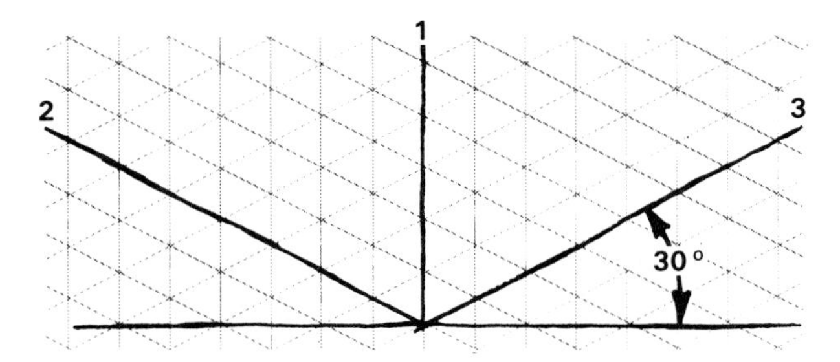

Figure 2.9. To begin an isometric sketch, make these three lines to set up the three axes (edges) of the sketch.

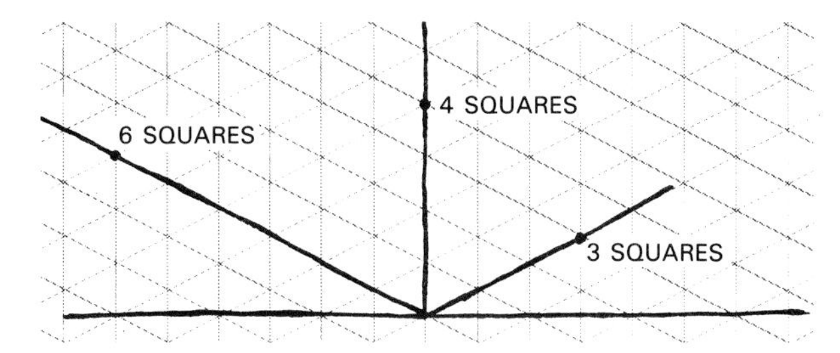

Figure 2.10. Next, establish the three dimensions of the isometric sketch.

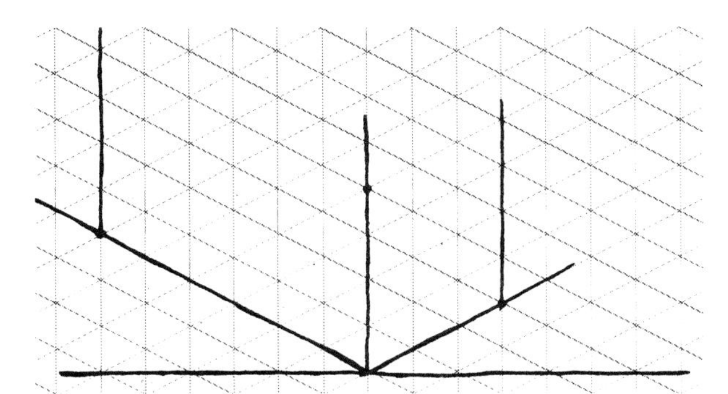

Figure 2.11. Add two more vertical lines to represent the visible vertical edges of the box.

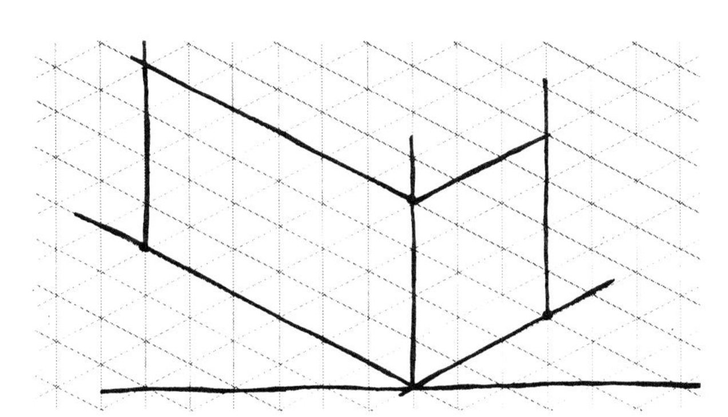

Figure 2.12. Next, begin to sketch in the top edges of the box.

Whatever the shape of the object to be drawn, it is easiest to begin by drawing a box. In most cases, however, you will have to remove parts of the box to create the shape, figure 2.14.

Sometimes you will want to add a piece. Could you make a sketch of a box with a small block added to one side? This method of isometric sketching may be used to draw a simplified house.

1. Lightly construct an isometric box as in figure 2.15. Use the dimensions, eight squares long, four squares wide, and six squares high.
2. Construct the basic shape of the house by removing a corner of the box, figure 2.16.

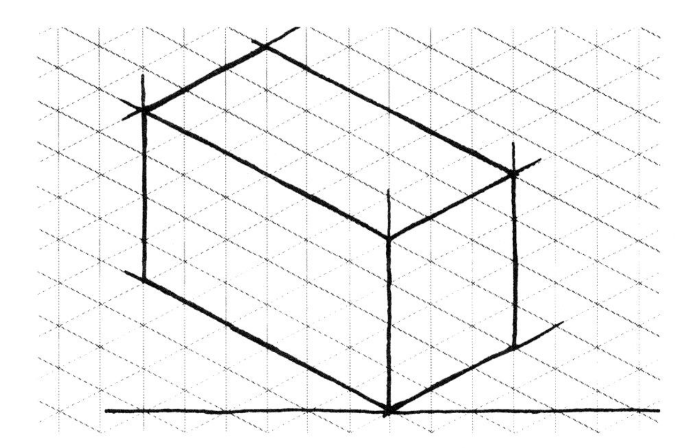

Figure 2.13. The box is completed.

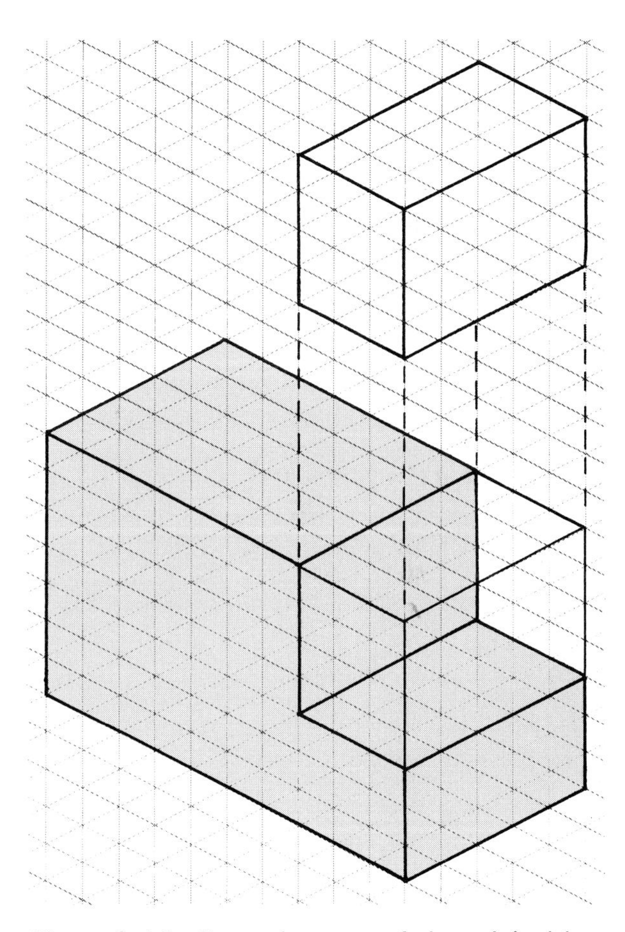

Figure 2.14. Removing part of the original box will create a new shape.

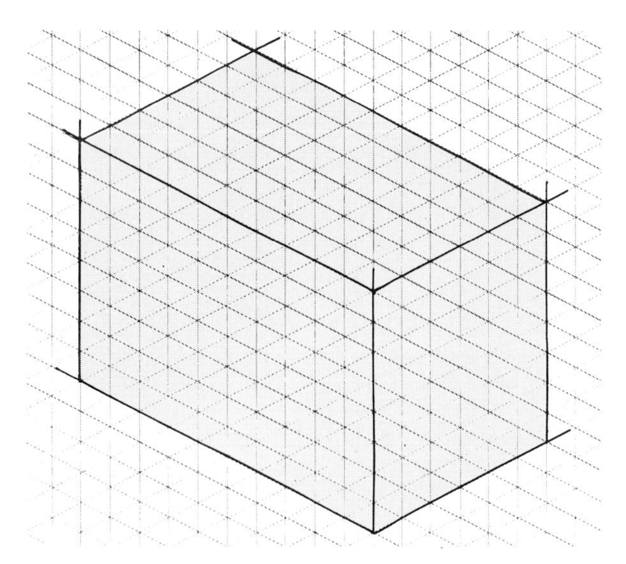

Figure 2.15. To draw a house, again make a box on an isometric grid.

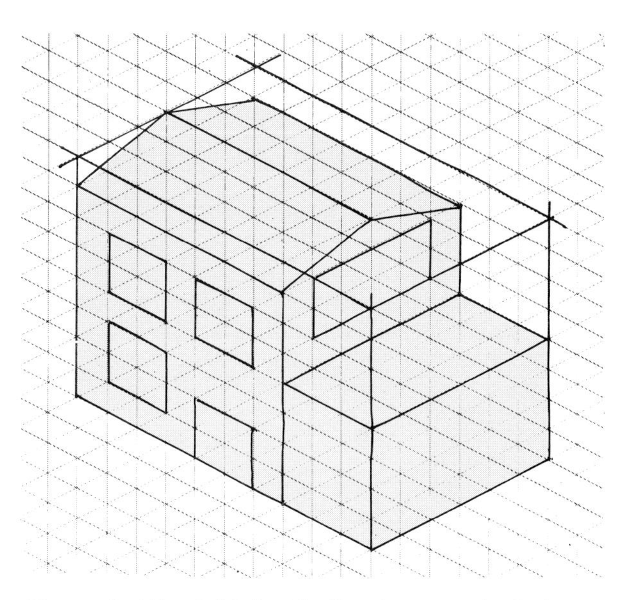

Figure 2.17. Add details like doors and windows.

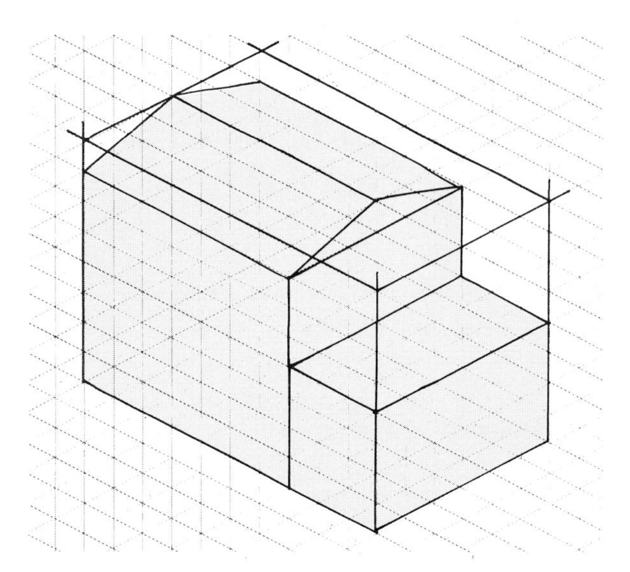

Figure 2.16. The house takes shape! See how points on the grid are used to draw in the roof lines.

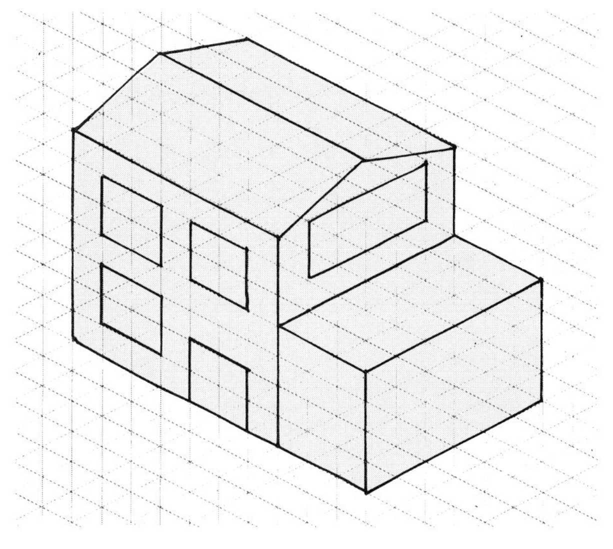

Figure 2.18. The sketch of the house is completed.

Add lines for the roof.

3. Add details, including windows and doors, figure 2.17.
4. Complete the line work by removing unnecessary construction lines. Darken the remaining lines to form the building's shape, figure 2.18.

While isometric paper makes sketching easy, it has one disadvantage. It leaves grid lines on the final drawing. These could confuse someone looking at your drawing. Designers often prefer to sketch on plain paper. To make a freehand isometric sketch of a rectangular block on plain paper, use the method shown in figure 2.19.

Perspective sketching. Look at the photograph of the railroad station, figure 2.20. What do you notice about the parallel lines of the tracks? They appear to converge (run together). What do you notice about the height of the lamp posts? Those farther away appear shorter. What do you notice about the

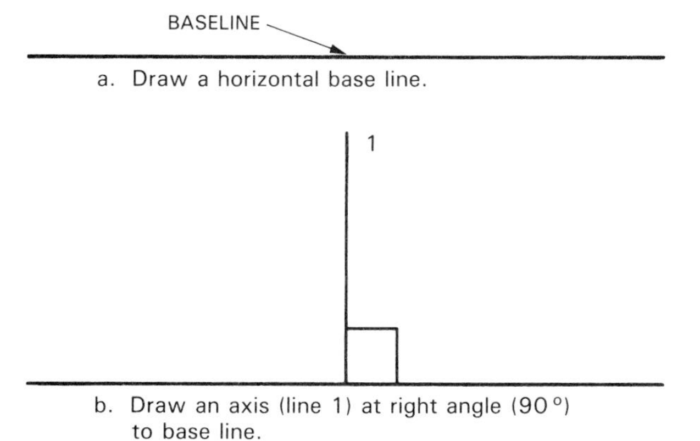

a. Draw a horizontal base line.

b. Draw an axis (line 1) at right angle (90°) to base line.

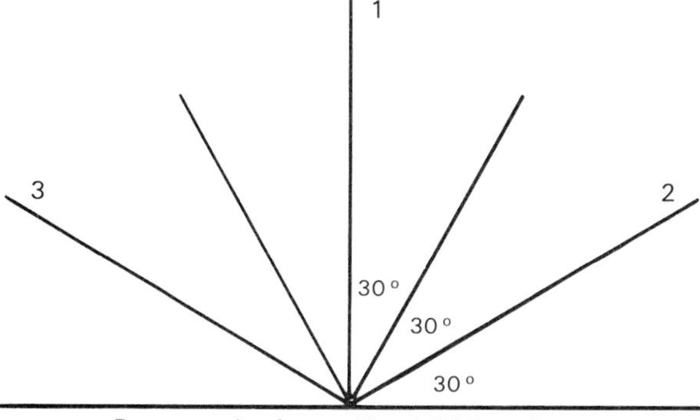

c. Draw a pair of axes (lines) 30° up from the horizontal. (You can judge this by dividing right angles into three equal parts).

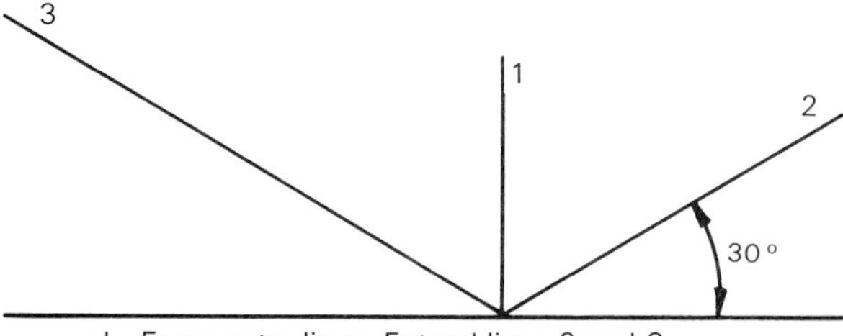

d. Erase extra lines. Extend lines 2 and 3.

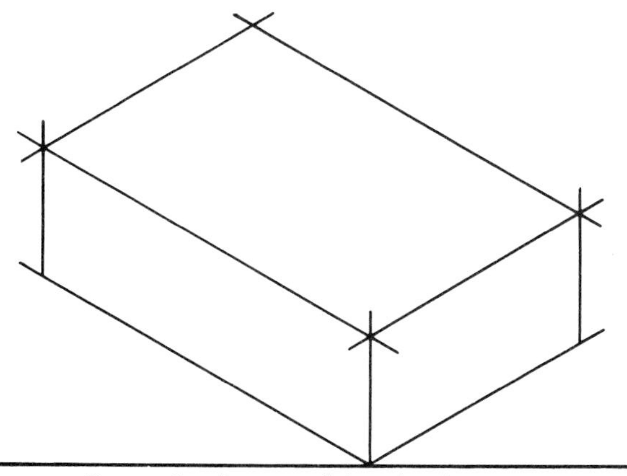

e. Draw other lines parallel to the three axes. This completes the box.

Figure 2.19. You can make a freehand isometric sketch by following these steps.

ECRITEK

Figure 2.20. In real life, objects at a distance seem narrower or shorter than they are close up.

width of the platform? The farther away it is the narrower it appears. Of course, railroad tracks don't converge, lamp posts don't get shorter, and platforms don't become narrower.

Perspective sketching provides the most realistic picture of objects. The sketches are drawn to show objects as we would actually see them. Parallel lines converge and vertical lines become shorter as they disappear into the distance. A perspective sketch of a block is drawn in the following way:

1. Draw a faint horizontal line, figure 2.21A. Think of this as representing the horizon. Mark two points, one at each end of the line. These are vanishing points (VP).
2. Draw the front vertical edge of the block, figure 2.21B.
3. Draw faint lines from each end of the vertical edge to the vanishing points. This is shown in figure 2.21C.
4. Draw vertical lines to represent the left- and right-hand edges of the block, figure 2.21D. The length of these vertical sides will be shorter than the real object.
5. Join the top of these vertical lines to the vanishing points. Darken the outline of the object as in figure 2.21E.

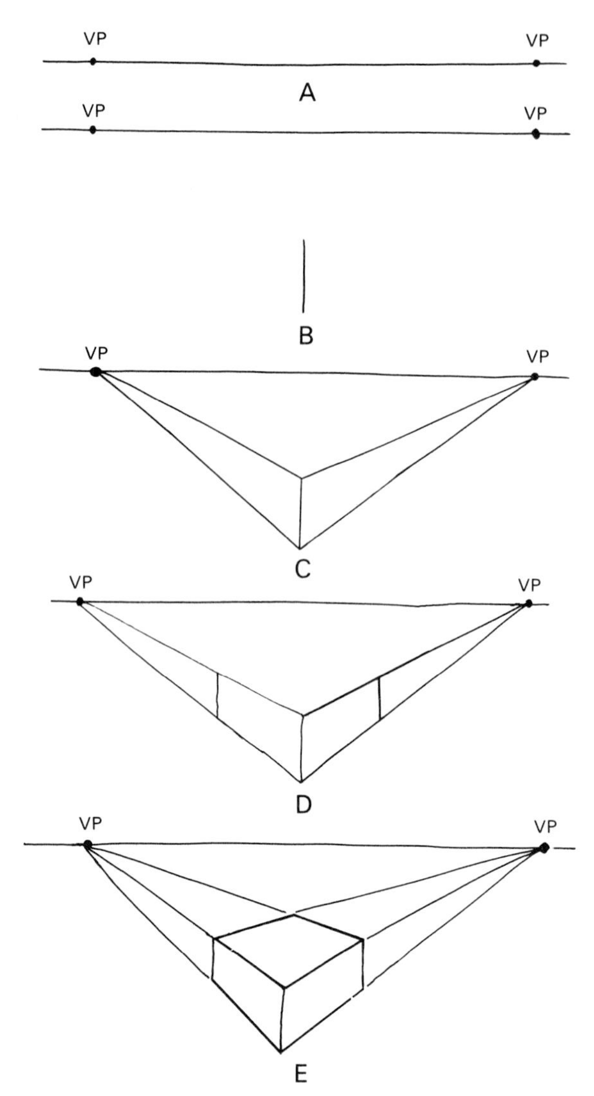

Figure 2.21. This is how you produce a perspective drawing. Notice how like a photograph it is.

Orthographic projection. You have seen how isometric and perspective sketches are simple methods of recording your ideas and communicating them to other people. They give a general idea of the shape and features of an object. Unfortunately, there are some disadvantages to isometric and perspective sketches. For example, they do not describe the shape of an object exactly because of distortion (change or true shape) at the corners. Neither do they provide complete information for the object to be made.

Orthographic projection overcomes both these problems. This kind of drawing shows each surface of the object ''square on,'' that is, at right angles to the surface. In this way, you see the exact shape, or view, of each surface. Complete information is usually given by drawing three **views:** front, top, and right side. To understand how a view is produced, imagine that you are the person in figure 2.22. Because you are looking at the object square on, you will only see the area that is colored red. Since this is the front of the object this view is called the front view.

To produce a top view, imagine you are above the object looking down on the top. The view you'd see is shown in blue. The right-side view, shown in green, is drawn by looking at the right side square on. These three views are always arranged as shown in figure 2.23.

To draw an orthographic projection of the house in figure 2.24, complete the steps described in figure 2.25.

NOTE: Work on squared paper. Remember to keep all lines in steps 1-3 as faint as possible.

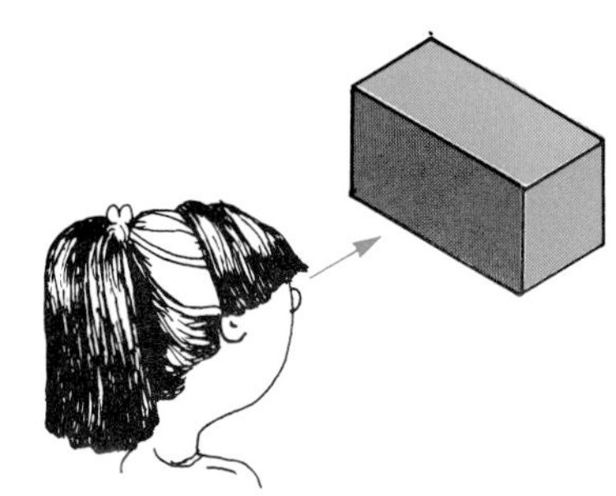

Figure 2.22. An orthographic view ''sees'' only one side of an object.

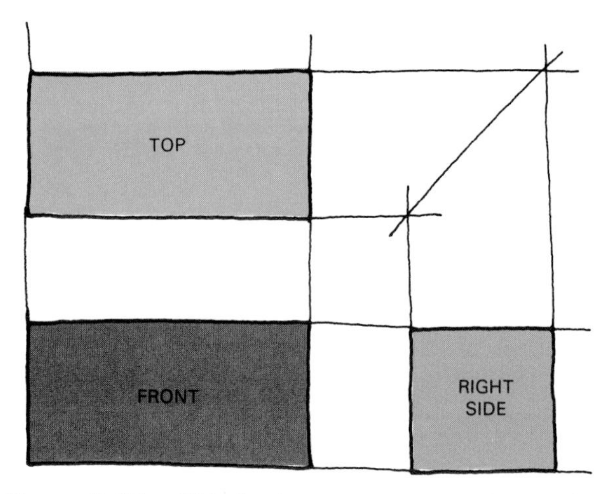

Figure 2.23. This is the proper arrangement for orthographic views of an object.

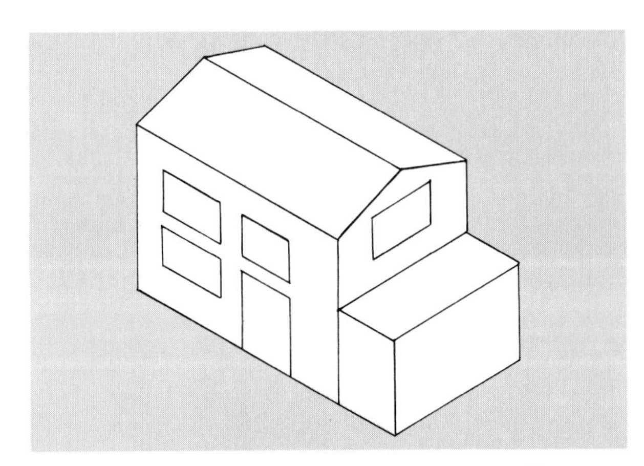

Figure 2.24. Isometric view of a house. The next drawing will show it as an orthographic drawing.

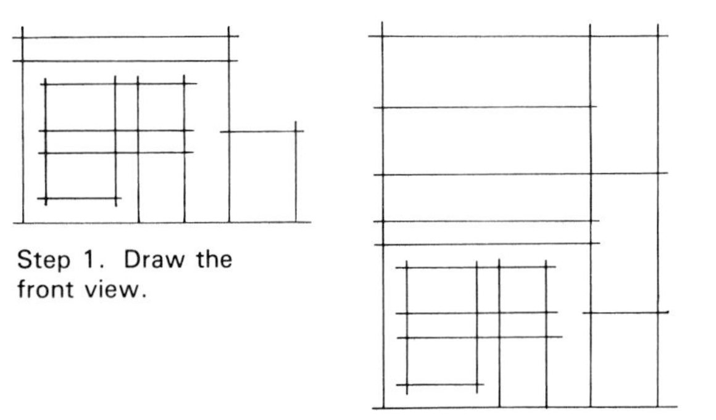

Step 1. Draw the front view.

Step 2. Project (extend) the vertical lines of the front view above the drawing. Draw the top view.

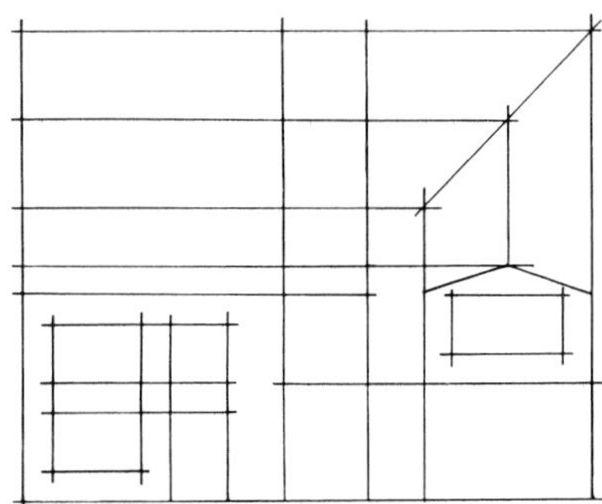

Step 3. Draw the projection lines as shown to complete the right-side view.

Step 4. Darken the outline of the object. Erase the projection lines, if you wish.

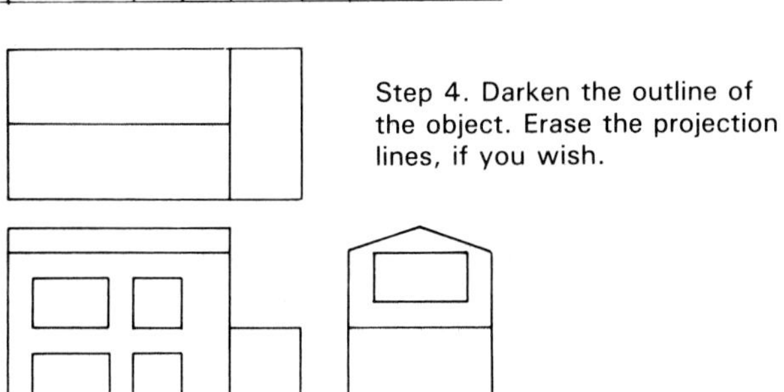

Figure 2.25. This is how orthographic drawings are developed.

□ DRAWING TECHNIQUES

The orthographic views shown in figure 2.25 were sketched on squared (grid) paper. This is a quick method. Its disadvantage, however, is that the grid could be confused with the lines of the drawing.

An alternative to grid paper is the use of plain paper and drawing instruments. The instruments most often used are the T squares, 45°, and 30°/60° set squares (drafting triangles), compass, and scale (ruler). The following are some techniques for drawing with these instruments:

1. As a general rule, when drawing lines with a T square, draw in the direction the pencil is leaning, figure 2.26.
2. When drawing vertical lines with drafting triangles, lean the pencil away from yourself and draw lines from bottom to top, figure 2.27.

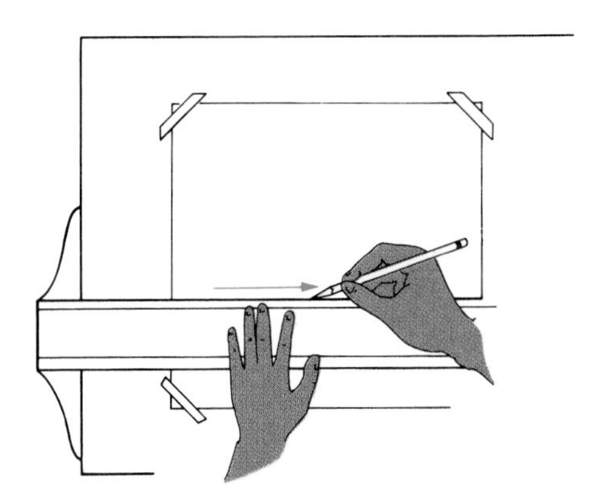

Figure 2.26. Always draw in the direction pencil is leaning.

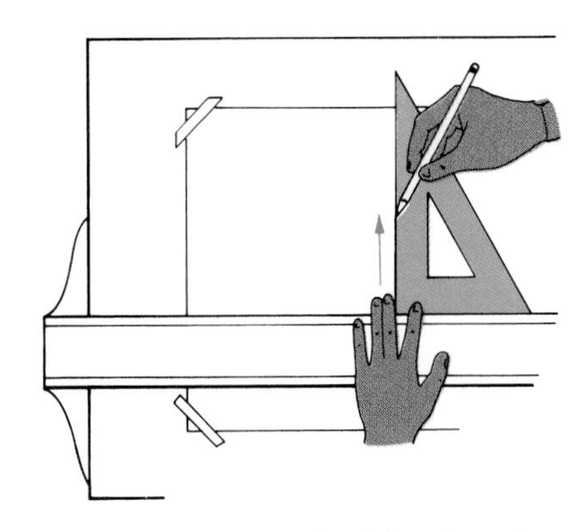

Figure 2.27. Proper method for drawing vertical lines using a triangle.

3. When using a 45° or 30°/60° triangle, draw lines in the directions shown by the arrows in figure 2.28.
4. Hold a compass between the thumb and forefinger and rotate clockwise. Lean the compass slightly in the direction of the rotation as you draw a circle, figure 2.29.

Alphabet of lines

There are a number of different types of lines used to produce orthographic drawings. Each line is used for a particular pupose and cannot be used for anything else. Look at the casting in figure 2.30. The alphabet of lines can be used to produce orthographic drawings of this casting. Figure 2.31 shows the rules for use of lines.

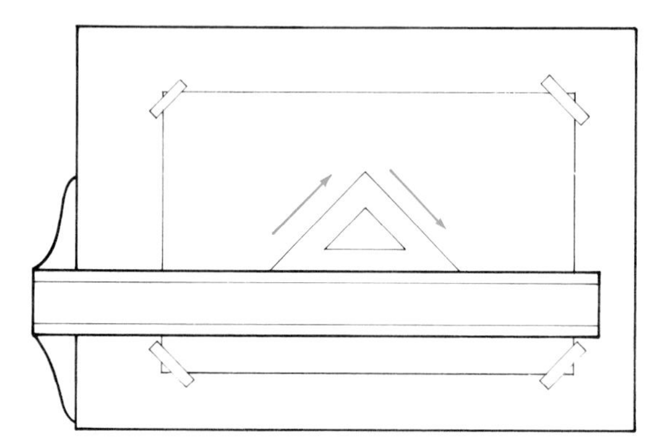

Figure 2.28. Proper method of drawing sloping lines.

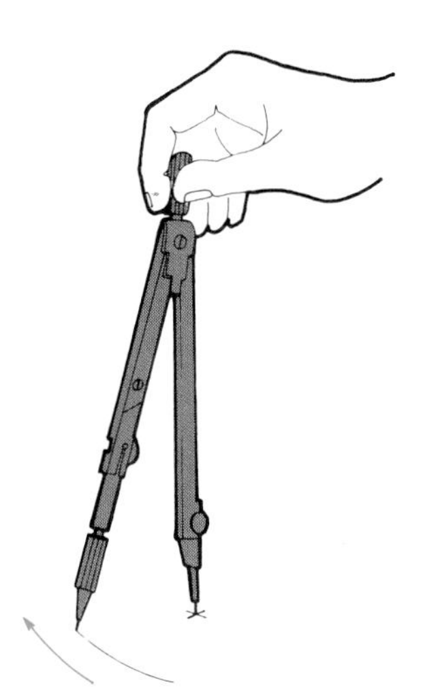

Figure 2.29. Proper method of using a compass. Draw circles or arcs lightly at first. Make repeated turns to darken the line.

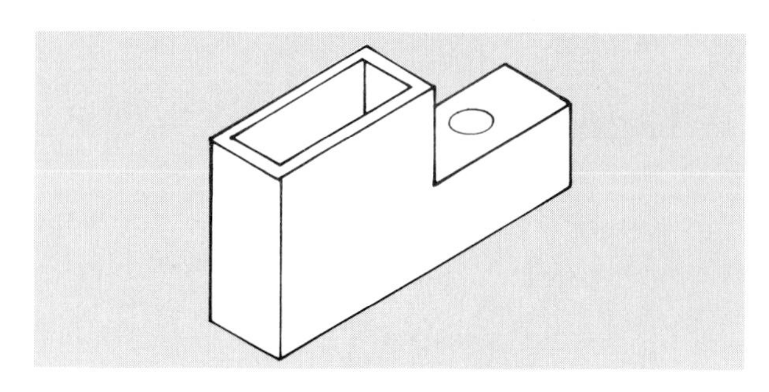

Figure 2.30. Drawing of metal casting.

Type of line	Example
Construction line Thin, faint lines used to start a drawing.	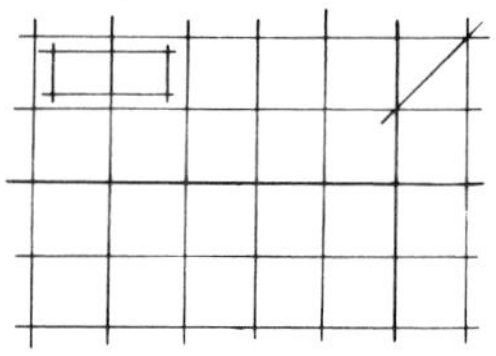
Object or visible line Darker, thicker lines used to show the outline of the object.	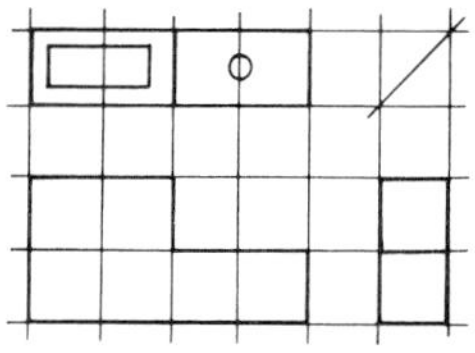
Hidden line Short and evenly-spaced dashes used to show hidden features.	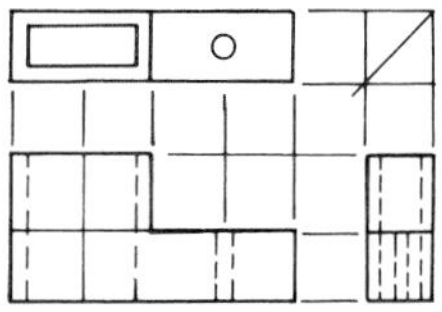
Centerline Alternating long and short dashes to show the centers of holes.	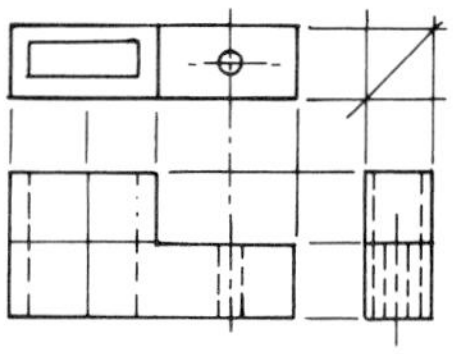
Extension and dimension lines Thin lines used to show the size of an object and its parts.	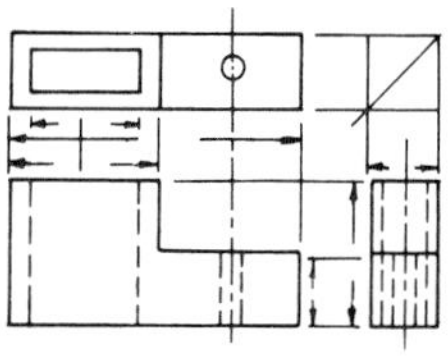

Figure 2.31. The alphabet of lines explains the types of lines and where they are used in drafting.

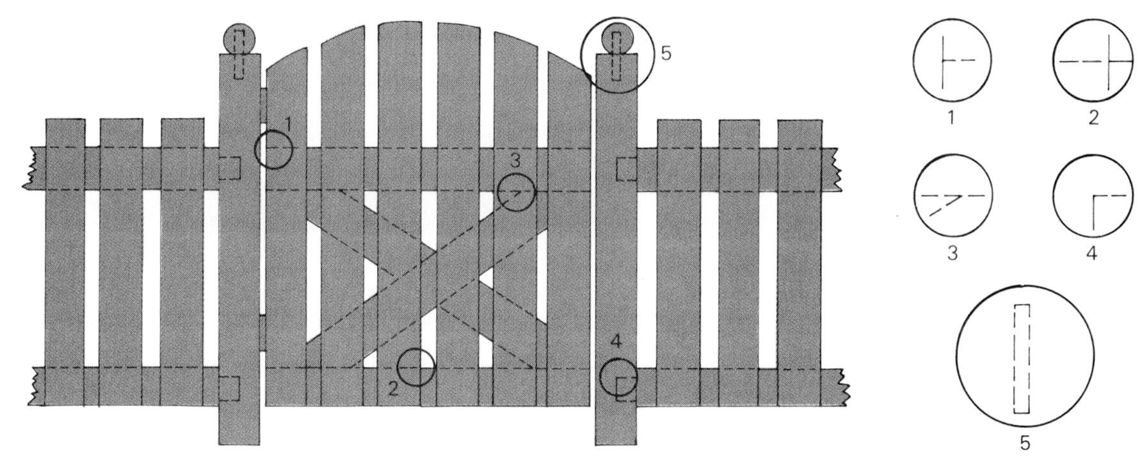

Figure 2.32. When you look at a board fence, parts of it are hidden by the boards. A drawing would show the hidden parts as short, evenly-spaced dashes.

Hidden lines are short, evenly-spaced dashes. They show the hidden features (lines or shapes) of an object, figure 2.32. Hidden lines almost always begin and end with a dash touching with the line where they start and end (1). This rule is not followed when the dash would continue a visible detail line (2). Dashes should join at corners (3) and (4). The dashes of parallel hidden lines that are close together should be staggered (5).

Dimensioning

Most drawings include two types of dimensions: overall dimensions and detail dimensions. To fully describe the size and shape of the view in figure 2.33A, you need two overall and two detail dimensions.

If a hole is added to this view, then you must add some dimensions as in figure 2.33B. The size of the hole and its location must be shown. What is important is to show the exact position of the center of the hole.

Using drawing techniques

The following method is used to produce a set of orthographic views using plain paper and instruments. You will also need to refer to the alphabet of lines, figure 2.31. You are going to draw the truck in figure 2.34.

1. Draw the front view using construction lines, figure 2.35.
2. Draw the top view directly above the front view, figure 2.36. Project all vertical lines upward as shown.

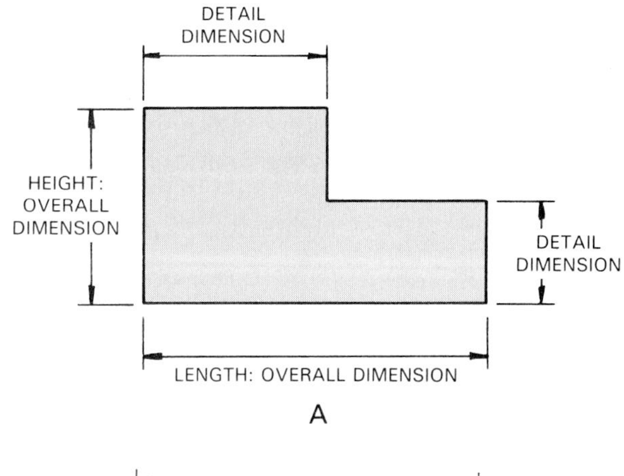

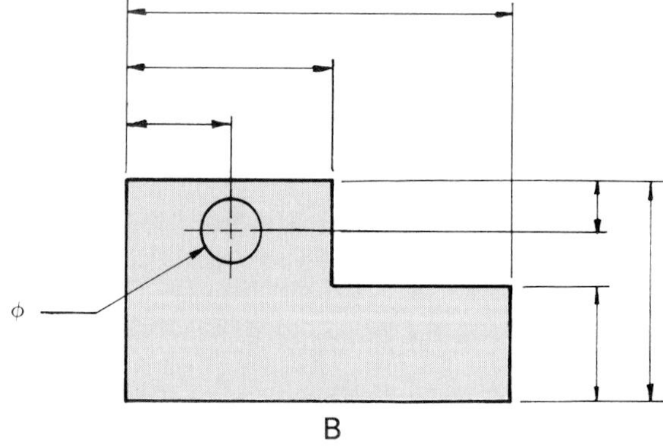

Figure 2.33. These two examples show overall and detail dimension lines.

3. Draw the edge of the right-side view (line a). Distances x and y should be the same. Project the lower edge of the top view (line b) to intersect line a. Project the lines of

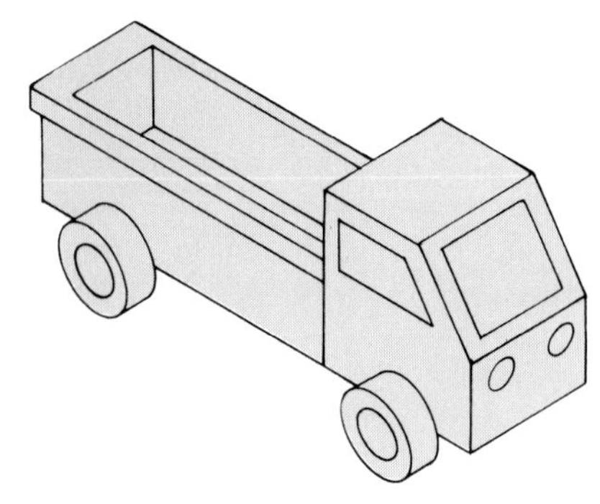

Figure 2.34. A toy truck in isometric view. Can you change the view to an orthographic drawing?

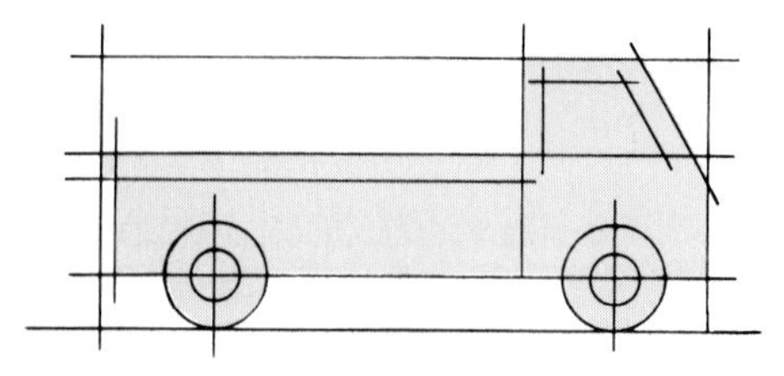

Figure 2.35. First step. Make a front view using construction lines. Take measurements from the isometric view in figure 2.34.

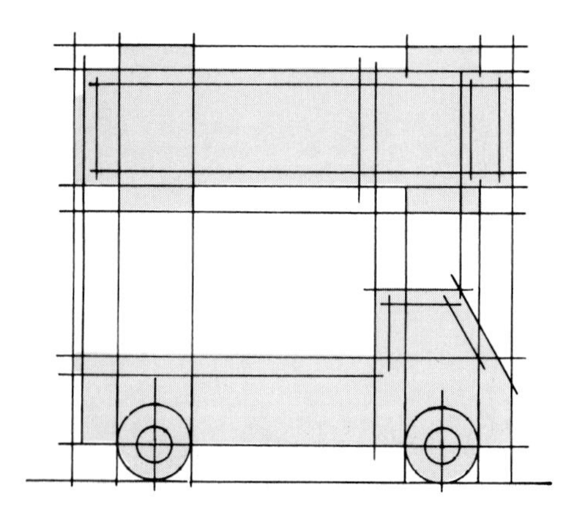

Figure 2.36. Extend the vertical lines to start the top view. Measure the drawing in figure 2.34 to get line lengths.

the front view to the right (lines c).

4. Draw a 45° line as shown in figure 2.38. Project horizontally the lines from the top view to the 45° line, then vertically to the right-side view.
5. Darken the object lines. Erase construction lines if necessary, figure 2.39.
6. Add the dimension lines and dimensions, figure 2.40.

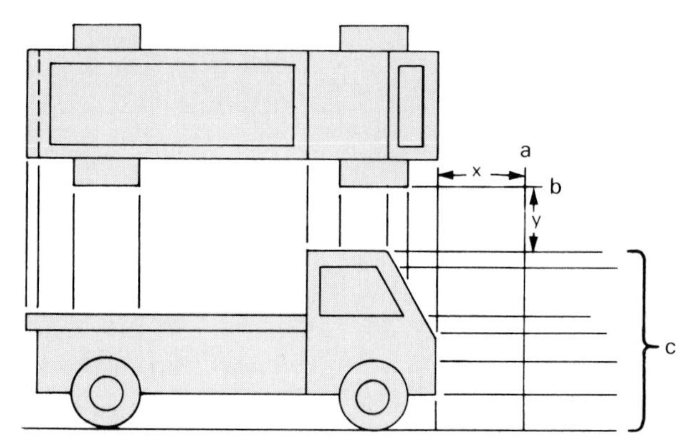

Figure 2.37. Getting ready to draw the right-side view. Lines a, b, and c have been drawn.

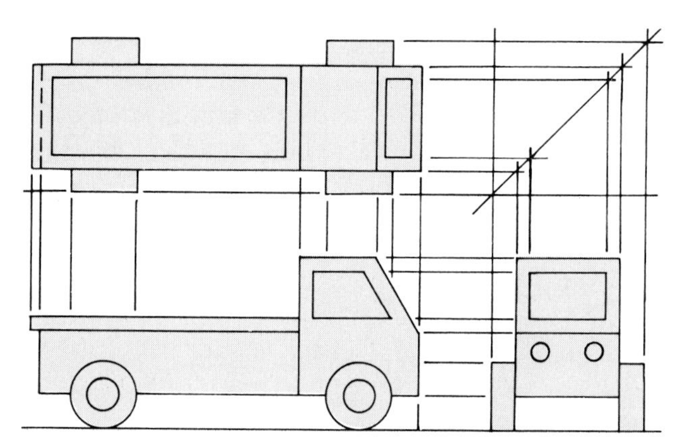

Figure 2.38. This drawing shows how to find the size of the front view.

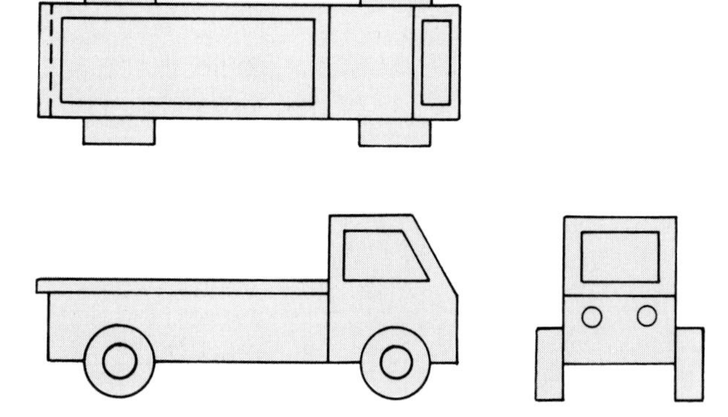

Figure 2.39. Orthographic drawing of the toy truck is almost completed.

Scale drawing

A **scale drawing** is one that is larger or smaller than the object by a fixed ratio. It is made when an object is either too large to fit

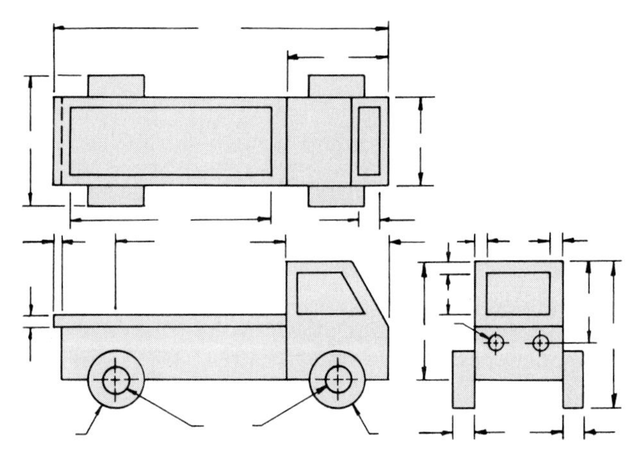

Figure 2.40. Final step adds dimension lines and dimensions.

onto the paper or too small to see the details. Examples of scaled drawings are: an architect's drawings of a building, a cartographer's (mapmaker's) map, and an electronic engineer's drawing of a printed circuit.

If you wanted to draw a full-size front view of the sailboat in figure 2.41, you would need a piece of paper larger than the sailboat. Full-size is a scale of 1:1. Each in. (or mm) of the drawing paper represents 1 in. (or 1 mm) of the actual object.

In a drawing one-half full-size (a scale of 1:2), each in. (or mm) on the drawing paper represents 2 in. (or 2 mm) of the actual object. Thus, the actual object would be twice the size of the views on the drawing paper.

If an object to be drawn is very small, it may be necessary to prepare drawings to a scale larger than full-size. Such a scale is referred to as an enlarged scale. In a drawing that is twice full-size (i.e., on a scale of 2:1), each 2 in. (or 2 mm) on the drawing paper represents 1 in. (or 1 mm) of the actual object. The parts of the compass shown in figure 2.42 are drawn twice their actual size.

□ COMPUTER-AIDED DESIGN

In the past, many people worked at making drawings. They used the tools just described. Today, however, fewer drafters (people who make drawings) use drafting boards. Instead, they make drawings using computers.

This new form of drawing is called CAD. It stands for computer-aided design. A typical

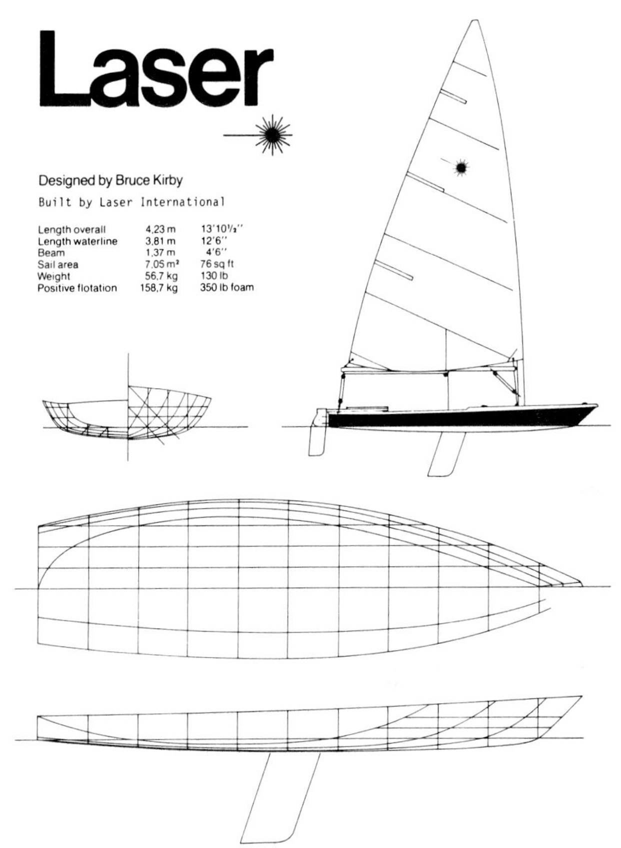

Laser
Designed by Bruce Kirby
Built by Laser International
Length overall 4,23 m 13'10½"
Length waterline 3,81 m 12'6"
Beam 1,37 m 4'6"
Sail area 7,05 m² 76 sq ft
Weight 56,7 kg 130 lb
Positive flotation 158,7 kg 350 lb foam

Figure 2.41. This is a scale drawing of a sailboat.

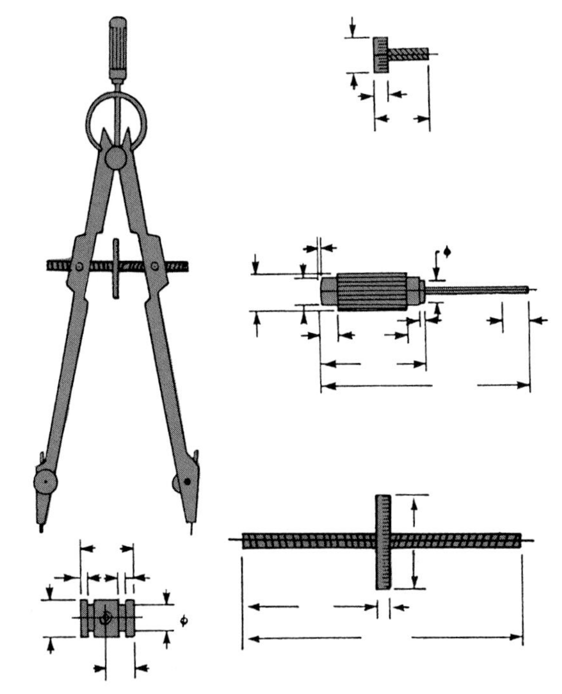

Figure 2.42. Which parts are scaled larger than their actual size?

CAD system has three types of devices or parts. These parts are:

1. Input device. It gives information or instructions to the computer.
2. Processor. This part carries out the instructions.
3. Output device. This part actually makes the drawing.

An input device, figure 2.43, sends a signal to the computer. Often the device is used to point to a particular part of a drawing. These kinds of input devices can be:

1. A hand-held light pen. The drafter points the pen directly at a spot on the computer screen. It sends a signal so the computer can find the spot on the screen.
2. Thumbwheels and joysticks. The drafter uses them to move a crosshair cursor around the screen. (The cursor is a ''light'' that can be moved around the screen.)
3. A tablet menu. This is a list of symbols and parts on the computer screen. The drafter can transfer the symbols and parts to a design with a stylus or mouse. Like the light pen, the stylus and mouse both send signals to control things on the computer screen.

Processing of inputs is the job of the CPU (central processing unit) of the computer. It carries out commands given to it by input devices. The designer can create the drawing details, add, delete (erase), change scale, rotate features, call up standard notes, title blocks, and other standard information. Components can be viewed in three dimensions, in color, and from various angles. Color graphics enable images to stand out and depth to become a reality.

Once the designer is satisfied with the image on the screen, he or she can produce a drawing on paper. This is called an output. It may be produced by a Xerox machine or by a plotter. This is a drafting machine that controls a drafting pen.

A computer-aided design system has many advantages over drawing by hand. CAD is an additional tool for the designer. It is like a template that helps you to draw more accurately and quickly. The computer works at high speed. The designer does not have to spend hours producing perfect line work and lettering. The CAD system makes them perfect the first and every time. CAD relieves the designer of repetitive (do over and over) tasks. This leaves more time for creative work. Still, CAD cannot replace the individual. It cannot think for the designer. Concepts and principles learned in drawing and design classes

TEC

Figure 2.43. The input devices on this CAD system are a keyboard and a digitizer.

continue to be important. Just as you need a knowledge of mathematical concepts when using a hand calculator, so the designer needs a knowledge of drawing and design when using a CAD system.

SUMMARY

When we tell people about our ideas we are communicating. All forms of communication use a code or symbols. Examples include hand signals, sounds, simple pictures, signs, and drawings.

Of these, the most important for the designer is drawing. Designers most commonly use three types of drawing: isometric and perspective sketches and orthographic projection.

Isometric sketches are used to record ideas quickly on paper. Perspective sketches show objects as you would see them. Neither isometric nor perspective sketches describe the shape exactly. Orthographic projection shows the exact shape of each surface of an object. Orthographic views can be drawn freehand on squared paper, or by using instruments. Most drawings today are made using computer-aided design (CAD).

Sometimes an object is too large to be drawn on paper, or too small to be seen clearly. Scale drawings are made when the object is either too large to fit onto the paper or too small to see all the details.

KEY TERMS

CAD
Communication
Construction line
Isometric sketching
Isometric paper
Object line
Orthographic projection
Perspective sketching
Scale drawing
Symbol
View

TEST YOUR KNOWLEDGE

Write your answers to these review questions on a separate sheet of paper.

1. List six ways in which people communicate ideas to one another.
2. Name the three types of drawing. Make small sketches to show each type.
3. A sketch is a drawing made ______.
   a. Using a straightedge.
   b. Quickly to communicate an idea.
   c. With the help of drawing instruments.
   d. In pencil only.
4. How many sides of a rectangular block are shown in an isometric drawing?
5. Which type of drawing provides the most realistic picture of an object?
   a. Isometric sketch.
   b. Plan view.
   c. Perspective sketch.
   d. Orthographic drawing.
6. A(n) ________ drawing describes the exact shape of each surface of an object.
7. The front view of an object is usually the view which ______.
   a. Shows its width.
   b. Best describes its shape.
   c. Shows its depth.
   d. Is the smallest.
8. List the most common instruments used to draw orthographic views.
9. What technique is used by a drafter or architect to draw a large object on a small piece of paper?
10. List the advantages of computer-aided design.

APPLY YOUR KNOWLEDGE

1. Create a symbol that can be used at a zoo to communicate the message: "Do not feed the animals!"
2. On isometric paper, draw a cube with sides measuring 12 squares. Remove pieces from the block to create a "swiss cheese" effect.
3. Design a logo (name or symbol) that you could use on your own personalized work sheets. Letters, geometric shapes, natural shapes, and simplified pictures are most appropriate for a logo.
4. Draw an isometric sketch, a perspective sketch, and an orthographic projection of a die.
5. Make an isometric sketch of either a tool you have used in the technology lab or an object used in the kitchen.
6. Select a day of the week. List all the ways that information is communicated to you during that day.

ECRITEK

Materials can be used to make individual decorative objects.

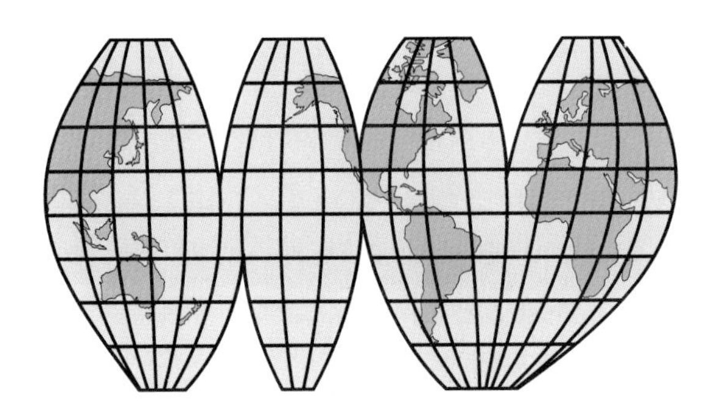

Chapter 3
Materials

OBJECTIVES

After reading this chapter you will be able to:
- O List the principal properties of materials.
- O Perform tests to determine the properties of selected materials.
- O Identify various materials and their characteristics.
- O Describe the major processes used to change raw materials into standard stock.
- O Select materials to meet the needs of a particular product.

Designing and building objects always involves the use of materials, figure 3.1. Concrete is used for walkways, cotton for clothing, fiberglass for insulation, copper for water pipes, and glass for bottles.

In fact, materials have been so important that some of the major periods of history have been named after materials. Perhaps you have heard of the Stone Age, the Bronze Age, and the Iron Age.

If we were to name the current period or "age" by the name of a material, it might be the "Age of Composites." **Composite** materials are combinations of different materials. Water skis used to be made of laminated wood. (Thin strips of wood were glued together.) Today they are made of a foam core (center) wrapped with layers of fiberglass and graphite reinforcement.

Designers need to know about the properties of many different materials. They use this knowledge to choose the most appropriate material for the object being designed and

A

B

D&S ADVERTISING, INC.

Figure 3.1. Composite materials are combinations of materials. A—Concrete is a mixture of stone, gravel, and cement. B—Plastic cores are used to support the copper windings of these coils.

built. Faced with a choice, designers must first learn how the product must function. Examples of questions they should ask are:

- Does product have to withstand heat?
- Is color important?
- Should it be heavy or light?
- How strong does it have to be?
- Must it withstand bad weather?
- Does it have to conduct electrical current?

These types of questions will narrow the choice of materials. The next step is to determine whether the material is available and affordable. Equally important are having the tools and knowledge to work the material.

STRENGTH	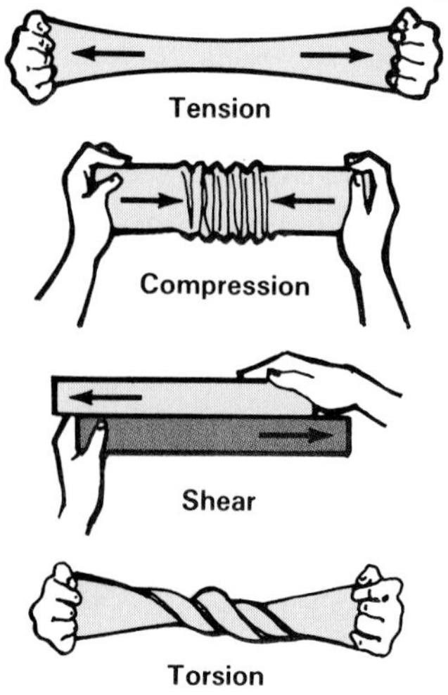 **Tension** **Compression** **Shear** **Torsion**	Strength is a material's ability to withstand a mechanical force. Tension is a pulling force. A material that resists being pulled apart has tensile strength. An elastic band holding a package together is in tension. Compression is a squeezing force. A material that resists being crushed has compressive strength. The concrete pillars for a bridge are in compression due to the mass of the materials and the traffic on the bridge. Shear is a sliding and separating force. A material that resists separation has shear strength. Torsion is a twisting force. A material that resists torsion has torsional strength. A screwdriver blade must resist torque when force is being applied to the screws.
ELASTICITY	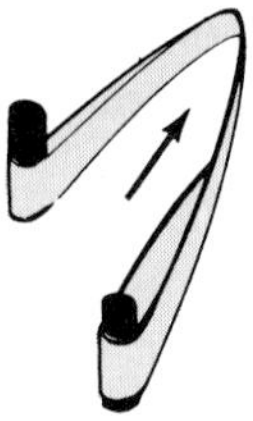	Elasticity is the ability to stretch or flex but return to an original size or shape. A material that resists elasticity has stiffness. A rubber band is elastic. A piece of glass has high stiffness.
PLASTICITY	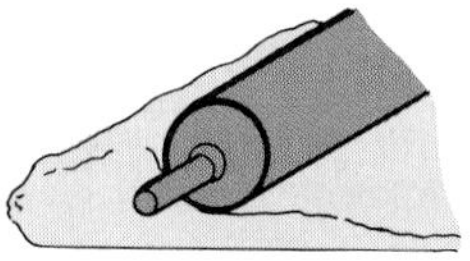	Plasticity is the ability to flow into a new shape under pressure and to remain in that shape when the force is removed. Plasticity can be measured in two ways: 1. Ductility is a material's ability to be pulled out under tension. Chewing gum is very ductile. 2. Malleability is a material's ability to be pushed (compressed) into shape. Potter's clay is very malleable. The opposite of plasticity is brittleness. Glass is very brittle.
HARDNESS		Hardness is the ability to resist cuts, scratches, and dents. Harder materials will wear less under use. Cutting tools such as knives, scissors, and drills should be hard. Diamonds are the hardest of all materials.
TOUGHNESS	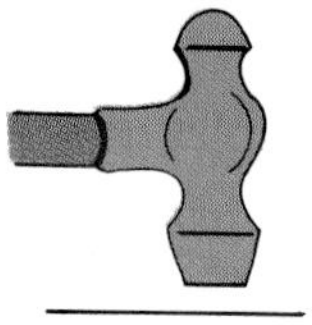	Toughness is the ability to resist breaking. A hammer head should be tough so as not to shatter when it strikes other materials.
FATIGUE	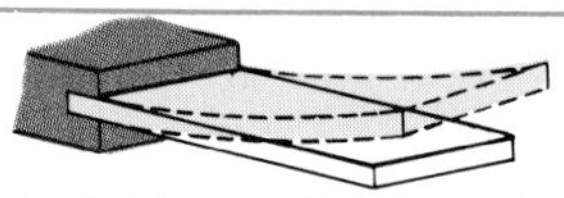	Fatigue is the ability to resist constant flexing or bending. A springboard must have high fatigue strength.

Figure 3.2. A description of mechanical properties. These are the most common ones.

■ PROPERTIES OF MATERIALS

Designers and engineers judge materials by their properties. These tell how the material can be expected to perform. Properties of materials can be grouped as:

- Physical.
- Mechanical.
- Thermal.
- Chemical.
- Optical.
- Acoustical.
- Electrical.
- Magnetic.

Physical properties

Physical properties give a material its size, density, porosity, and surface texture. You can describe any material or product by physical properties. Consider an eraser, for example. It may be 2 1/2 in. long, 1 in. wide, and 3/8 in. thick (64 x 25 x 10 mm). It is not very dense. Therefore, its mass is small. Its surface is smoother than a pine board. However, it is not as smooth as glass.

Mechanical properties

Mechanical properties are the ability of a material to withstand mechanical forces. An elastic band will stretch and return to its original shape. A diving board will spring back. The head of a hammer withstands sharp blows.

The common mechanical properties, as shown in figure 3.2, are:

- Strength.
- Elasticity.
- Plasticity.
- Hardness.
- Toughness.
- Fatigue.

Thermal properties

Thermal properties control how a material reacts to heat or cold. Materials will generally expand when heated and shrink when cooled. They will also conduct heat.

Sometimes the expansion of metals causes problems. On a hot day, railroad tracks may expand and buckle. However, the **thermal expansion** (expansion caused by heat) of metals can be useful. The flashing lights on a Christmas tree are an example. A strip of brass is joined to a strip of invar. (Invar is an alloy or mixture of iron and nickel.) The two are clamped at one end. The brass expands more than the invar. Thus, a change in temperature will cause the bimetal strip to bend. This movement can be used to open and close an electrical circuit. When the bimetal strip is cool the lights are on. When the strip heats up, it bends, figure 3.3. The lights go off.

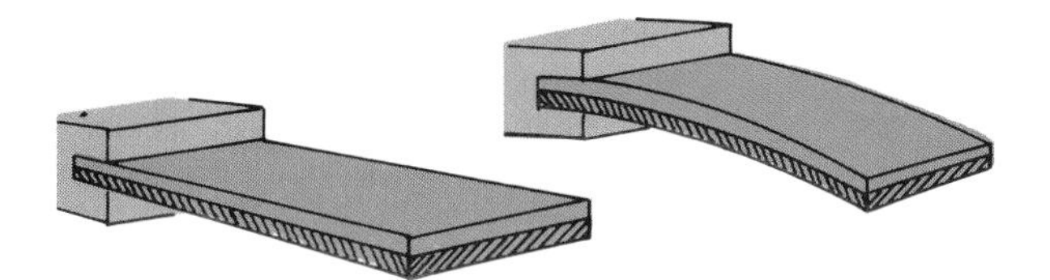

Figure 3.3 A bimetal strip can be used as a switch. Thermal expansion makes it work.

Thermal conductivity is a measure of how easily heat flows through a material. All metals conduct heat. Some do it better than others. Copper is a good conductor of heat. The copper bottom of a frying pan quickly conducts heat from the stove element to the pan. The metal pan then conducts this heat to the food.

Thermal insulators are materials that do not conduct heat well. Nonmetallic materials are generally thermal insulators. Plastic and wood handles on saucepans prevent heat being conducted (moved) from the hot metal to your hand. A cooler used for camping may have a casing filled with polyurethane (plastic) foam to keep out heat. Fiberglass batts are used to insulate walls and ceilings in order to reduce heat losses in a home, figure 3.4.

ECRITEK

Figure 3.4. Insulating materials slow down heat trying to move through walls of homes.

Chemical properties

A material's chemical properties affect how it reacts to its surroundings. Steel rusts. Glass becomes pitted. Plastics become etched and brittle. These are all the result of a chemical reaction. The material is changed by its environment. The reaction is called **corrosion**.

Probably the most familiar example of corrosion is the rust on a car body. What is corrosion? When a material contacts both air and water there is a chemical change:

Iron + oxygen + water = iron oxide (rust).

Sometimes the water contains dissolved chemicals, such as the salt used on roads to melt snow or the salt present in sea spray. Then rusting occurs much faster. Since water and air cause certain metals to rust, we can prevent rusting by covering the metal with paint.

For centuries, guns and tools have been coated with oil and grease. Paint, varnish, and enamel have been widely used to protect ships, trains, cars, and bridges. Nails and heating ducts are galvanized (coated with zinc). Food cans are plated (coated) with tin. Many decorative objects are electroplated. (A coating of nickel, chromium, copper, silver, or gold is applied to their surface.)

Optical properties

Optical properties are a material's reaction to light. Materials react to light in several important ways. One has to do with how well they transmit light that strikes them. Some materials cannot transmit light at all. When a material stops light, we say it is "opaque." A roller blind in your bedroom should be made of an opaque material. Translucent materials, waxed paper and stained glass, for example, allow some light to pass through. However, you cannot see clear images through them. Other materials allow all light to pass through. These materials are transparent. Clear glass windows are an example.

The second optical property of a material is color. The color of a material affects its ability to absorb or reflect light. (Light is the visible part of the sun's energy.) Light, as well as other kinds of radiant energy, is reflected by shiny, smooth surfaces. It is absorbed by dark, dull surfaces. A car with black upholstery is far more uncomfortable on a hot day than one with a white interior.

The ability of a material to absorb heat can be useful. The pipes of a solar panel are painted black so the panel will absorb more heat from the sun.

Acoustical properties

Acoustical properties in a material control how it reacts to sound waves. (Sound waves are pressure waves that are carried by air, water, and other materials. They are what the ear "hears.") All sound energy is produced by vibrations. Sound energy will travel through some materials. For example, a piece of string tightly stretched between two tin cans will carry a voice message over a short distance. Materials used in most musical instruments also transmit and amplify (increase) sound.

The speed of sound in a material depends on the spacing of the **molecules** and how easily the molecules move. (Molecules are groups of atoms. **Atoms** are the smallest possible particles of matter.) Sound travels faster in aluminum than in pine. This is true only because the molecules in aluminum are closer together. They transmit sound energy more easily. See figure 3.5.

Materials vary in their ability to absorb sound. For example, acoustical tiles or heavy carpeting both absorb sound. The sound waves become trapped in the air pockets of the material. Hard materials such as the walls of a canyon reflect sound. Call out a name and it will bounce back at you as an echo.

Figure 3.5. Guitar strings vibrate. They set air inside the guitar sound box vibrating.

Electrical properties

Some materials will conduct electricity. Others will not. This is controlled by electrical properties. Materials that will carry an electric current are called **conductors.** Those that will not are called **insulators.**

Metals are good electrical conductors. Some are better than others. One of the best is copper. It is used in cables that supply electricity to lights, appliances, and machines.

Wires carrying electrical current must be insulated. They are covered with materials that are poor conductors. Insulators made from ceramics or plastics are used.

Between these two extremes of good conductors and good insulators are a third type of material. This material is called a **semiconductor.** It allows electricity to flow only under certain conditions. Silicon and germanium are two important semiconductors. They are used in the production of transistors. Semiconductor materials are also used in devices that detect heat or light. Two examples are fire alarms and light meters for cameras.

Magnetic properties

A **magnetic** material will be attracted to a magnet. The most common magnetic materials are iron, nickel, cobalt, and their alloys. Most other materials, such as wood, plastic, and glass, are nonmagnetic. How a material reacts to magnetism is known as the material's magnetic properties.

■ TYPES OF MATERIALS

Many materials exist in nature. These are called natural or **primary materials.** In this group are:

- Trees.
- Animal skins.
- Clay.
- Crude oil.
- Iron ore.

Very seldom are these materials used in their natural state. Usually they are changed so that they are more useful. Some examples:

- Trees are cut into boards or planks.
- Animal skins are treated to make leather.
- Clay is baked into pots.
- Crude oil is refined into gasoline.
- Iron ore is processed to produce steel.

Other materials are created by people. These are called synthetic. Examples include:

- Plastic.
- Glass.
- Cement.

For convenience, you can classify most solids under five groups. These are: woods, metals, plastics, ceramics, and composites.

Woods

There are two families of wood, **softwoods** and **hardwoods.** Softwood trees are coniferous, figure 3.6. Coniferous trees retain their needlelike leaves and are commonly called evergreen trees. Three examples of softwood trees are cedar, pine, and spruce. Hardwood trees have broad leaves which they usually lose in the fall, figure 3.7. They are also known as deciduous trees. Examples are birch, cherry, and maple.

NFB

Figure 3.6. A coniferous tree. The name means "cone bearing."

Figure 3.7. Hardwoods come from deciduous trees. Most such trees lose their leaves in the fall.

The terms softwood and hardwood refer to the botanical origins of woods, not their physical hardness. For example, balsa wood is botanically a hardwood, yet it is physically very soft.

Like all living things, trees need nutrients. They must have them to live and grow. Their roots absorb water and mineral salts. Carbon dioxide is absorbed through the leaves. A process known as photosynthesis uses energy from sunlight to convert the carbon dioxide, water, and minerals. They become changed into the valuable nutrients needed by the trees, figure 3.8.

If you look at the cross section of a tree trunk, figure 3.9, you will see the various parts of the tree. Bark, the outer layer, protects the tree. Beneath the bark is a very thin, single-cell layer called the cambium. Here new wood cells are created, adding to the sapwood. The center part of the trunk is formed of the older, dead wood. This is called the heartwood. Water and minerals move through the trunk by the medullary rays.

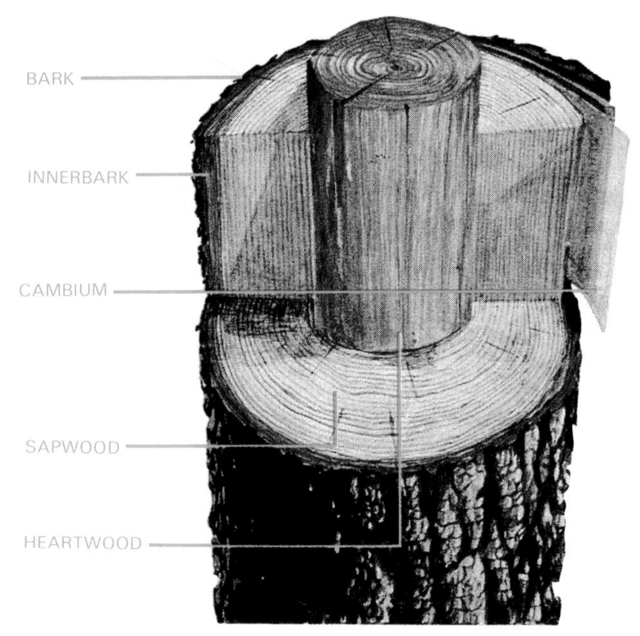

Figure 3.9. Cross section of a tree trunk shows parts of a tree. (The dark lines running outward from the center are medullary rays.)

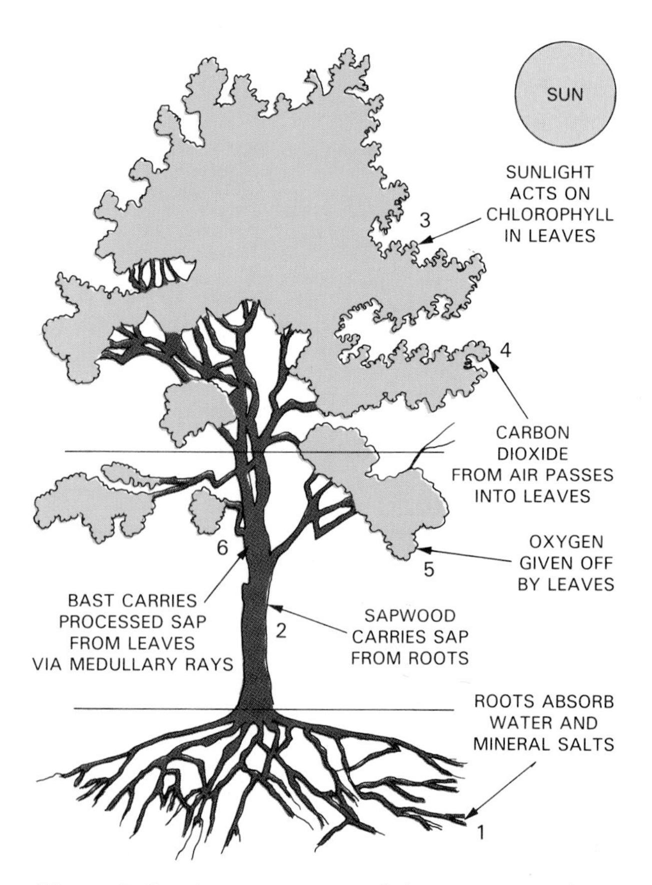

Figure 3.8. A tree uses sunlight, water, carbon dioxide, and minerals to "feed" itself.

The internal structure of wood is cellular, figure 3.10. The cells conduct sap, store food, and provide the support system for the tree. During the spring when there is plenty of water available, larger cells are produced. They form the spring growth. During the dryer summer season the cells are smaller, heavier, and thicker walled.

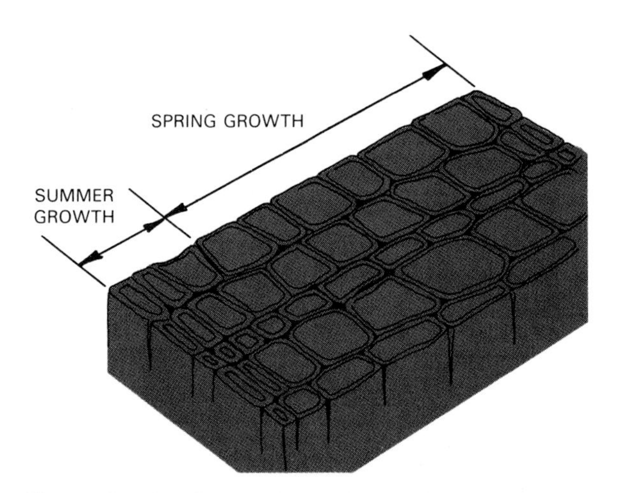

Figure 3.10. Cells of wood look like this under a microscope.

	SOFTWOOD			HARDWOOD		
Wood	SPRUCE	PINE	CEDAR	BIRCH	MAPLE	CHERRY
Resistance to decay	slight	moderate	great	slight	slight	great
Strength	medium	low	extremely low	extremely high	extremely high	high
Color	white	white to creamy yellow	reddish-brown	light brown	light brown	golden brown
Uses	• building construction • floorboards • packing cases • newspaper • pit props	• indoor joinery • matchsticks • telegraph poles	• building construction • paneling	• plywood • chairs • furniture	• pool cues • furniture	• chairs • other furniture

Figure 3.11. These are the characteristics (how they look or act) and uses of common types of wood.

The characteristics and properties of typical softwoods and hardwoods are summarized in figure 3.11. These qualities are important if wood is to be put to best use.

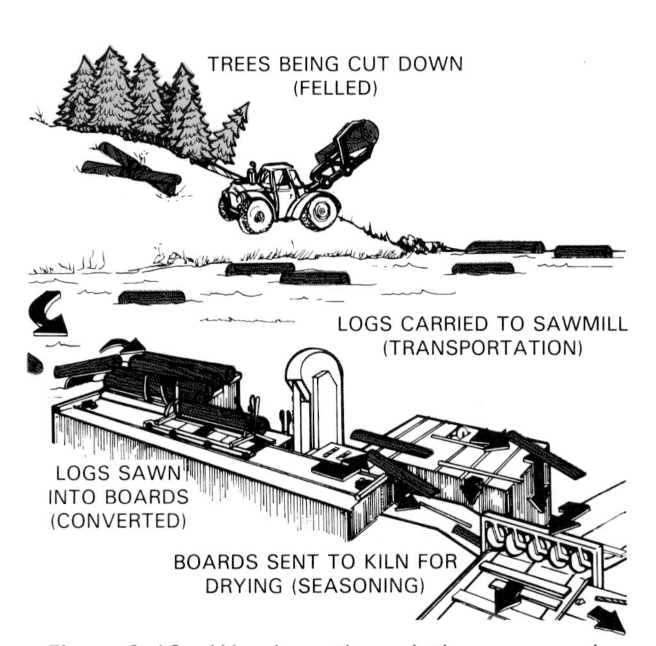

Figure 3.12. Woods go through these processing steps from standing trees to finished lumber.

To change the living tree into usable pieces of wood the tree must be cut down, carried to the sawmill, sawn into boards, and dried in kilns. These steps are shown in figure 3.12.

Wood is seasoned to reduce its water content to about the same as the air surrounding it. Traditionally, this has been done by stacking the wood outside. This allows the moisture to evaporate. The process usually takes a year or more. Nowadays, to speed up the process, the wood is dried in a special oven called a kiln. Even kiln dried wood will still expand or contract as the moisture content in the air changes with the seasons.

Drying out causes the wood to shrink unevenly. This leads to warping as shown in figure 3.13. Expansion, contraction, and warping are major problems. They are most troublesome when large pieces of solid wood are needed to make, for example, a tabletop. A way to overcome these problems is to join narrow boards edge to edge, figure 3.14. However, this solution is time-consuming and costly.

Today, the usual solution is to use manufactured boards. The most common are plywood,

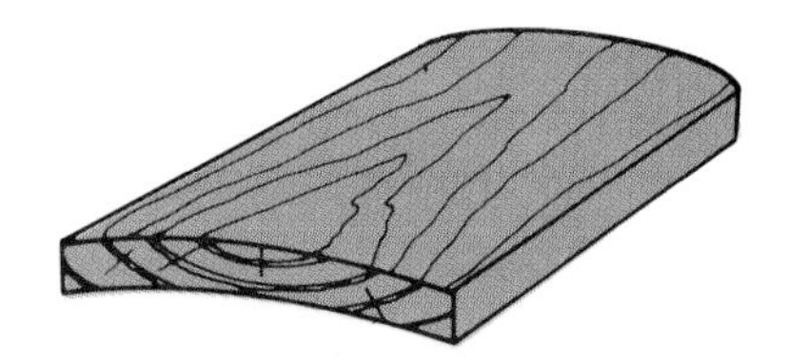

Figure 3.13. Uneven expansion and contraction causes boards to warp.

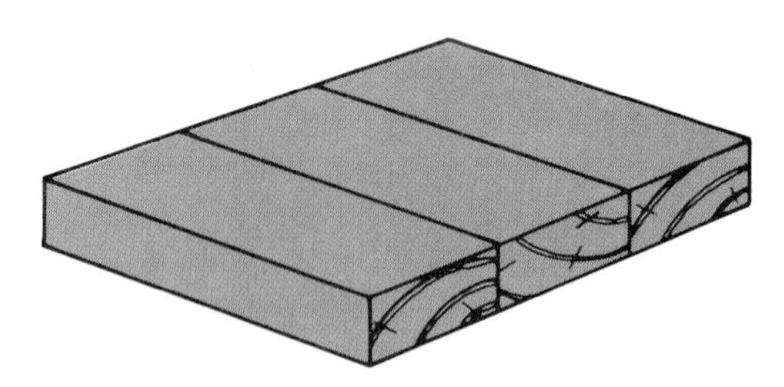

Figure 3.14. Woodworkers join narrow boards edge to edge to reduce warpage.

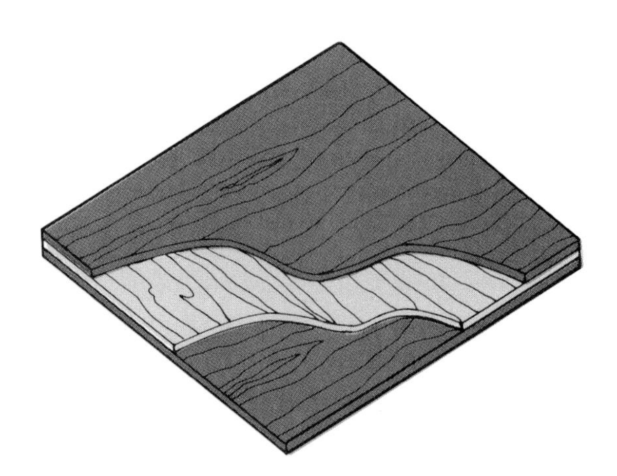

Figure 3.15. Plywood is always made with an odd number of plies (layers). Can you guess why?

blockboard, particleboard, and hardboard.

Plywood. Plywood is the strongest manufactured board and is usually the most expensive. It consists of an odd number of veneers (thin sheets of wood) glued together so that the grain in one is at right angles to the grain in the layers above and below it. See figure 3.15. Gluing the veneers in this way prevents the wood from twisting and warping.

The veneers are cut from a log. The log is mounted on a huge lathe-like machine, figure 3.16. A large blade cuts the wood, unrolling it like a reel of paper.

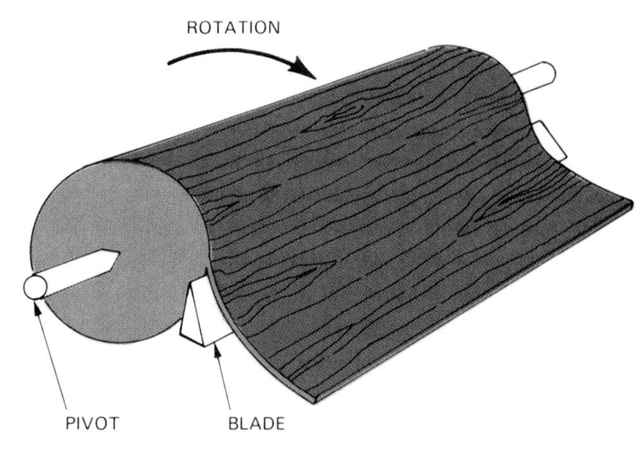

Figure 3.16. Rotary cut veneer is "peeled" from a log with a sharp blade.

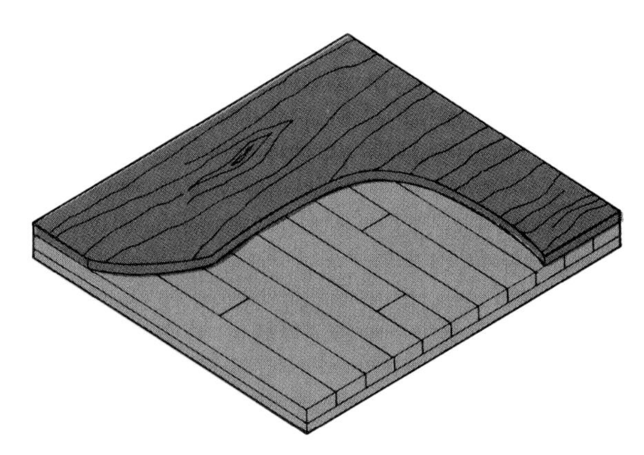

Figure 3.17. Blockboard always has a core (center) of solid pieces of softwood.

Blockboard. Blockboard consists of a core of softwood strips faced with a veneer on each side, figure 3.17. The grain of the face veneer is at right angles to the direction of the core strips. Blockboard is strong and durable, but expensive.

Particleboard. Particleboard consists of a core of wood chips. They are bonded together with an adhesive and pressed into a flat sheet. It can be left in this form or faced with a more expensive veneer, such as mahogany or teak. Particleboard is rather brittle, difficult to join, but cheap. It is used in the manufacture of less expensive furniture, figure 3.18.

Hardboard. Hardboard is made from wood fiber. These fibers are treated with chemicals. Then they are reformed into sheets using heat and pressure. Hardboard is usually smooth on one side and textured on the other.

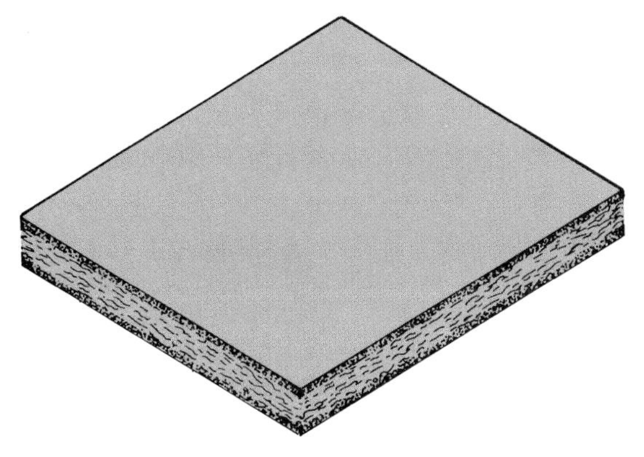

Figure 3.18. Particleboard is made from wood chips. They are glued and then pressed into a sheet. Sometimes the sheet is faced with a veneer.

CHRISTOPHARO

Figure 3.19. Three kinds of hardboard. Sometimes a decorative face is fixed to one side.

Hardboard can be tempered. It is made by impregnating standard hardboard with resin and heat-curing. Tempering improves water resistance, hardness, and strength. Hardboard is made in thin sheets and is cheap. It is used to cover hidden parts of cheap furniture or for drawer bottoms. Figure 3.19 shows several types.

Metals

Metals are inorganic materials (never were living). There are two families of metals: **ferrous** and **nonferrous.** The word ferrous is from the Latin word ferrum. It means "iron." Thus, any metal or alloy that contains iron is a ferrous metal. Metals or alloys that do not have iron as their basic component are called nonferrous metals.

When magnified, the internal structure of metals can be seen to be crystalline (made up of small crystals). The crystals are made of atoms arranged in boxlike shapes. Figure 3-20 shows how the atoms might be arranged.

In the body-centered cubic structure, the atoms arrange themselves into a cubic structure with an extra atom at its center. Chromium, molybdenum, tungsten, and iron at room temperature have this structure.

In the face-centered cubic structure, figure 3.21, each face of the cube has an additional atom at its center. Copper, silver, gold, aluminum, nickel, and lead have this structure.

Crystals, figure 3.22, are the basic units of metals. As molten metal cools from the liquid

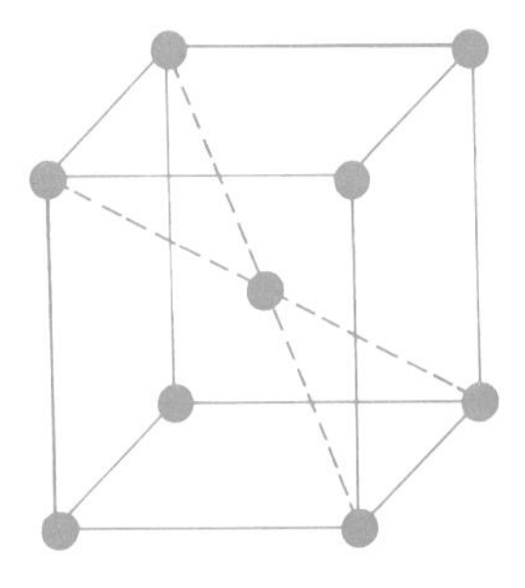

Figure 3.20. Body-centered atoms. This arrangement is found in ferrous (iron) metal. (The dots represent atoms.)

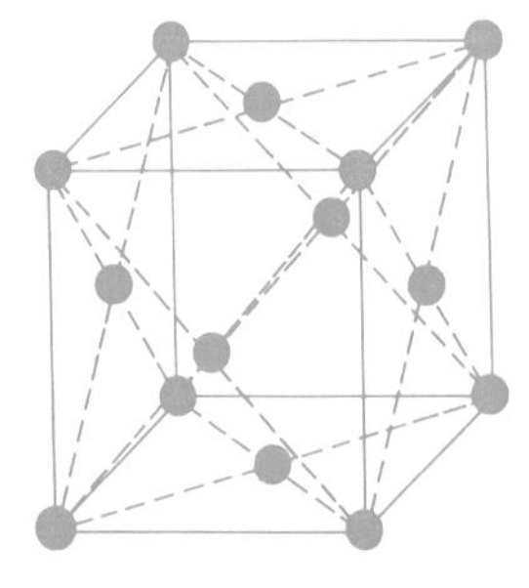

Figure 3.21. This is the structure of the atoms of nonferrous metals. Each face of the structure has an atom in its center.

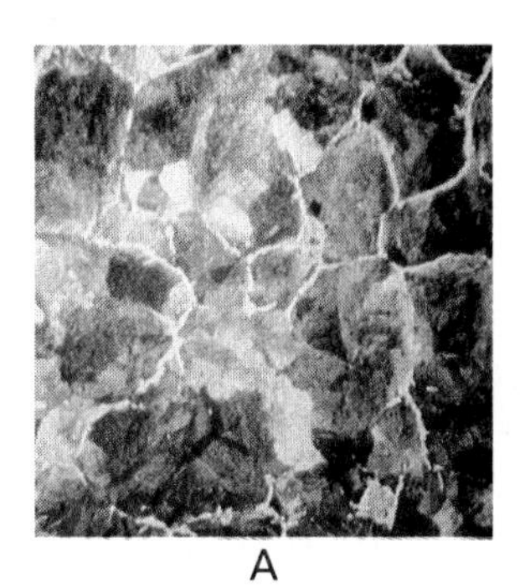

A

B

ALCAN

Figure 3.22. Photographs of magnified samples of metals. A—Cast iron. B—Aluminum.

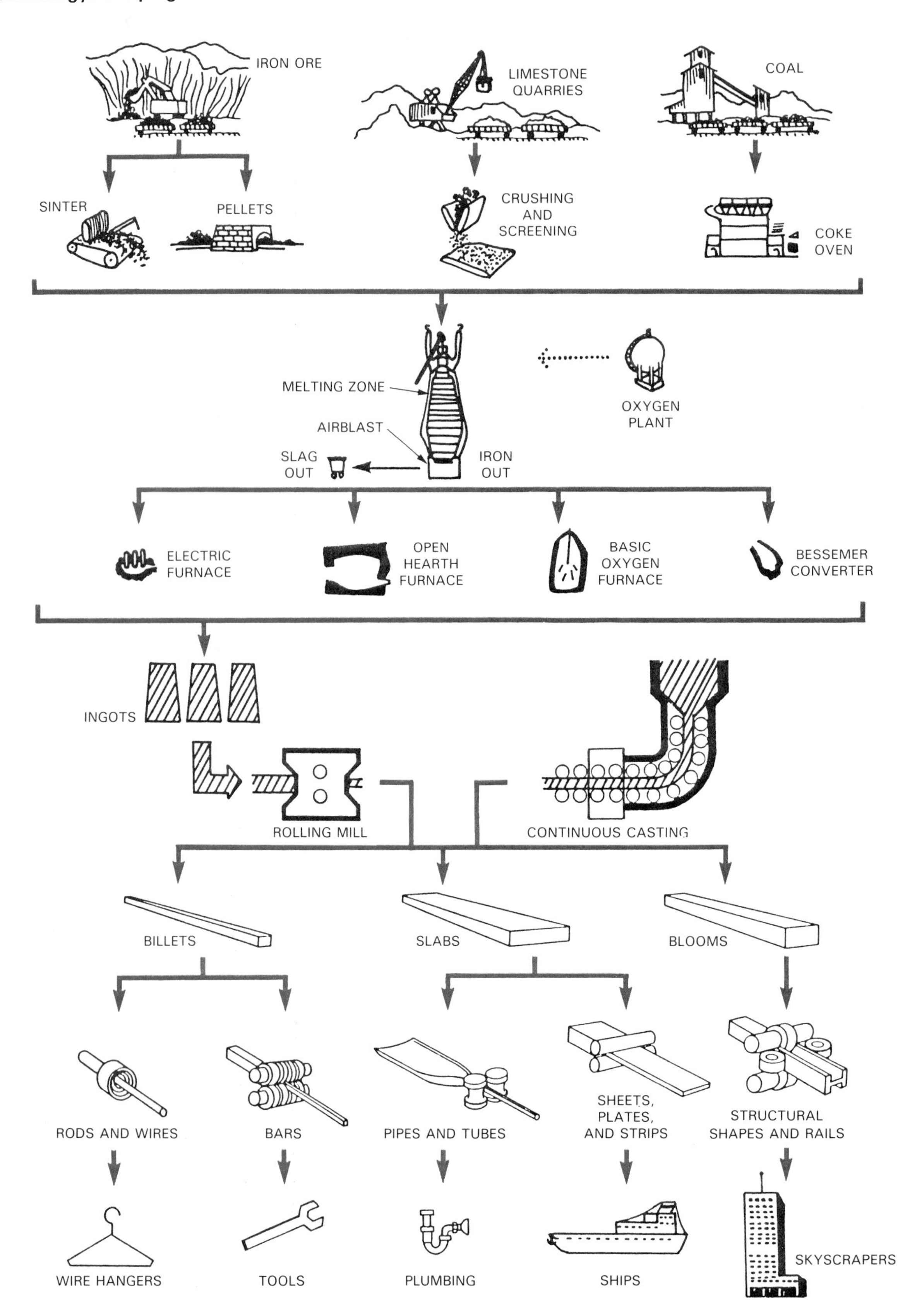

Figure 3.23. Processing iron ore into steel requires many steps.

state, atoms bond themselves together permanently, forming crystals. The crystals pack themselves together like the pieces of a jigsaw puzzle.

All crude (unprocessed) metals are found buried in the ground. Most are in the form of ore. This means that they are mixed with rock and other impurities. The ore is extracted from the ground by mining or quarrying. It is then crushed. The waste earth and rock is removed. The remaining ore is sintered (formed by heat) into pellets.

Iron ore, limestone, and coke (made from coal) are required for making iron. The iron ore coke, and limestone are dumped into the top of a blast furnace. The mixture is heated to 2912 °F (1600 °C). The burning coke and very hot air melt the iron ore and limestone. The limestone mixes with the ashes and waste rock. It forms a waste called slag. The molten iron sinks to the bottom of the furnace. The iron and slag are tapped (drawn) off separately. The iron is poured into large containers. It is ready to become steel. Figure 3.23 shows the entire process.

Metals are rarely used in their pure state. They are often too soft and ductile (bend too easily). Instead, they are often alloyed (mixed

	METAL	IMPORTANT CONTENT	MELTING TEMP.	RESISTANCE TO CORROSION	CHARACTERISTICS	COLOR	USES
FERROUS	Cast iron	93% iron 3% carbon	2200 °F (1204 °C)	poor	-hard -brittle -heavy	dark-gray	-bodies of machine tools -engine blocks -bathtubs -vises -pans
FERROUS	Mild steel	99% iron 0.25% carbon	2500 °F (1371 °C)	very poor	-stronger and less brittle than iron -can be easily joined by welding	gray	-girders in bridges -tubes in bicycle frames -nuts and bolts -car bodies
FERROUS	High-Carbon (tool steel)	0.60-1.30% carbon	2500 °F (1371 °C)	very poor	-not easy to machine, weld, or forge	gray	-cutting tools -drill bits -self-tapping screws -wrenches -railroad rails
NONFERROUS	Aluminum	base metal	1220 °F (660 °C)	high	-light, soft, and malleable -conducts heat and electrcity -very difficult to solder or weld	gray	-cooking utensils -foil -siding -window frames -aircraft
NONFERROUS	Tin	base metal	450 °F (232 °C)	excellent	-soft -nontoxic -shiny -most often used as an alloying agent	silver	-rolled foil -tubes and pipes -tinplate -ingredient of solder -galvanizing
NONFERROUS	Copper	base metal	1980 °F (1083 °C)	high	-tough and malleable -good conductor of heat and electricity -expensive -easily joined by soldering	reddish-brown	-electrical wiring -cables -water pipes

Figure 3.24. Characteristics and uses of ferrous and nonferrous metals.

Alloy	Major Components	Uses
Brass	Copper and zinc	Electrical fittings, locks, hinges
Bronze	Copper and tin	Bells, castings, sculptures
Pewter	Tin and lead	Tableware, tankards
Solder	Tin and lead	General solder uses a 50-50 mixture. Electrical solder uses a 60-40 mixture.
Stainless steel	Steel and chromium	Cutlery, sinks, surgical instruments

Figure 3.25. Alloys are mixtures of two or more metals. Alloying makes a metal more useful. It combines good characteristics of the different metals used.

together), see figure 3.25. Alloying increases strength or hardness. It may also inhibit rust, change color, and electrical or thermal conductivity. For example, mixing 10 percent aluminum with copper produces an **alloy** with approximately three times the strength of pure copper.

Plastics

Synthetic plastics include two basic categories: **thermoplastics** and **thermosets.** To understand how a plastic is made, you must first remember that all matter is composed of minute particles called atoms. An atom is the smallest part of an element. For example, carbon, hydrogen, and oxygen are all elements. Atoms can combine with one another to form molecules. These molecules form the fundamental building blocks of a plastic material. The molecules are joined together in chains, figure 3.26.

The scientific name for plastic is **polymer.** Polymeric materials are basically materials that contain many parts. ''Poly'' means many and ''mer'' stands for monomer or unit. A polymer is a chain-like molecule made up of smaller molecular units.

Thermoplastics. The chains in a thermoplastic material, figure 3.27, remain separated from one another even after heating. The material can be repeatedly softened by heating and hardened by cooling. Any time heat is applied the material becomes soft. If pressure is applied the material will take on new shape. The thermoplastic material may be formed into a new shape as many times as desired.

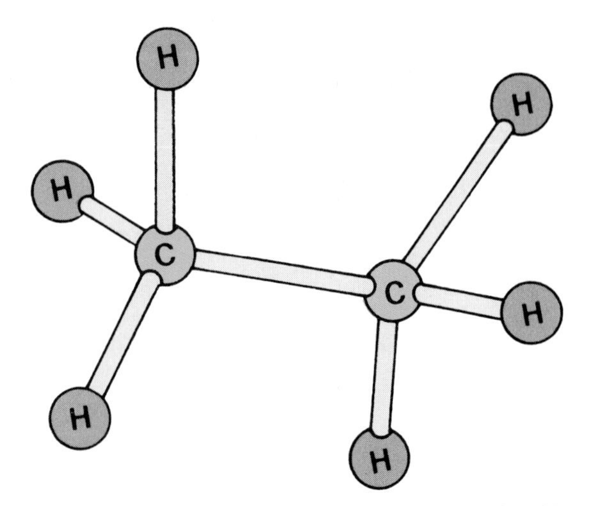

Figure 3.26. This is how a molecule of ethane looks. It has two atoms of carbon and six of hydrogen.

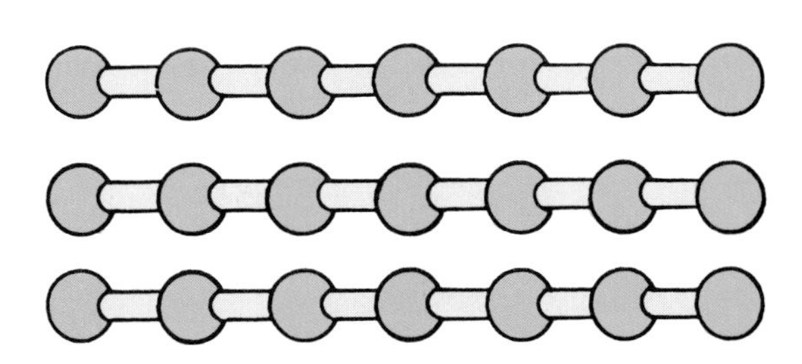

Figure 3.27. The molecular structure of a thermoplastic material. The ''chains'' of atoms have no links to one another.

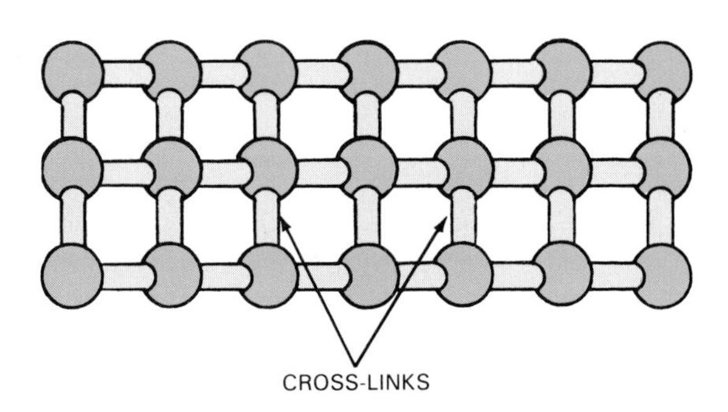

Figure 3.28. Molecular structure of a thermoset. Note the cross-links between chains.

Thermoplastics are not used to make objects that must resist high temperatures. Examples of thermoplastics include acrylic, polyethylene, polyvinyl chloride (PVC), nylon, and polystyrene.

Thermosets. Thermoset plastics, figure 3.28, can be heated only once. This is at the time they are made. Do you know how they got their name? "Thermo" means heat, and "set" means permanent. Heat causes the material to develop cross-links between the chains. The individual chains, once connected, form a rigid structure. The plastic is no longer affected by heat or pressure. Its shape has become permanent.

Articles made from thermosetting materials have good heat resistance. Examples of thermosets include polyester resins (often reinforced with glass fiber), phenolic resins, urea resins, melamine resins, epoxy, and polyurethane.

Plastics are synthetic, manufactured materials. The first plastics were made from natural substances such as animals, insects, and plants. Currently, most plastics are made from crude oil. Coal and gas are also used.

Today, there is no single material called plastic. Chemists can alter the "mix" each time to create plastics that look and behave very differently from each other, figure 3.29.

Plastics have many important properties. These include:

- Ability to be colored.
- Ease of molding.
- Flexibility or rigidity.
- Good electrical or thermal insulation.
- Light mass (weight).
- Low cost.
- Resistance to rot and corrosion.
- Strength.

The characteristics and uses of typical thermoplastics and thermosets are summarized in figure 3.30.

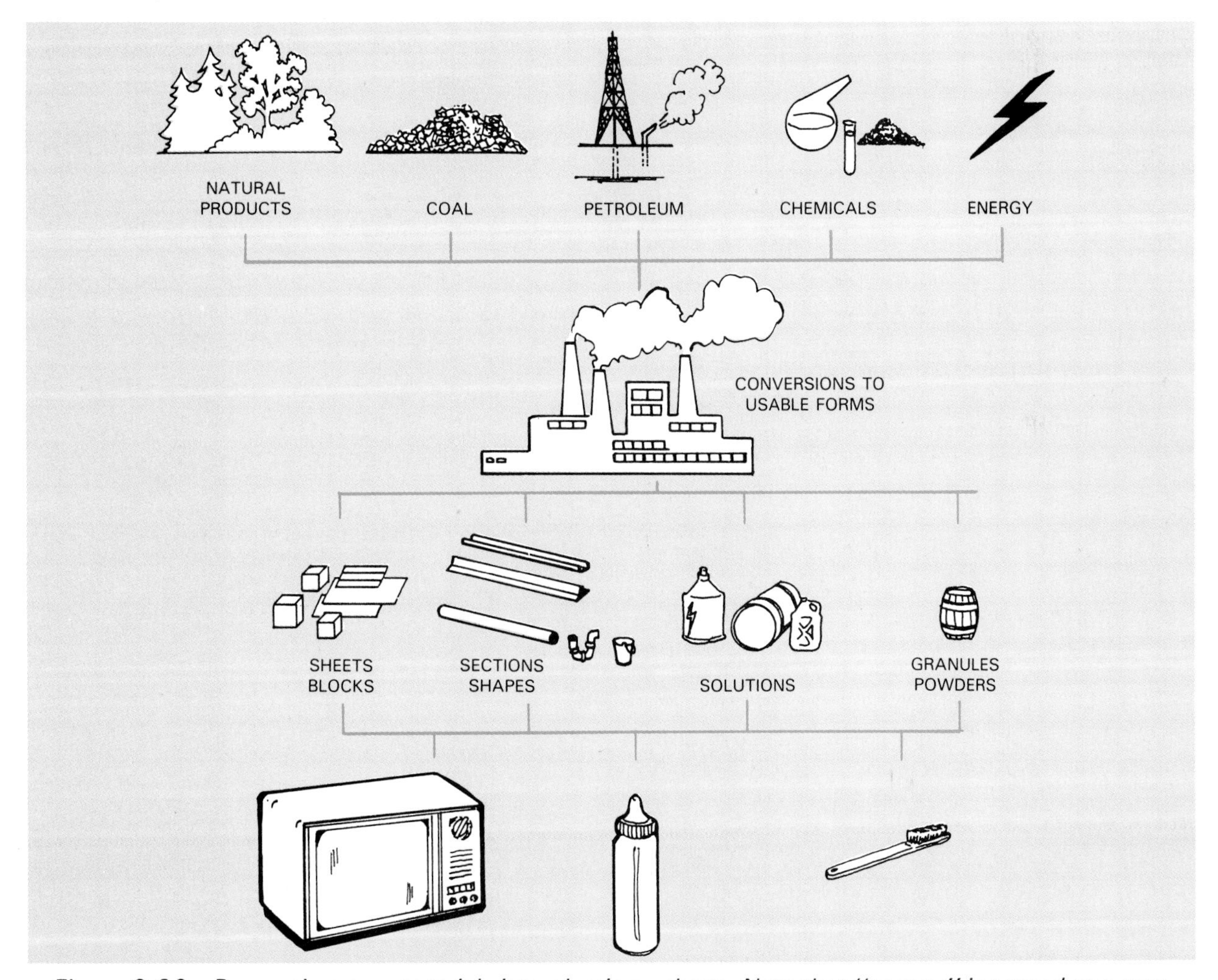

Figure 3.29. Processing raw materials into plastic products. Note that "energy" is named as a raw material. Why?

	PLASTIC	CHARACTERISTICS	USES
THERMOPLASTICS	Polyethylene (low density)	-fairly flexible -soft -cuts easily and smoothly -floats -''waxy'' feel -not self-extinguishing -transparent when thin -translucent when thicker -can be dyed various colors	-detergent squeeze bottles -plastic bags -electrical wire covering
	Polyvinyl chloride	-rigid or flexible -transparent -fairly easy to cut -smooth edges -sinks -self-extinguishing	-water pipes -records -raincoats -soft-drink bottles -hoses
	Polystyrene	-opaque -usually white -tends to crumble on cutting -very buoyant in water -burns readily, not self-extinguishing -very lightweight	-packing materials -insulation -ceiling tiles -disposable food containers
THERMOSETS	Polyester resin	-stiff -hard, solid feel -difficult to cut -brittle -burns readily	-repair kits -car bodies -boat hulls -garden furniture
	Urea formaldehyde	-opaque -usually light in color -stiff, hard, solid feel -sinks -flakes on cutting -burns with difficulty -good heat insulator	-light-colored domestic electric fittings (e.g., plug tops, adaptors, switch covers) -waterproof wood adhesives
	Melamine	-opaque -usually light in color -stiff -hard, solid feel -flakes on cutting -sinks -burns with difficulty -good heat insulator -resists staining	-tableware -surfaces for counters, tables, and cabinets -expensive electrical fittings

Figure 3.30. Note the characteristics and uses of plastics.

Ceramics

The word **ceramics** is from the Greek *Kermos* meaning ''burnt stuff.'' The important characteristics of ceramics are hardness, strength, imperviousness to (blocks the passage of) heat, resistance to chemical attack, and brittleness.

Today, the term ceramics covers a wide range of materials. Abrasives and cement are ceramics, as are window glass and porcelain enamels on bathroom fixtures.

Most ceramics are thermosetting materials. Once they have been processed and hardened, they cannot be made soft and pliable again.

The one exception is glass. It can be continuously reheated and reshaped. It is, therefore, thermoplastic in nature.

One of the great advantages of ceramics is the abundance of the raw materials used to make them: silicon and oxygen. Silicates (including sand, clay, and quartz) are the most common minerals on earth.

The two ceramic materials most familiar to you are glass and cement. Their manufacture is shown in figures 3.31 and 3.32.

Ceramics' characteristics, figure 3-33, are:

- Strong and resistant to attack by nearly all chemicals.
- Withstand high temperatures.
- Lack malleability (are stiff, brittle, and rigid).

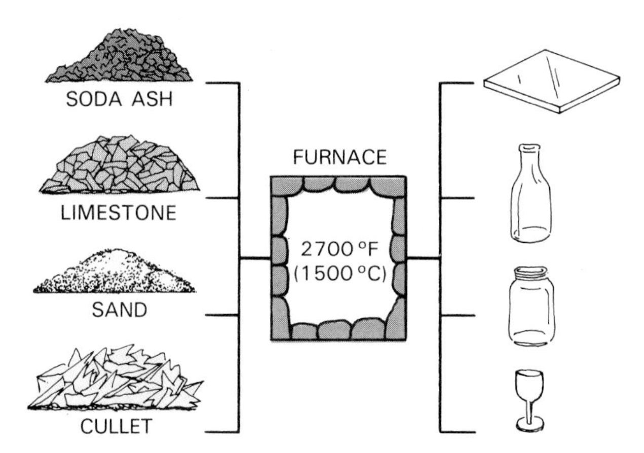

Figure 3.31. Turning raw material into glass.

- Very stable (not likely to change shape because of heat or weather).
- High melting point.
- Hardest of all engineering (solid) materials.
- Raw materials are available world-wide, and are consequently low in cost.
- Withstand outdoor weathering from the sun, moisture, environmental pollutants, and dramatic temperature changes.
- Poor electrical and thermal conductivity.
- Poor thermal shock resistance.

Composite materials

When two or more materials are combined, a new material, known as a **composite,** is formed. For instance, concrete is a composite. It combines cement, sand, and gravel. Fiberglass is a composite of glass fibers and plastic resin.

Combining materials produces a new material. This composite has improved mechanical or other properties. Each material in the composite keeps its own properties. Combining them also adds new properties.

There are three types of composites: layered, fiber, and particle. Each will be explained.

Layered composites consist of laminations like a sandwich. Thin layers of material are tightly bonded.

How much do you know about the tires that support you when you ride in a car or bus, or

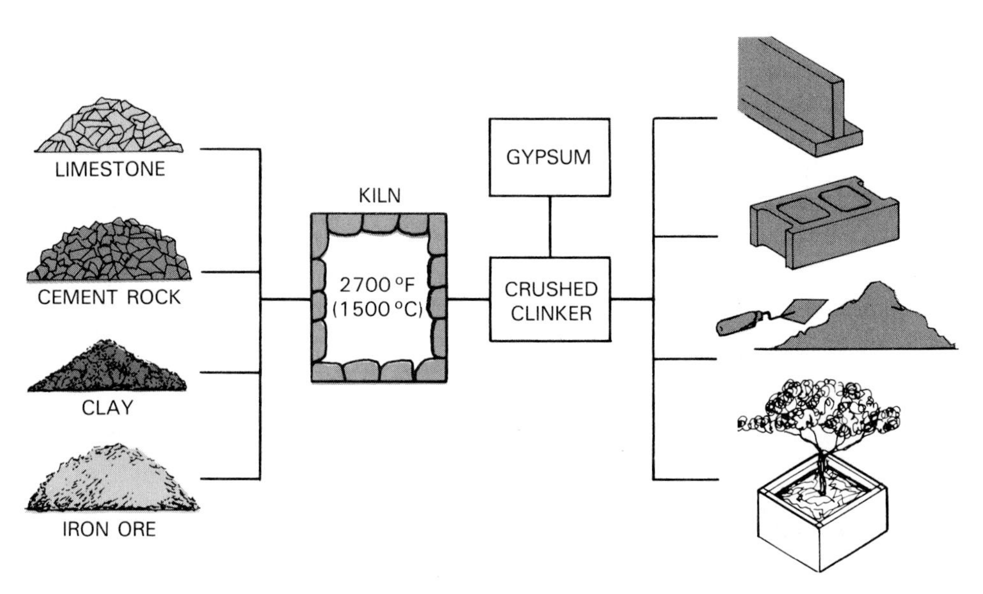

Figure 3.32. Cement is a common ceramic material. This is how it is made.

TYPE	CHARACTERISTICS	USES
Stone	-hardness varies from soft sandstone to hard granite -generally able to withstand effects of rain, spray, wind, frost, heat, and fire -color varies -surface appearance varies -resistant to corrosion -opaque	-building material -abrasives
Clay	-becomes plastic when mixed with water -has a bonding action on drying	-house bricks -tiles -earthenware -stoneware -china -porcelain
Refractories	-stability at high temperatures -can withstand pressure when hot -can withstand thermal shock -highly resistant to chemical attack	-lining of furnaces to produce glass, cement, metals, bricks -insulation on spark plugs -tiles on space shuttle -firebrick
Glass	-transparent -poor in tension -transmits visible light well -opaque to ultraviolet light	-fibers for insulation -fibers for filters -windows and doors -heat resistant cookware -light bulbs -drinking glasses -lenses
Cement	-correctly called Portland Cement -made from limestone (80%) and clay (20%) heated in kiln to form a clinker that is ground to a fine powder -very durable -can withstand frequent freezing and thawing -can withstand wide range of temperatures	-the ingredient of concrete that binds the crushed rock, gravel, and sand together

Figure 3.33. Ceramics have characteristics that make them useful for many products and building materials.

land in an airplane? Tires are also layered composites, figure 3.34.

Another common use of layered composites is in the construction of hollow core doors, figure 3.35. Sheets of plywood are glued to spacers of cardboard and solid wood.

The wings of some aircraft are made of layered composites. The center of the composite is an aluminum honeycomb. This honeycomb is covered with layers of fiberglass to form a sandwich, figure 3.36.

Fiber composites consist of short or long fibers of a material such as glass or carbon embedded in a matrix of another material such

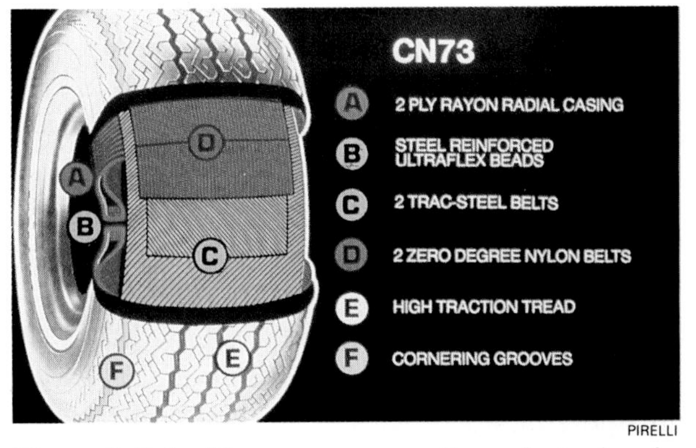

Figure 3.34. Tires are composites of steel, rubber, plastics, and fibers.

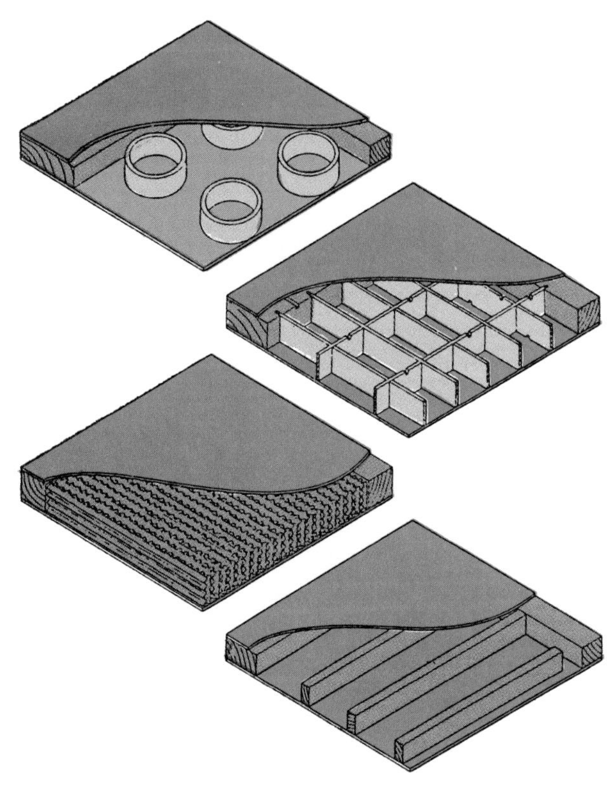

Figure 3.35. Sample sections of hollow core doors. Layers are made up of plywood, cardboard, and solid wood pieces.

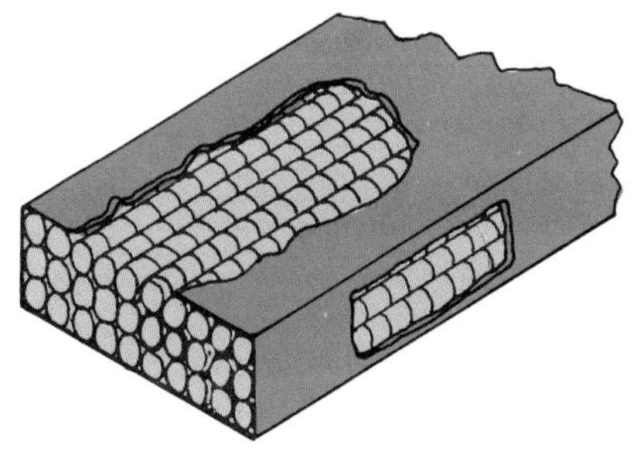

Figure 3.37. Fibers in a matrix. The matrix could be a liquid such as resin. It is poured over the fibers and left to harden.

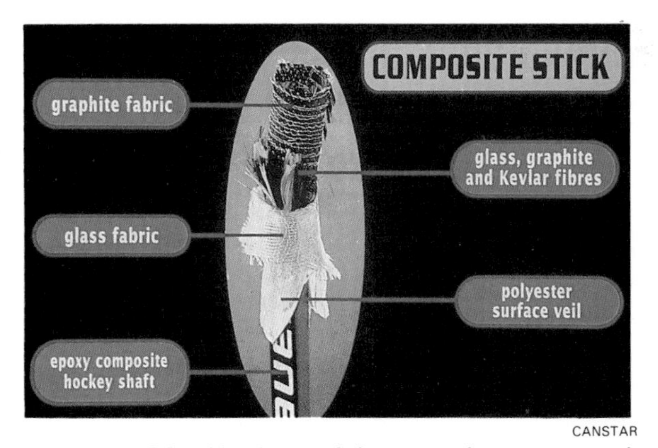

CANSTAR

Figure 3.38. Hockey sticks must have strength and flexibility.

CANADAIR

Figure 3.36. This airplane wing is also a composite. The aluminum center looks like a honeycomb. It is covered with sheets of fiberglass.

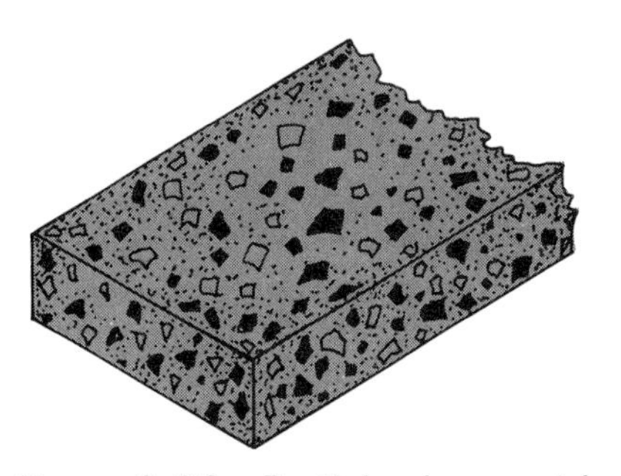

Figure 3.39. Particles in a matrix.

as resin, or other plastic, and metal, figure 3.37. One use of a fiber composite is shown in figure 3.38.

Particle composites, figure 3.39, consist of particles held in a matrix. The most common particle composites are concrete and particleboard, figure 3.40. Concrete is a mixture of gravel, sand, and a matrix of cement. Particleboard is a mixture of wood chips in a resin matrix.

A MORRISON KNUDSEN CORP.

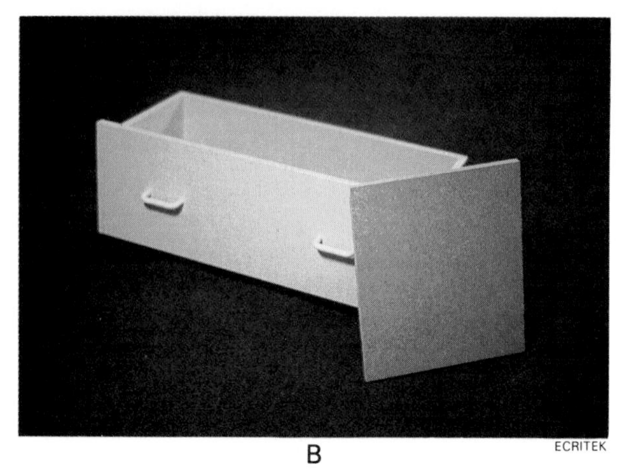
B ECRITEK

Figure 3.40. A—Concrete, a composite, is being poured into a form. B—This drawer is made of another composite, particleboard.

SUMMARY

Designers need to know about the properties of different materials. The properties of a material will determine its suitability for the object to be designed and built. Materials may be chosen for their physical, mechanical, thermal, chemical, optical, acoustical, electrical, or magnetic properties.

For convenience most materials can be classified into five groups. These are woods, metals, plastics, ceramics, and composites. Within each group there are many varieties. Woods may be subdivided into softwoods and hardwoods. Metals include ferrous and nonferrous. Plastics are either thermoplastics or thermosets.

Most ceramics are thermoset materials, except glass which is a thermoplastic. A composite is formed when two or more engineering materials are combined. Composites are increasingly important in the production of the objects we use today.

KEY TERMS

Alloys
Atoms
Ceramic
Composite
Compression
Conductors
Corrosion
Elasticity
Ferrous
Hardness
Hardwood
Insulators
Magnetic
Molecules
Nonferrous
Plasticity
Polymer
Primary material
Semiconductor
Shear
Softwood
Tension
Thermal expansion
Thermoplastics
Thermosets
Torsion
Toughness

TEST YOUR KNOWLEDGE

Write your answers to these review questions on a separate sheet of paper.

1. Why have some major periods throughout history been named after materials?
2. Name and define each of the seven properties of materials.
3. Set up a chart like the one below. Place each of the following materials in the appropriate column in the chart: polystyrene, cast iron, cedar, copper, glass, maple, aluminum, pine, cement, melamine, spruce, polyethylene, birch, porcelain, clay, steel.

Category of Material			
Woods	Metals	Plastics	Ceramics

4. List the characteristics of hardwood and softwood trees.
5. What happens to a tree once it has been transported to a sawmill?

6. Sketch the construction of the following manufactured boards: plywood, blockboard, and particleboard.
7. What is the difference between a ferrous and a nonferrous metal?
8. Is pure iron an alloy? Explain your answer.
9. Give three examples of both ferrous and nonferrous metals.
10. Copy the chart below and fill in the blank spaces.

Metal	Typical Use
Copper	
Aluminum	
High-carbon steel	
Cast iron	

11. The scientific name for plastic is ________.
12. Sketch the molecular chains of thermoplastics and thermosets.
13. Describe the fundamental difference between thermoplastics and thermosets in terms of the way their molecular chains are formed.
14. Plastics are generally manufactured from ______.
    a. Crude oil.
    b. Wood chips.
    c. Mineral ores.
    d. Animal by-products.
15. Copy the chart displayed in the next column. List four objects in your home that are made of plastic. In your opinion, what particular properties make plastic the most appropriate material for each of the objects you have chosen? Refer to figure 3.30 in your text.

Object	Important Characteristics
detergent squeeze bottle	flexible, transparent, lightweight
1.	
2.	
3.	
4.	

16. The two ceramic materials most familiar to you are ________ and ________.
17. State three advantages and three disadvantages of ceramic materials.
18. What is a composite material?
19. Using sketches and notes, describe the differences between layered, fiber, and particle composites.
20. List six objects in your home made using composite materials. For each object, state the type of composite and the materials used.

APPLY YOUR KNOWLEDGE

1. Choose three objects in your home that are made of different materials. State whether the materials used are appropriate. Explain your answer.
2. Describe the major processes used to change (a) a tree, and (b) iron ore, into standard stock.
3. State one advantage and one disadvantage of five different materials.
4. Collect pictures of objects that are made of (a) layered, (b) fiber, and (c) particle composites. Label the diagrams to show the materials used in each composite.
5. Choose one property of materials, e.g., conduction or elasticity. Collect samples of five different materials. Design and build an apparatus to test the materials for the property you have chosen.

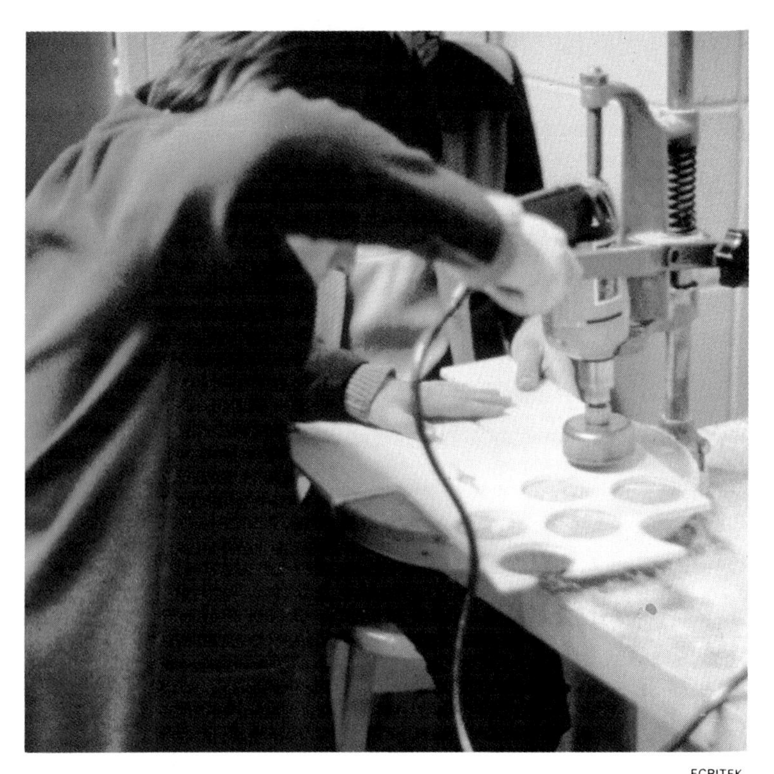

ECRITEK

Cutting disks from plywood is an example of processing materials.

ECRITEK

Accurate machining using a lathe produces parts that fit together precisely.

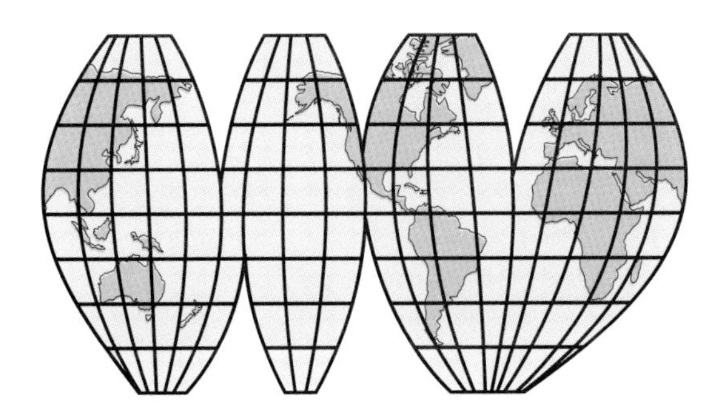

Chapter 4

Processing Materials

OBJECTIVES

After reading this chapter you will be able to:

- O Demonstrate responsible and safe work attitudes and habits.
- O Design and make simple objects using wood, metal, and plastic materials.
- O Identify the correct tools and processes to be used to shape a material.
- O Select the correct method for joining materials.
- O Select and apply an appropriate finish to a material.

■ SAFETY

As you work through your technology course, you will use various materials. You will learn to select the correct hand and machine tools, and practice safe techniques for each material. Tools can be dangerous unless you learn to use them correctly.

In addition, you will build models and prototypes. Each must be built with **safety** in mind. When testing models or prototypes, or when you are involved in other experimental work, use great care. You must ensure that you, your friends, and the equipment are not harmed. The following notes cover some of the safety rules you should observe as you work through your Activity Book.

Materials and processes

- Remove loose clothing (coats, jackets, and sweaters).
- Roll sleeves above the elbows.
- Remove all jewelry.
- Tie back long hair.
- Put on safety glasses before switching on a machine.
- Report all accidents, including minor cuts and scratches, to your teacher.
- Report unsafe conditions such as damaged or worn equipment.
- Place rags containing oil, gasoline, paint, or solvents, in approved metal containers.
- Use the right tool for the job, and use it correctly.
- Keep tools in proper condition and store in a safe place when not in use.
- Make all adjustments to a machine with the power off.
- Do not leave a machine until it has come to a complete stop.
- Avoid using dangerous materials such as lead or asbestos.

Structures

- Ensure that the structure will withstand the loads imposed on it.
- When testing a structure by loading it with weights, be prepared for the structure to collapse: protect yourself.
- When using a hot wire cutter to cut Styrofoam®, avoid noxious fumes by working in a well-ventilated area.

Machines

- Test all mechanisms. Move the parts by hand before connecting them to a power source.
- Check to ensure that all parts of a mechanism are correctly attached to the supporting structure.
- Test vehicles only when you have sufficient space to permit the vehicle to move without danger to anyone.

Energy

- Ensure that all parts are fully insulated against leakage of electric current.
- If possible, use a 1.5 V cell or a 9 V battery as a power source for projects.
- When a 120 V power source is used, the project must be correctly grounded.

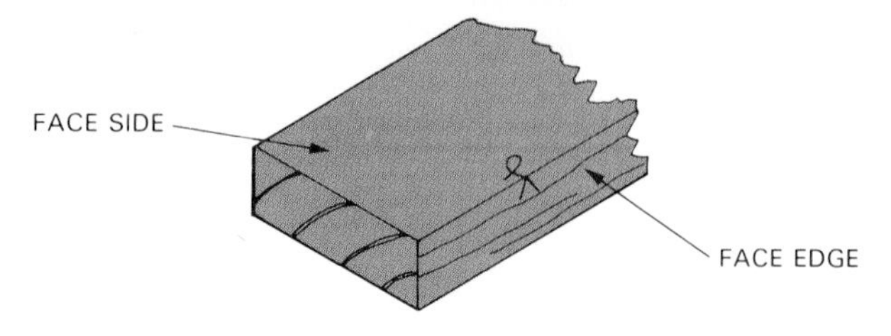

Figure 4.1. Mark lumber on both the face side and the face edge.

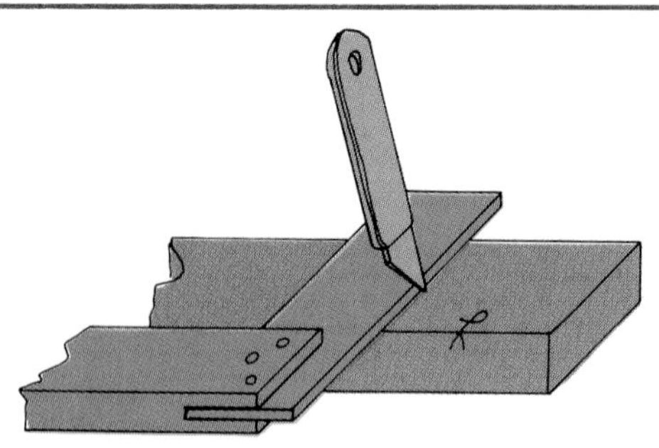

Marking wood to length
Use a try square.
Hold the handle firmly against the face edge of the wood.
Mark a line with a marking knife.
Always use the outside edge of the square.

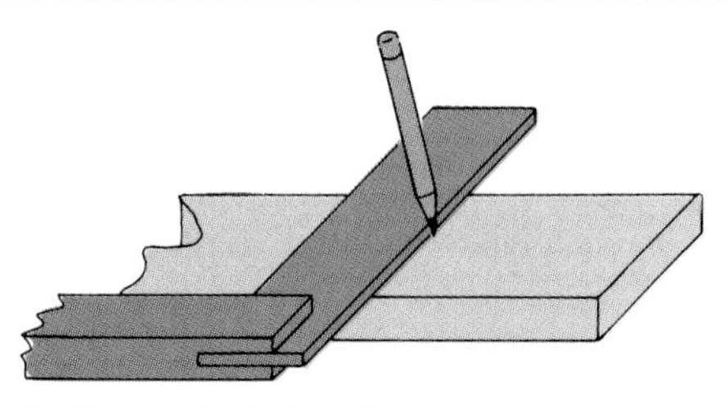

Marking metal to length
Coat the metal with marking blue.
Use an engineer's square.
Hold the handle firmly against the straight edge of the metal.
Mark a line using a scriber.
Always work on the outside of the square.

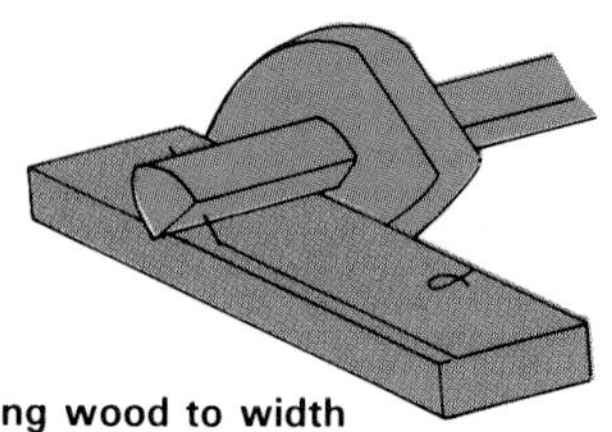

Marking wood to width
Use a marking gauge.
Press the stock of the gauge firmly against the wood.
Tilt the gauge in the direction you will push it.
Practice on scrap wood (the gauge is a difficult tool to use).

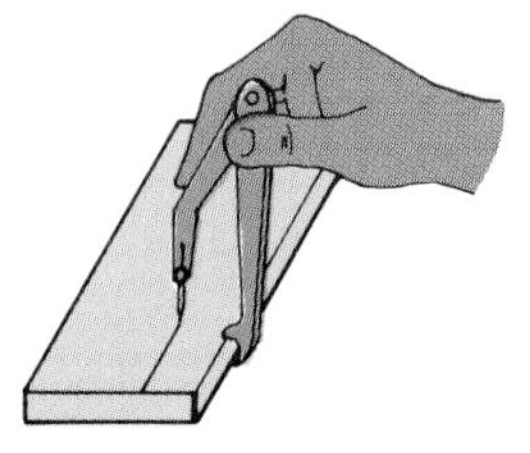

Marking metal to width
Use odd-leg caliper.
Press the stepped leg of the caliper against the straight edge.

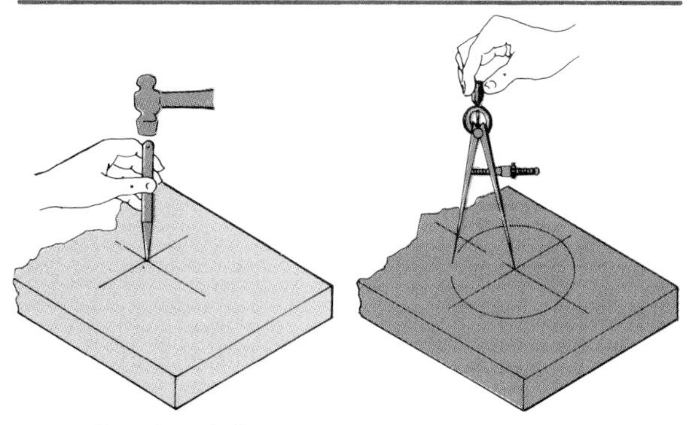

Drawing circles
Mark the center of the hole with a center punch on metal and an awl on wood.
Use a compass on wood, and on the protective paper of plastic.
Dividers are used on metal.

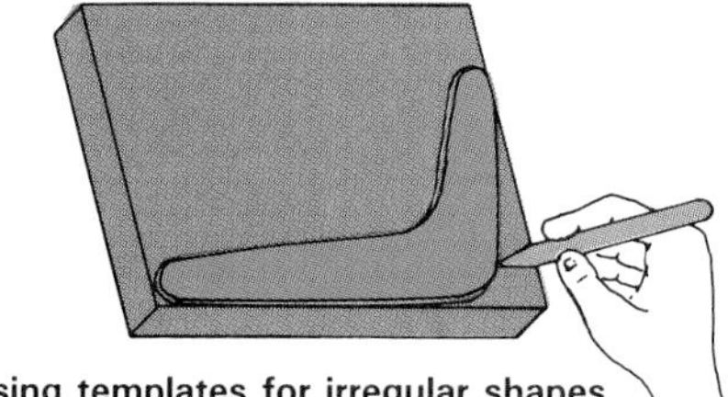

Using templates for irregular shapes
Draw and cut out the shape in paper or cardboard. This is called a template.
Hold the template to the material and draw carefully around it.
When drawing on wood take careful note of the grain direction.
When many pieces of the same shape are to be made, use masonite to make the template.

Figure 4.2. Tools for marking out. Read the instructions for using them.

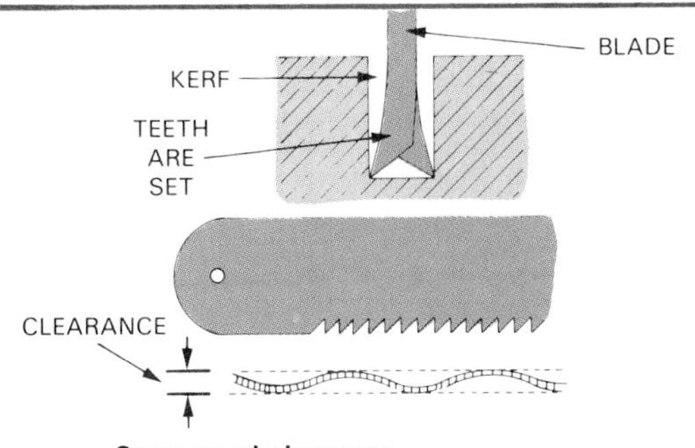

Saws need clearance
Teeth are bent in alternative directions; this is called set.
The kerf made by the teeth is wider than the blade thickness.
Teeth usually point away from the handle and material is cut on the push stroke.
The teeth on hacksaw blades are too small to be set. Clearance is achieved by stamping a wavy edge on the blade.

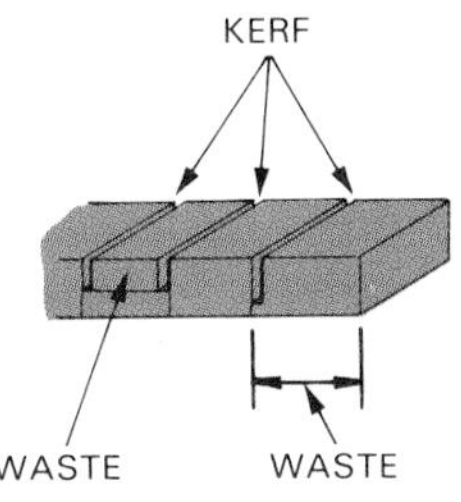

Sawing techniques
Never cut on the line; the kerf is made touching the line but on the waste side.
Stand with your hand and arm in line with the saw cut.
Use your thumb at the side of the blade when you start the cut.
Always use the full length of the blade.
At the end of the cut support the work underneath.

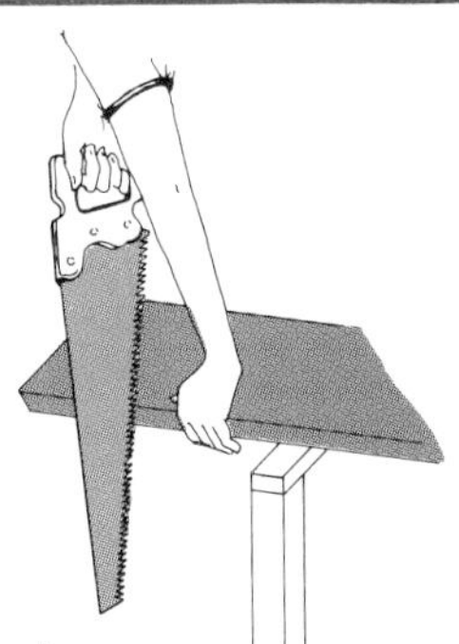

Using a hand saw
Used for first cutting wood to approximate size.
There are two types: cross cut and rip.
A cross cut saw has finer teeth and is used for cutting across the grain.
A rip saw has larger teeth and is used for cutting parallel to the grain.

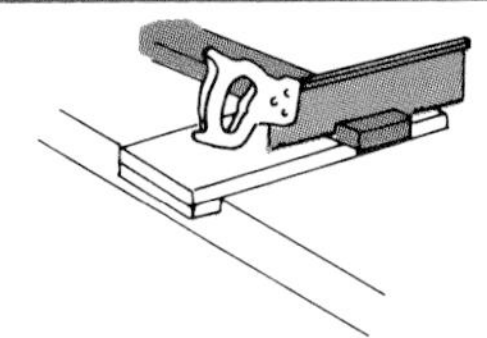

Sawing wood accurately
A tenon saw is used to make straight, accurate cuts in wood.
A bench hook, held firmly in the vise, should always be used.

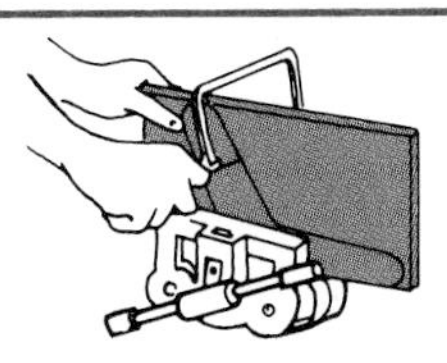

Coping saw
Used to make curved cuts through wood and plastic.
Teeth point towards the handle: cuts on the pull stroke.

Hacksaw
Used to make straight cuts through metal and plastic.
Make sure at least three teeth are in contact with the material all the time.
Use small teeth for hard materials and large teeth for soft materials.

Junior hacksaw
Small and inexpensive.
Useful for cutting thin metals and light sections.

Abrafile (rod saw)
Used to make curved cuts through metal, plastic, and ceramics.
The blade is like a file and is held in a hacksaw frame.
The cutting edge is made of tungsten carbide particles bonded to a steel rod.

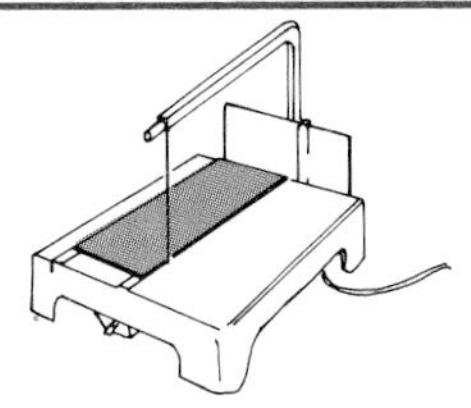

Hot wire cutter (an alternative to sawing)
Heated wire cuts straight and curved shapes in rigid foam plastic.
Must be used in a well-ventilated area as fumes are produced.

Figure 4.3. These tools are used for cutting operations.

■ SHAPING MATERIALS

When you have designed your product and have chosen the material, you are ready to make the product. You will do this following several carefully thought out steps.

Marking out

The first step in making a product is called **marking out.** Marking out involves measuring material and marking it to the dimensions on your drawing. You should do this carefully for two reasons. First, your marks must be ac-

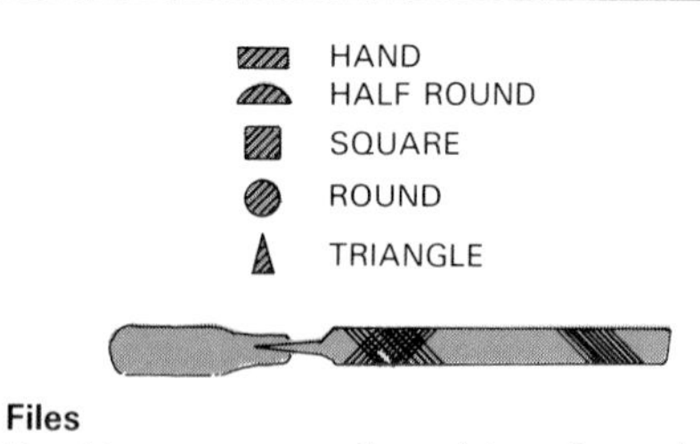

Files
Used to remove small particles of metal and plastic.
Double cut files remove metal faster but make a rougher surface.
Single cut files produce a smooth surface.
Each shape is available in many sizes and degrees of coarseness.
Always use a file with a handle.

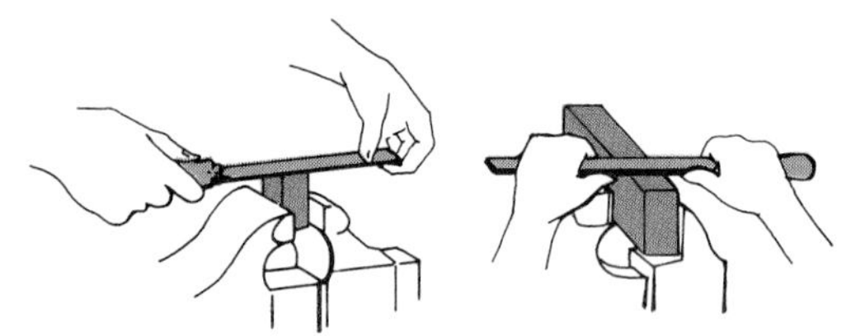

Using a file
For normal filing, hold the file at each end; it only cuts on the forward stroke (cross filing)
To produce a very smooth finish, push the file sideways (draw filing).

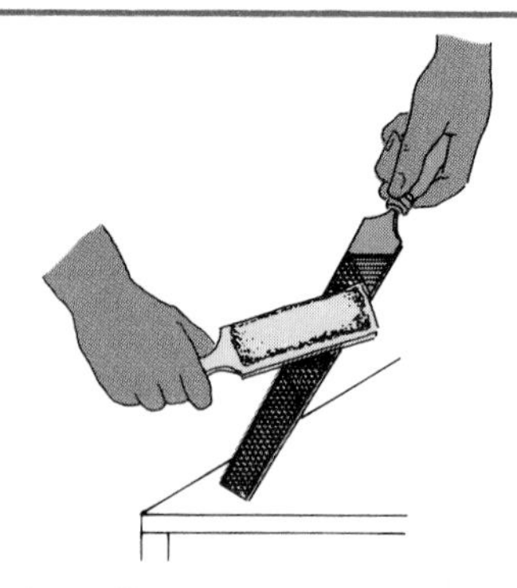

Cleaning a file
Small pieces of metal sometimes stick in the teeth of the file; this is known as pinning.
A special wire brush called a file card is used to clean the file.
Keeping the file clean is particularly important when filing plastic.

Rasps
Similar to coarse files.
Used for rough shaping (e.g.), free-form, sculptured shapes.

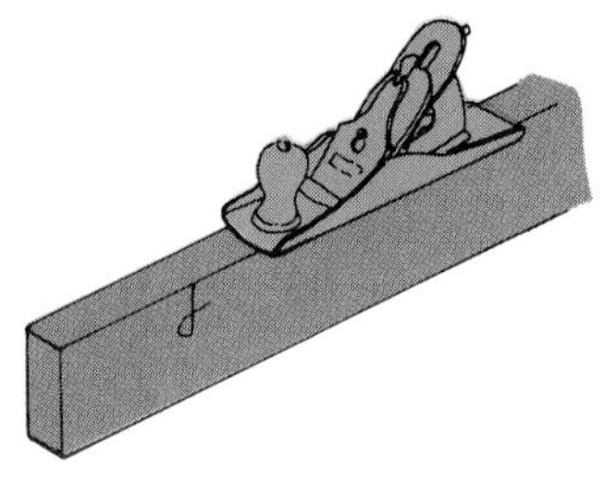

Planes
Used to remove a thin layer of wood (shaving).
A short plane, called a smoothing plane, is used on short pieces of wood.
A longer plane, called a jack plane, is used on longer pieces of wood.
Always plane in the direction of the grain.

Surforms
Are held like files and rasps but cut like planes.
Cut wood quickly but leave a rough surface.

Figure 4.4. Filing and planing tools are used to shape and smooth materials.

curate so that the pieces fit together. Second, materials are expensive. Making mistakes wastes time and money.

Most marking out starts from a straight edge. To make a straight edge on wood a plane is used. Plastic and metal are filed. When wood is used, this edge is called the face edge and the best side is the face side, figure 4.1. Lines should be easily seen. To be accurate, they must be thin. They should be marked as shown in figure 4.2.

Sawing

Sawing removes material quickly. All saws have a row of teeth. They chip or cut away the material. Like all tools, the part that cuts must be harder than the material being cut. There are special saws for cutting wood and for cutting metal, figure 4.3. Plastics and composites mostly use saws designed for cutting wood and metal.

Filing and planing

Small amounts of material may be removed by **filing** and **planing,** figure 4.4. Files are mainly used on metal. A special type of file, called a rasp, is used on wood. Various types of planes produce smooth, flat surfaces on wood.

Most filing is done while the work is held in a vise. In straight or cross filing you would push the file across the work straight ahead or at a slight angle. Never run your fingers over a newly filed surface. Sharp burrs on the workpiece may cause a severe cut.

Planing also requires that the wood piece be held in a vise. When properly adjusted, a plane should take off fine shavings from the piece of wood. Planing removes the small ridges left by the power planer. This saves having to remove them with sandpaper—a slow process.

Shearing and chiseling

Shearing and **chiseling** are other techniques used to shape material. These tasks require shears and chisels. Shears, also called snips, cut thin metals. Chisels are designed to cut wood or metal, figure 4.5.

Snips are designed for various kinds of cuts. A straight snips cuts straight lines and large curves. For cutting curves and intricate

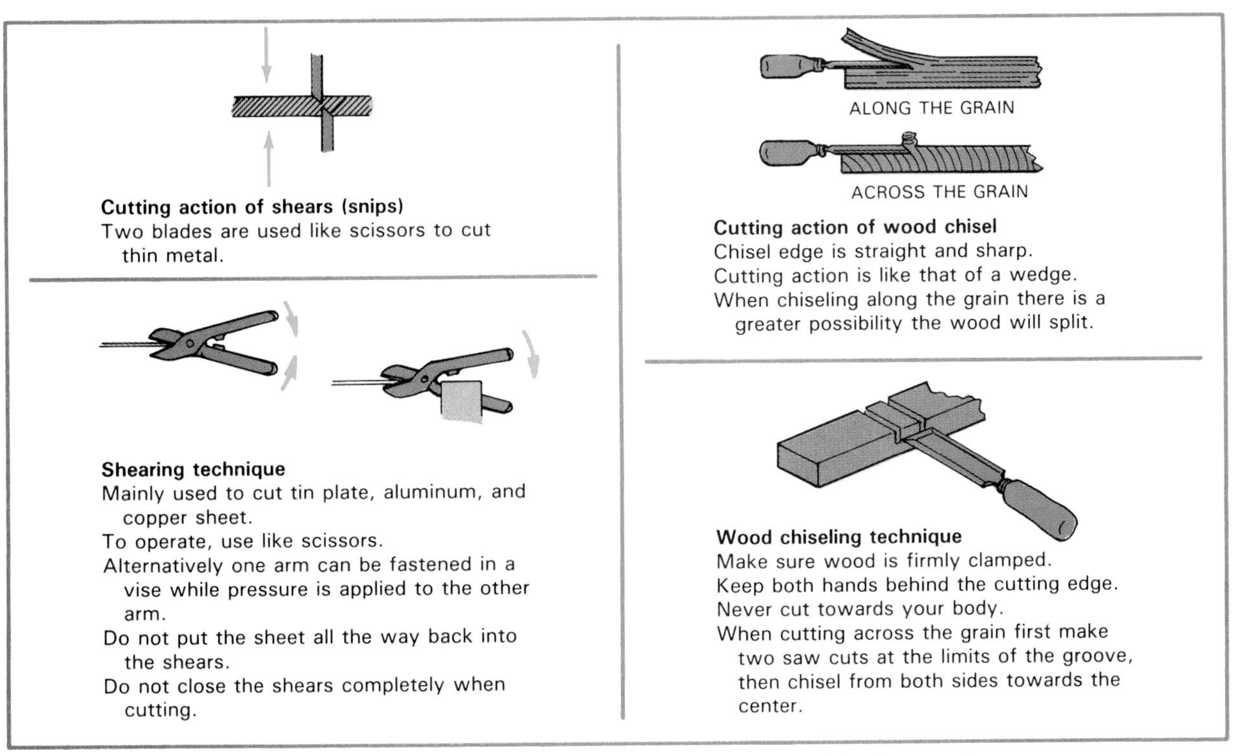

Figure 4.5. These are common shearing and chiseling tools.

designs an aviation snips or a hawk-billed snips are best.

Chisels must be designed for the type of material on which they are to be used. Wood chisels have very sharp cutting edges so they will cut rather than tear the wood. Metal-cutting chisels have thicker, tougher cutting edges. Refer to figure 4-5 once more. It shows how shears and chisels are to be used.

Drilling

Drilling is a process used to make holes in wood, plastic, metal, and other materials. A drill cuts while turning. Twist drills for cutting metals are made of carbon steel and high-speed alloy steel. Twist drills are also used to bore holes in woods and plastics. Figure 4.6 shows some types of drills and how you are to use them.

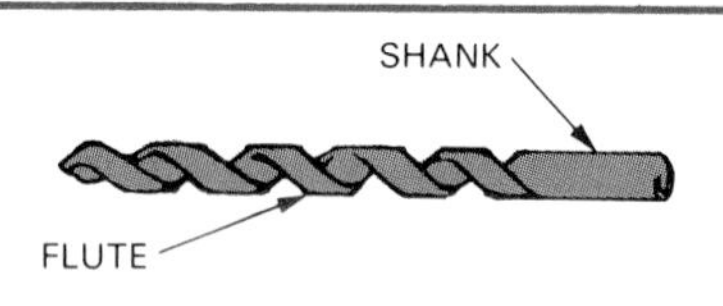

Cutting action of drill
Drills cut by rotating a cutting edge into a material.
A twist drill has a cutting edge, a spiral groove (the flute) to release the chips, and a straight shank to hold the drill in a chuck.

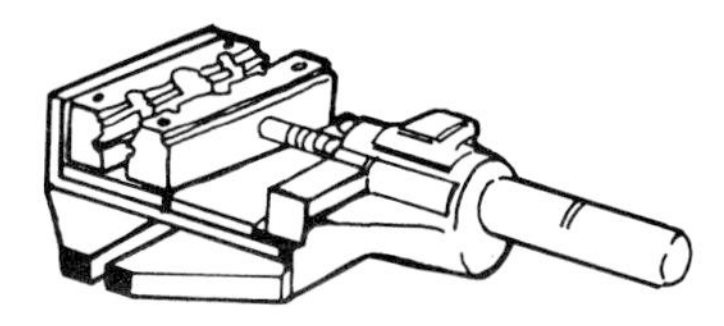

Drilling technique
Mark the point you want to drill.
Use a center punch on metal.
Never hand hold work to drill; hold work in a vise or a clamp.
Place waste wood under the work to prevent damaging the bench after the drill has gone through; this also prevents the material cracking away.
When drilling deep holes remove the drill from the hole from time to time to avoid clogging and overheating.

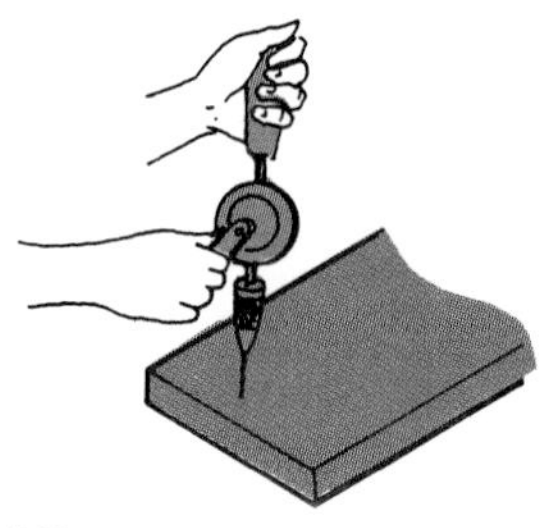

Hand drill
Concentrate on keeping the drill vertical.
Turn the handle at a steady speed, trying not to wobble the drill.

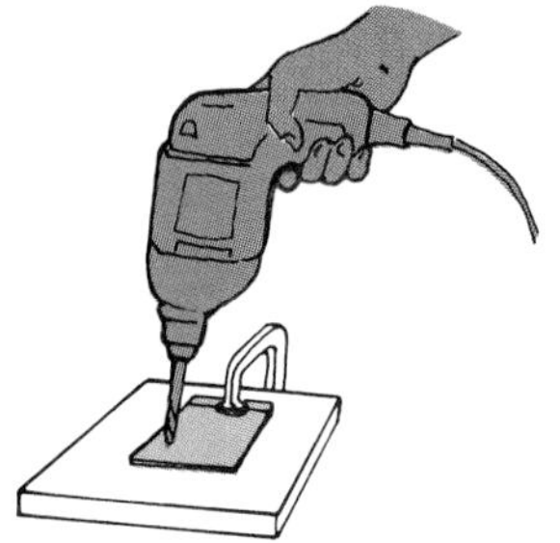

Portable electric drill
Clamp small pieces of material in a vise.
Center punch the location of the hole.
Place a small piece of wood on the underside to prevent splitting.
Do not bend the drill sideways or you will snap the drill bit.

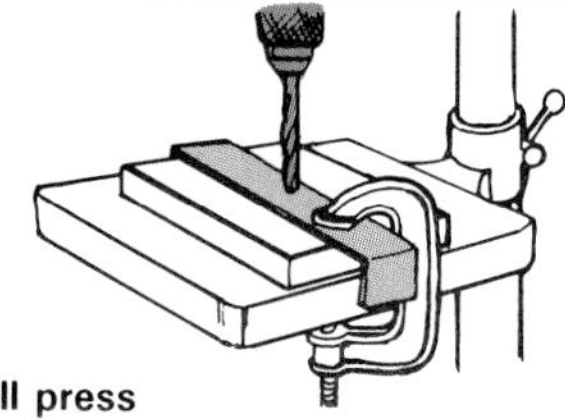

Drill press
Remember that work is always held in a machine vise or clamp.
Always wear eye protection.
The speed can be adjusted: the larger the drill bit, the slower it should turn.
Place a piece of scrap wood under the workpiece to protect the table.

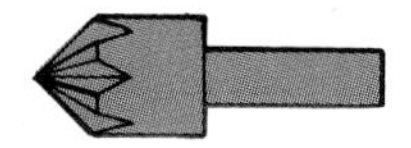

Countersink drill
Used to open out the end of a hole so that a flathead screw will fit flush with the surface.

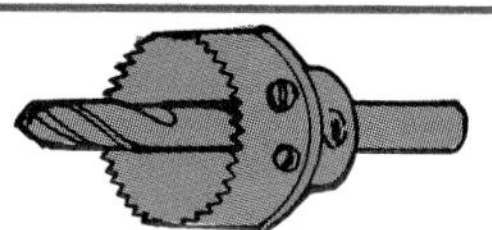

Holesaw
Used to drill large holes in wood up to 3/4 in. (18 mm) thick.
Very useful for making discs or wheels.

Figure 4.6. Hole-drilling tools and how to use them.

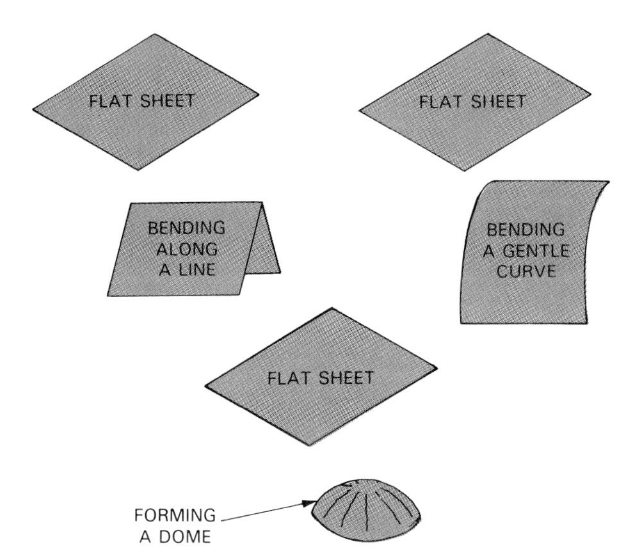

Figure 4.7. Sheet material will bend or curve easily in one direction. Bending in several directions at the same time is a problem.

Bending and forming sheet materials

Bending and **forming** are two different processes. Bending sheet material is like folding paper along a straight, sharp crease. It is also quite easy to bend it along a gentle curve.

However, forming sheet material is more difficult. Take a flat sheet of paper and try to form it into a dome. That is much harder, isn't it? The sheet must bend and curve in many directions. It creases and buckles, figure 4.7.

Bending wood

Being a stiff material, wood does not bend easily. One way to create a curved shape from wood is to cut the curve out of a thick, solid block as shown in figure 4.8A. The trouble is, the curved piece would break easily because of the short grain on the curved sections. The part made this way would be weak. But there is another way: laminating. It avoids the problem of short grain.

Laminating wood. Laminating means gluing together several veneers (thin sheets of wood). These can be easily bent, glued, and held in a mold until the glue dries. The steps for laminating are:

1. Make a mold, figure 4.8C. The two parts must fit together exactly.
2. Veneers vary in thickness. Calculate the number you will need to get the right thickness of laminate.
3. Clamp all of the veneers together without glue. If they don't bend easily, dampen the veneers and leave them clamped in the mold overnight, figure 4.8D and E.
4. Completely cover the surfaces of the veneers with glue. Do not glue the outside surface of the top and bottom pieces or they will be permanently attached to the mold! Use a resin adhesive that becomes rigid when it sets. Avoid using contact cement or PVA which remain rubbery. Once the glue has dried, the laminated wood will hold its shape. It is unable to spring back to its old shape. It will be held by the glue.

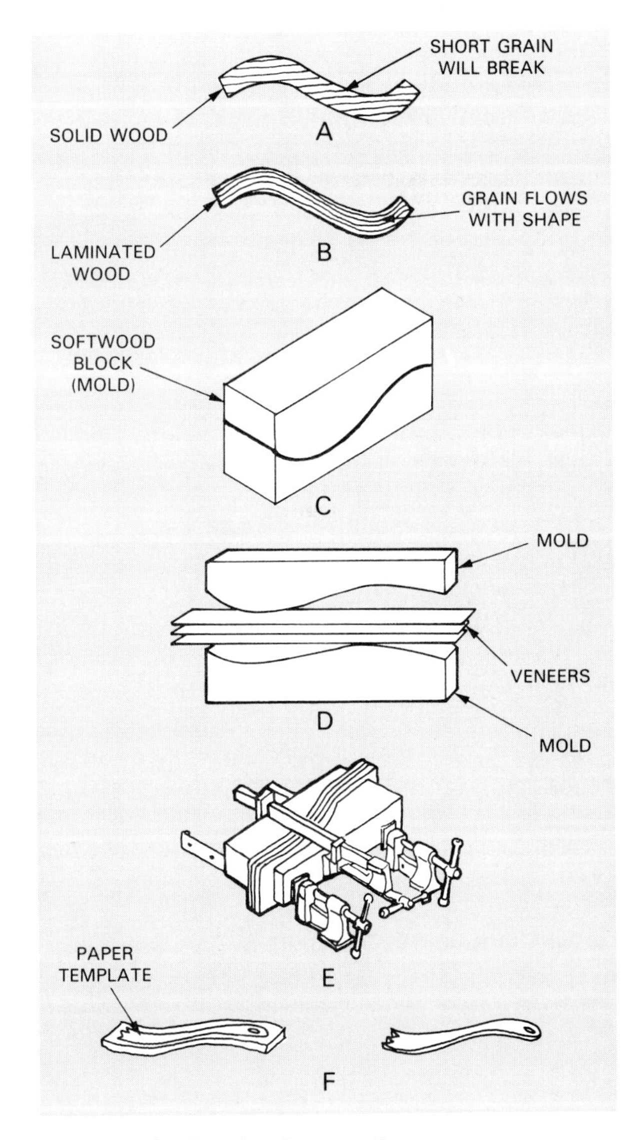

Figure 4.8. Laminating produces stronger parts than sawing the curved shapes out of solid wood.

5. To prevent the veneers sticking to the mold, wrap them in wax paper or a thin plastic sheet.
6. Use a thin rubber sheet between the mold and the veneers. It will take up any irregularities (roughness) in the mold surface.
7. Use bar clamps, figure 4.8E, and squeeze the mold together until the glue is forced out along the edges of the veneer.
8. As shown in figure 4.8F, make a template of the shape you want. Either glue it to the laminate or mark around it. Use it as a pattern for cutting out the shape.

Bending sheet metal

Sheet metal can be bent and folded into three dimensional objects. A pan is a good example. Before you begin to shape the metal you must work out a development. (A development is a pattern. It shows where the sheet metal must be cut and folded to make the object, figure 4.9A.) The development is marked on the sheet metal. You then use tin snips to cut the shape.

Bending along straight lines. Straight bends follow these steps:

1. Place the marked sheet metal into folding bars. The bend line should be touching the top edge of the bar.
2. Fasten the folding bars in a vise.
3. If the metal is wider than the vise jaws, add a C-clamp.
4. Use a mallet with a rawhide or nylon head. Bend the metal over the bar, figure 4.9B.
5. For some shapes, you may find a block of wood more useful than folding bars.

Bending strip steel. Lengths of mild steel not more than 1/4 in. (6 mm) thick may be bent fairly easily.

1. Clamp the metal vertically in a vise.
2. Hammer the metal from one side to bend it to the needed angle.

Jigs. Jigs are useful tools for bending metal.

- Small diameter rods can be bent on a peg jig.
- Metal tubing can be bent using the jig shown in figure 4.9D.

Bending and forming plastics

Thermoplastics can be bent or formed when heated to between 300° and 400°F (150° and 200°C). Heat the plastic sheet or rod along the line of the bend. The narrower the heated line, the sharper the bend will be. Refer to figure 4.10A.

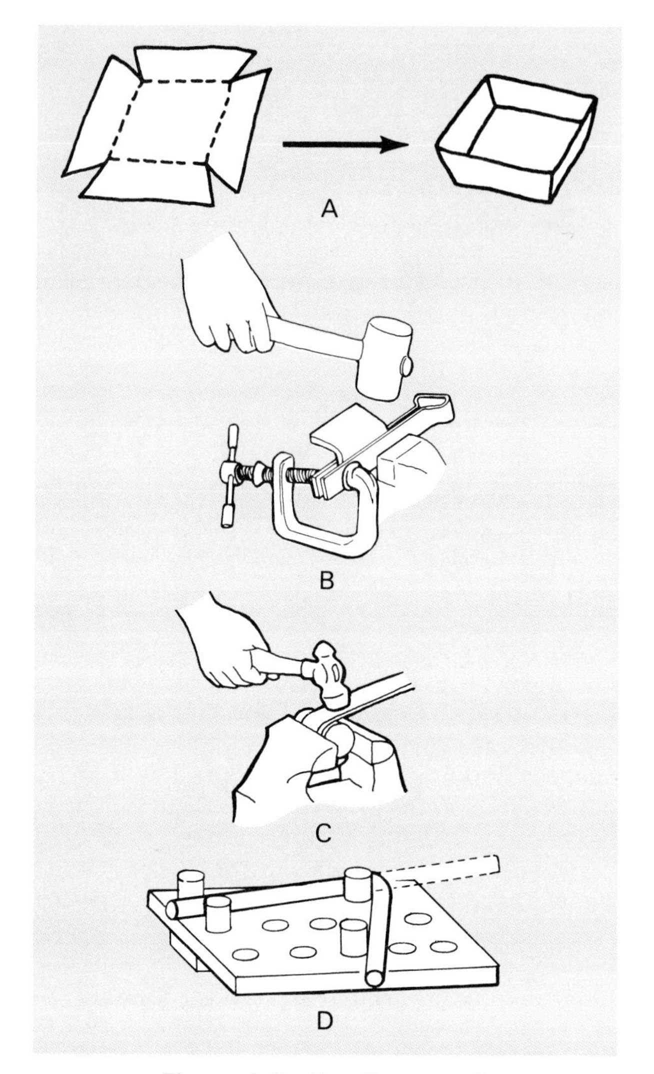

Figure 4.9. Bending metal.

Using a strip heater. This type provides heat along a straight, narrow line. The result will be a sharp bend.

To use the strip heater:

1. Place the sheet of plastic on the heater. The bend line must be exactly over the heat element.
2. For safety, wear gloves to protect hands from the hot plastic.
3. Heat both sides of the sheet.
4. Bend the plastic to the required shape. For a 90-degree bend, press the plastic into a mold, figure 4.10B.

5. To produce a sharper bend, press a second mold into the corner, figure 4.10B. Hold until cool.

NOTE: Wooden molds must be covered with cotton or felt material. This prevents the grain from marking the plastic.

Forming in a mold. This process calls for a two-part mold. To form the plastic:

1. Heat the acrylic sheet in an oven until pliable (bends easily).

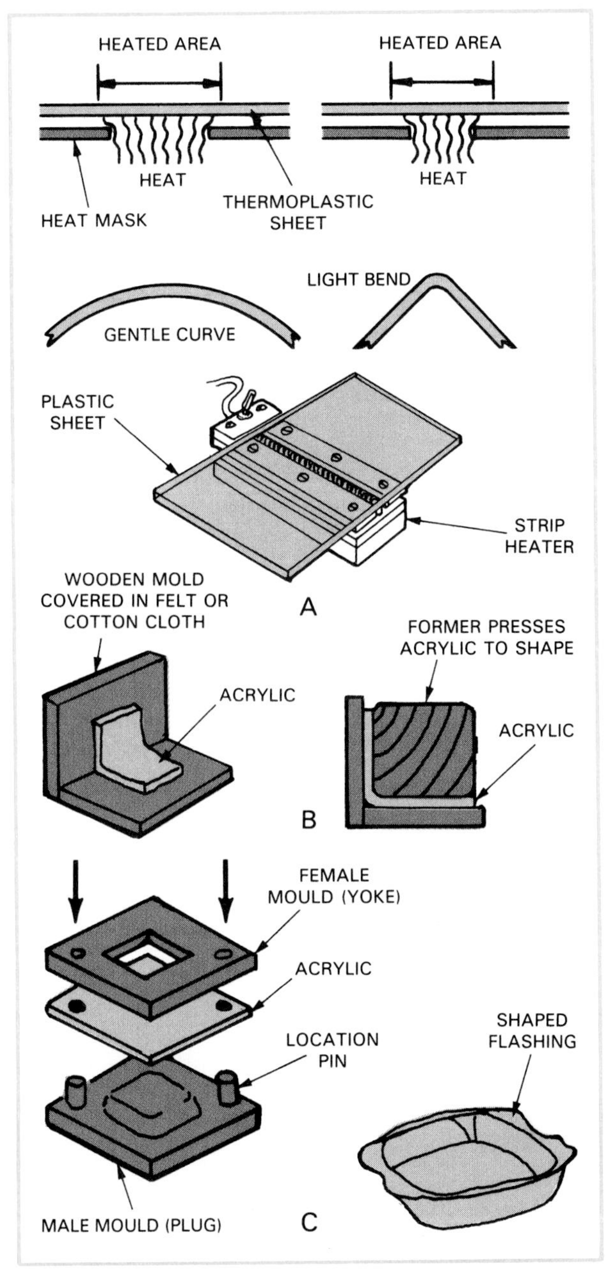

Figure 4.10. Forming and bending plastic.

2. Place it over the plug. Press the yoke down on top, figure 4.10C.
3. Allow the plastic to cool before separating the mold.

Casting and molding

Pouring liquid or plastic material into a mold (container) to shape it is called **casting** and **molding.** Do you realize that every time you make ice cubes you are actually casting? You are making a solid shape by pouring water into a tray (mold). The liquid takes the shape of its container.

Casting is a method of making shapes that are almost impossible to produce by sawing, drilling, or filing. We use three basic materials for casting and molding. These are metals, plastics, and ceramics. When the material is poured into the mold, the process is called casting. When the material is forced into the mold, it is called molding.

Metals become liquid when heated above their melting point. As they cool, they solidify (become solid).

Plastics are available in a liquid form. These liquids set hard through chemical action.

Ceramics are materials such as silica (sand), clay, and concrete. Silica must be melted to make glass and other products. Clays and concrete are not melted. They are mixed with liquid. Then they are poured into a mold.

Casting metals. Use the following steps to cast metal:

1. As shown in figure 4.11A, produce a pattern. (A pattern is just like the finished product.) Usually it is made of wood. But it could be made of some other easily worked material.
2. Place the pattern on a flat surface. On the same surface, place a molding box, figure 4.11B.
3. Pack the mold material around the pattern. The mold should be made from a material that will not burn as the hot metal is poured into it. Sand is often used because it is inexpensive.
4. Fill the molding box with sand. Tamp it tightly around and over the pattern. Fill the box completely.
5. Cover the box with another board and turn it over. Remove the board that was on the bottom.

6. Carefully remove the exposed pattern, figure 4.11C.
7. Pour molten metal into the cavity formed by the pattern, figure 4.11D.
8. Allow the casting to cool and solidify. Then remove it, figure 4.11E.

Casting plastics. Casting plastics has one big advantage over casting metal. The resins can be cast at room temperature.

Small articles, such as paperweights, can be cast. If you wish, you can also embed decorative objects in them. Follow these seven steps:

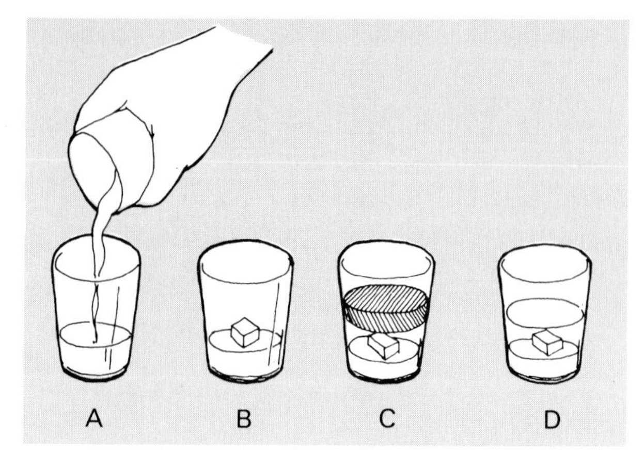

Figure 4.12. Steps for casting plastics.

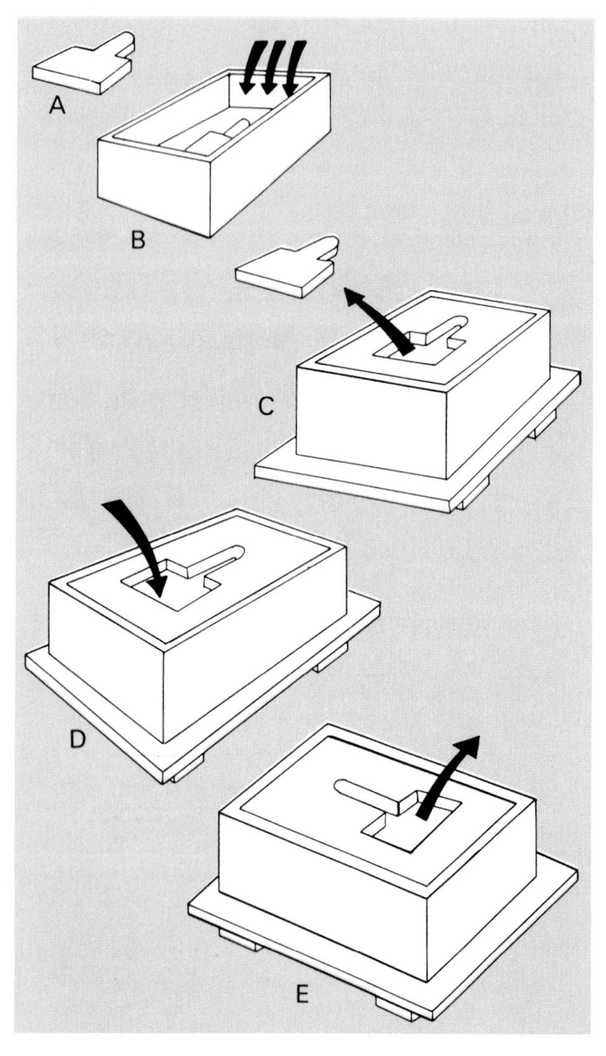

Figure 4.11. How to cast metals. A—Make a wooden pattern. B—Place pattern in a molding box. Pack sand around it until box is full. C—Place a flat wood cover on the box. Turn it over. Remove the pattern carefully. D—Pour molten metal into the cavity. E—When cooled and solid, remove the casting.

1. Use a smooth mold (container). It will produce a smooth surface on the casting. Waxed drinking cups work well. Never use polystyrene (Styrofoam®) cups. The resin will dissolve them!
2. Measure the amount of resin you will need. Add the recommended amount of catalyst (hardener). Mix thoroughly.
3. Pour a layer of the mixed resin into the mold. Leave it to harden. This will form the top layer of the casting, figure 4.12A.
4. Place the decorative object (coin, insect, or stamp) on the hardened layer of resin, figure 4.12B. (REMEMBER TO PLACE THE OBJECT FACE DOWN.)
5. Pour more resin around and over the object, figure 4.12C. You can use clear resin throughout or you can add pigment to the last layer. This forms the base of the object.
6. When the resin has hardened, figure 4.12D, remove the casting from the mold.
7. Smooth rough edges and surfaces with wet or dry sandpaper. Then polish the casting with a polishing paste.

Molding fiber reinforced plastic. Fiberglass canoes, racing car bodies, and crash helmets may all be made from plastic resin reinforced with glass fiber. The reinforcing glass fiber makes the shells very tough, while the thermosetting resin creates a hard surface. To make a fiberglass reinforced product:

1. Paint on a release agent over the surface of the mold, figure 4.13.

Figure 4.13. Molding fiberglass on a form, such as this boat, is called a "lay up."

2. Mix polyester resin and a catalyst (hardener) in recommended proportions.
3. Brush on a gel coat of polyester resin.
4. Add a layer of fiberglass and coat it with more resin.
5. Use a roller to make each layer take up the exact shape of the mold and remove air bubbles.
6. Add other layers of resin and fiberglass to produce the required thickness.
7. Allow the assembly to cure (set hard). Then remove it from the mold.
8. Color may be added to the resin.

	MECHANICAL	CHEMICAL	HEAT
WOOD	Nails Joints for gluing Screws Nuts and bolts KD (knock-down) fasteners Wedges Hinges	Glues Adhesives	
METAL	Rivets Nuts/bolts/screws KD fasteners Hinges	Adhesives	Weld Braze Solder
PLASTIC	Rivets Nuts/bolts/screws KD fasteners Hinges	Solvents Cements	Weld

Figure 4.14. Joining wood, metal, and plastics. Nuts and bolts and screws are used for all three.

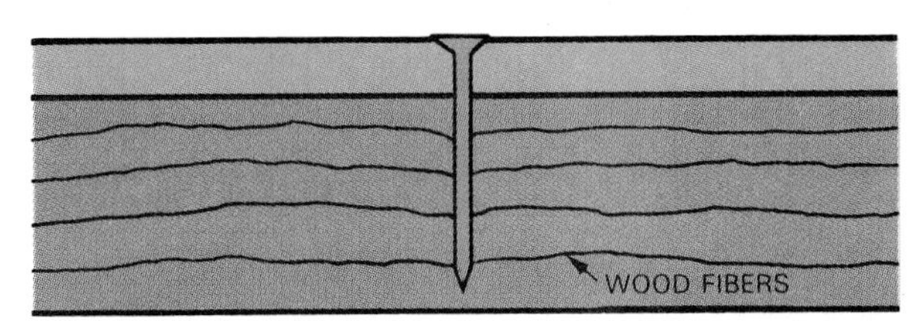

Figure 4.15. What holds the nail in the wood?

■ JOINING MATERIALS

There are many ways of joining materials. Figure 4.14 lists some common choices for wood, metal, and plastic. You will notice that some methods, such as nuts and bolts, are good for all three. Others, soldering, for example, may be used only for metal.

Mechanical joining

Mechanical joining is the use of physical means to assemble parts. It can be done in one of two ways. One method is to use hardware such as nails, screws, or special fasteners. Another method is to shape the parts themselves, so they interlock.

Nails. You may have used nails on a project. They provide one of the easiest ways to join two pieces of wood. Nails hold the wood by friction between the wood fibers and the nail, figure 4.15. Nails can be shaped many different ways. Each shape serves a special purpose, figure 4.16.

Remember these general points when using nails:

- Whenever possible, hold one of the pieces to be nailed in a vise, figure 4.17A.
- Always nail through the thinner piece into the thicker piece.
- Avoid bending the nail. Strike it squarely with the face of the hammer.
- When using finishing nails, drive the nail below the surface using a nail set, as shown in figure 4.17B.
- Stagger the nails. If you place them in a straight line you may split the wood along the grain, figure 4.17C.
- To remove nails, use a claw hammer, figure 4.17D. Always use a block of waste wood to protect the surface of the wood.

Screws. Screws have a greater holding power than nails, figure 4.18. However, just

Type of Nail	Uses
Common	Structural or other heavy work where head will be exposed.
Finishing	Finishing work where nail head should not be exposed.
Spiral	Building construction. Twisted shank causes nail to thread itself into wood, increasing its holding power.
Drywall	Fastening gypsum board to wood frames.
Concrete	Fastening to concrete. Nail is hardened to prevent bending.
Roofing	For wood, asphalt, and other roofing materials. Usually coated with zinc or galvanized to make rust resistant.

Figure 4.16. Nails are made in many different shapes to serve special fastening purposes.

like nails, they rely on the friction between the wood fibers and the screw for their strength. When two pieces of wood are held together the screw has only to grip the second piece. The two pieces are pulled together by the head

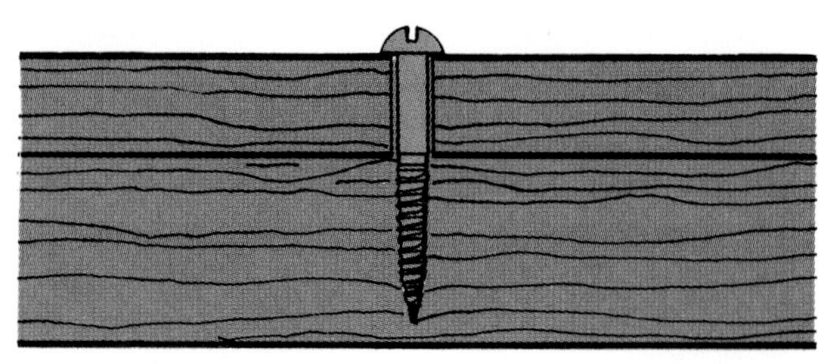

Figure 4.18. Wood fibers grip the threads of a screw. This gives the screw holding power.

of the screw and the grip of the screw thread. Screws can be withdrawn without damaging the material. They can be removed more easily than nails.

Wood screws are used for fastening wood to wood and metal to wood (hinges to a door). They are also used for fastening all types of hardware to furniture (handles to doors).

In choosing the correct wood screw, you must decide the:

- Shape of head and type of slot.
- Length.
- Thickness or gauge.
- Material.

There are three shapes of screw heads, figure 4.19.

- Flat head screws—used when the head of the screw must be flush with or below the surface of the wood.
- Oval head screws—used when the holding power of a flat head screw is needed. The screw head will show for decoration.

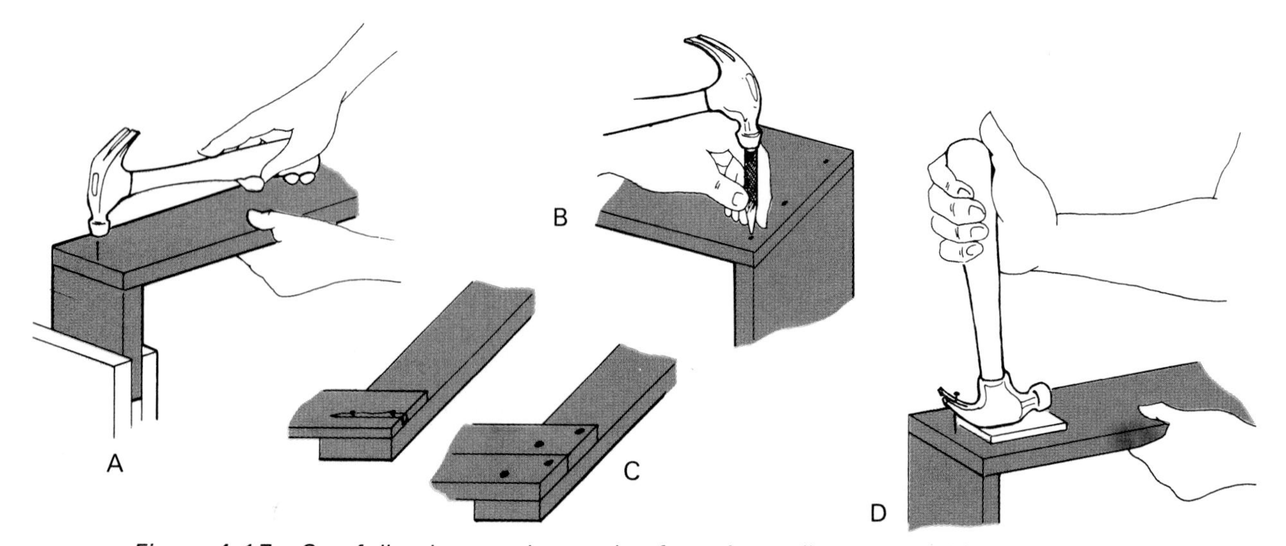

Figure 4.17. Carefully observe these rules for using nails to assemble wooden parts.

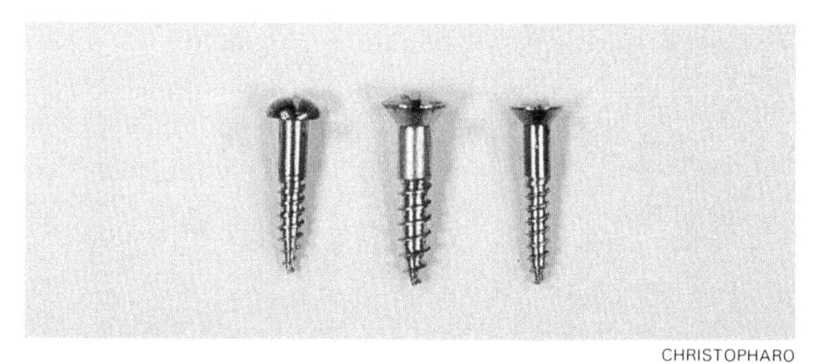

CHRISTOPHARO

Figure 4.19. Types of wood screws are named for the shapes of their heads.

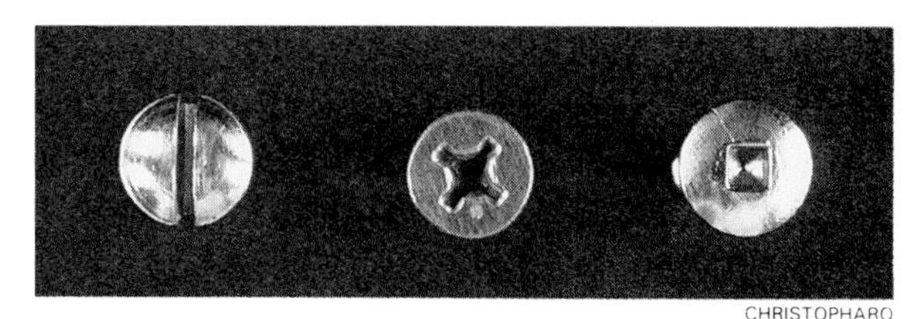

CHRISTOPHARO

Figure 4.20. These are the common types of screw heads. Left to right: Straight slot, Phillips, and Robertson.

- Round head screws—used when the object being fastened by the screw is too thin to be countersunk.

There are three common head styles or types:
- Straight slot.
- Phillips.
- Robertson.

The types are shown in figure 4.20.

Screw lengths range from 1/4 to 6 in. (6 mm to 150 mm). The thickness of a wood screw is called its "gauge." Gauge is expressed as a number. Screw gauges range from 0 to 24.

Most screws are made of steel. They are very strong, but they can rust. Brass screws do not rust, but they are weaker than steel. Sheet metal screws, figure 4.21, are used to join sheet metal, plastics, and particleboard.

Figure 4.22 shows the following three steps to fastening wood parts with a wood screw:
1. Hold the two pieces together. Drill a pilot (guide) hole the length of the screw.
2. In the top piece, drill a clearance hole. This is a hole the same diameter as the screw shank (unthreaded part of the screw below the head).

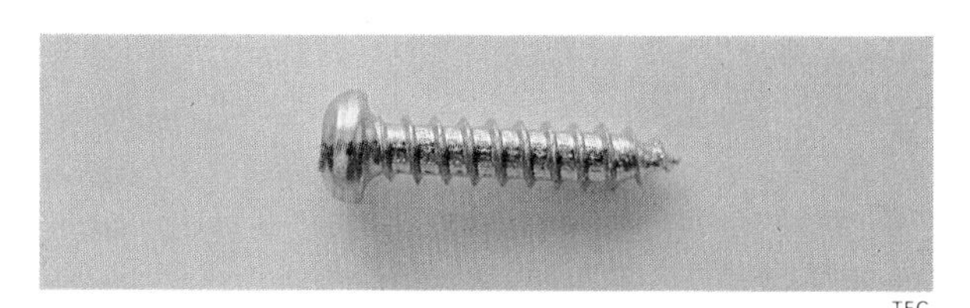

TEC

Figure 4.21. Sheet metal screws have threads that go all the way to the head.

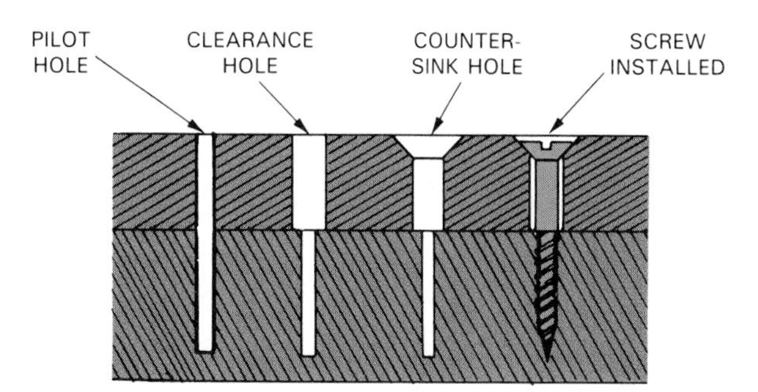

Figure 4.22. Special holes must be drilled for fastening wood with screws.

3. If you are using a flathead screw or oval head screw, countersink the hole.

Now you are ready to install the screw. The diameter of pilot holes and clearance holes used most frequently, are shown in figure 4.23.

Gauge No. of screw	Diameter of shank	Pilot hole	Clearance hole
4	7/64 (3 mm)	5/64 (2 mm)	7/64 (3 mm)
6	9/64 (4 mm)	3/32 (3 mm)	9/64 (4 mm)
8	11/64 (5 mm)	7/64 (3 mm)	11/64 (5 mm)
10	3/16 (5 mm)	1/8 (4 mm)	3/16 (5 mm)
12	7/32 (6 mm)	9/64 (4 mm)	7/32 (6 mm)

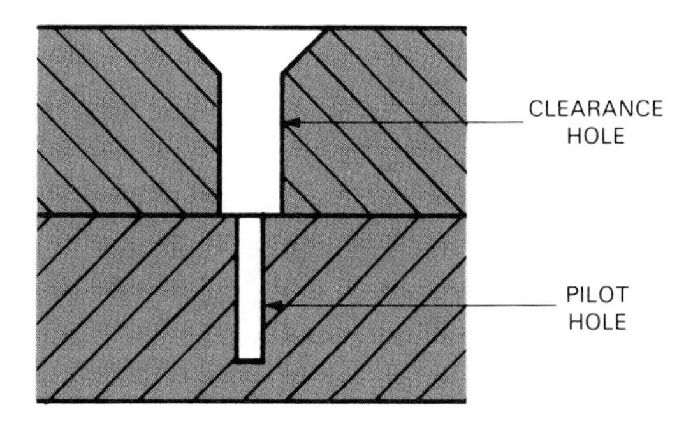

Figure 4.23. These pilot and clearance holes are recommended for wood screws.

Screwdrivers. The two most common types of screwdrivers are the straight slot (standard) and Phillips head. A third type is the Robertson. Use the largest screwdriver convenient for the work. More power can be applied to a long screwdriver than a short one. Also, there is less danger of it slipping out of the slot. The tip of the screwdriver must fit the slot correctly, figure 4.24.

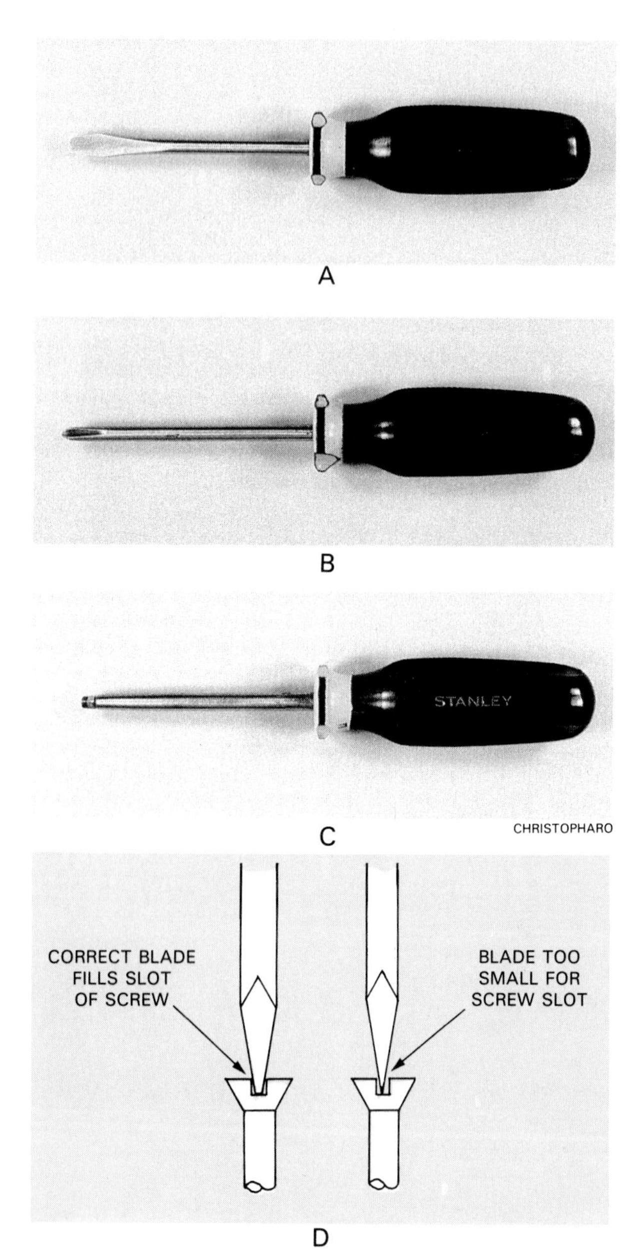

Figure 4.24. Screwdriver types. A—Straight-bladed or standard. B—Phillips. C—Robertson. D—Choosing the correct size standard screwdriver.

Nuts and bolts. Nuts and bolts fasten metal, plastic, and sometimes wood parts together. They are quite different from wood screws. Bolt threads do not depend on gripping the fibers of the material. Bolts go completely through a drilled clearance hole. (This is a hole large enough for the bolt to be pushed through.) A nut threads onto the bolt end. Tightening the nut squeezes the parts and holds them.

Washers are often used under the bolt head and the nut. This protects the surfaces. It distributes the load (pressure of squeezing) over a larger area, too. Lock washers prevent nuts from accidentally loosening due to vibration.

Sometimes, threads are cut into the hole in the second piece of material, figure 4.25. This takes the place of the nut. Joints fastened with nuts and bolts can be taken apart and reassembled.

You need to make a number of decisions to choose the right nut and bolt. You have to decide on:

- Length.
- Diameter.
- Shape of head.
- Thread series. Some are coarse; others are fine. There is also a standard metric thread.
- Material.

Some of the choices are shown in figure 4.26. Can you name any of the three shown?

You tighten most nuts and bolts with wrenches. Machine screws are tightened with screwdrivers or Allen wrenches. A combination wrench is shown in figure 4.27. It has one open end and one box end. The box wrench is usually preferred because it does not slip.

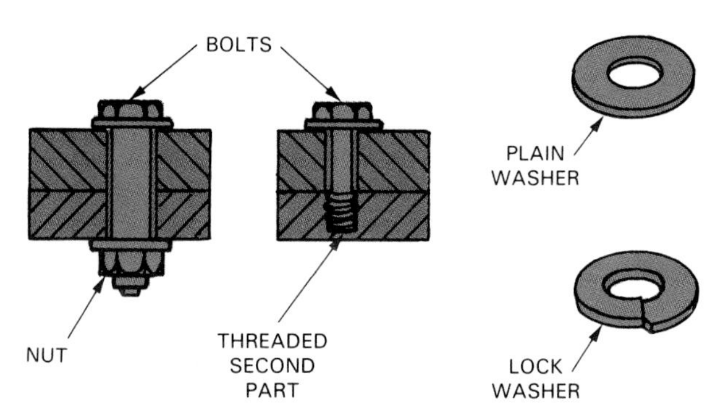

Figure 4.25. Two systems for fastening metal parts with bolts.

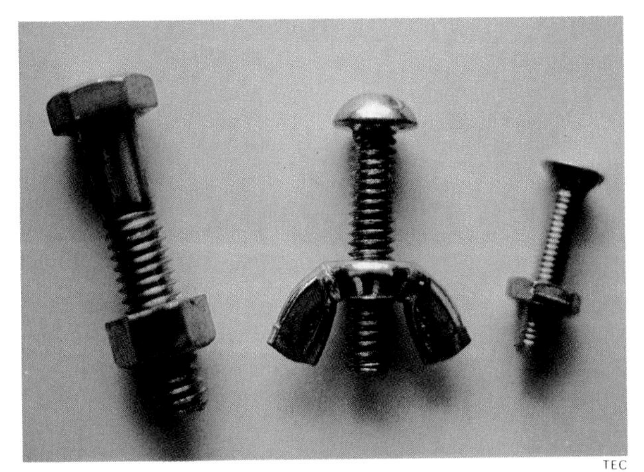

TEC

Figure 4.26. Bolts and nuts are made with many different shapes for special purposes.

TEC

Figure 4.27. This is known as a combination wrench

The open end is useful, however. Sometimes there is not enough room for a box end. The open end can be used where clearance is a problem.

An adjustable wrench fits a range of nut sizes. As figure 4.28 shows, always pull on the wrench handle. Pushing can be dangerous. If the wrench should slip, you would almost always injure your knuckles.

Rivets. Like nuts and bolts, rivets squeeze two or more pieces of metal or plastic together. They are either solid or pop type.

Solid rivets are usually made of mild steel. They may have round or flat heads. The four steps for installing a round head rivet are shown and described in figure 4.29.

To use solid rivets you must be able to reach both sides of the rivet. This is not always possible. Pop rivets are designed to be used

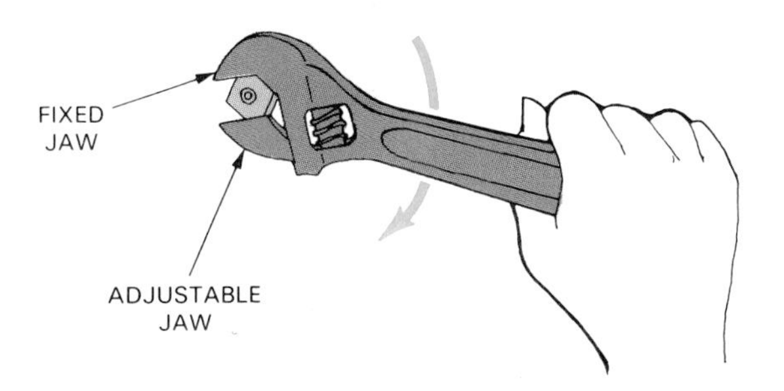

Figure 4.28. An adjustable wrench. Always pull against the fixed jaw.

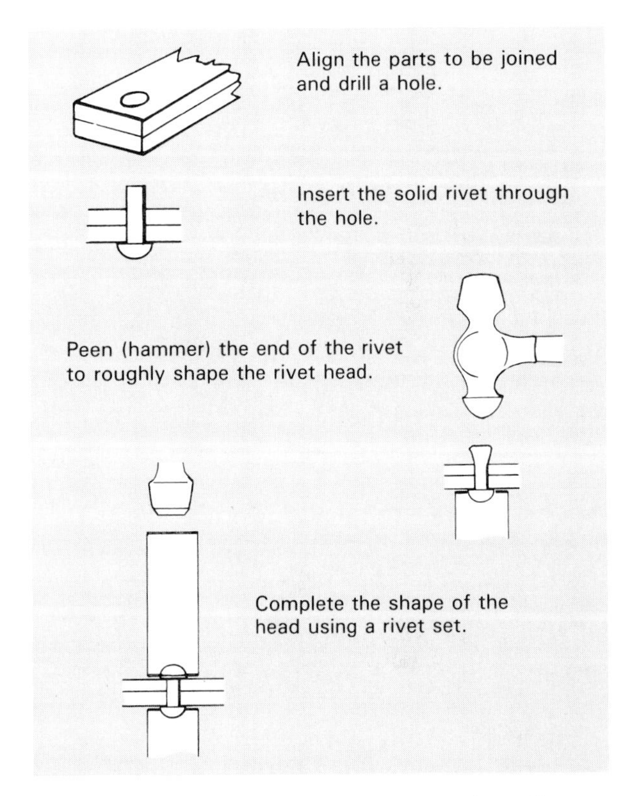

Align the parts to be joined and drill a hole.
Insert the solid rivet through the hole.
Peen (hammer) the end of the rivet to roughly shape the rivet head.
Complete the shape of the head using a rivet set.

Figure 4.29. This is the way to install a solid rivet

from one side. They are made of a hollow aluminum head with a steel pin through it. To use them:

1. Drill a hole in the parts large enough to receive the pop rivet.
2. Push the pop rivet through the hole.
3. Slip the rivet gun over the pin.
4. Squeeze the handle to pull the pin back. This creates the rivet head on the back (concealed) side.
5. Continue squeezing until the pin breaks off, figure 4.30.

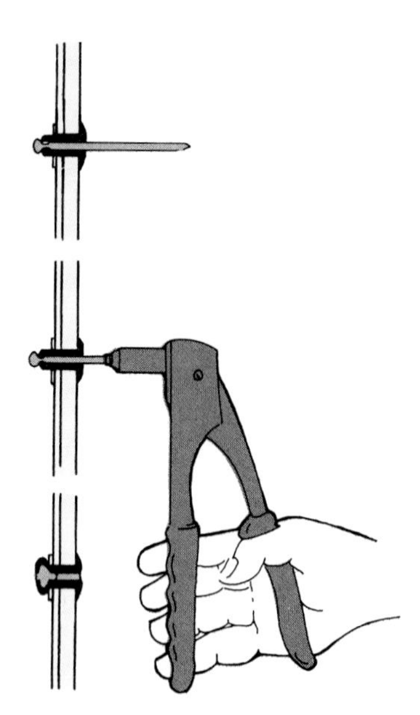

Figure 4.30. Using a pop riveter. Top. Insert rivet in predrilled hole. Middle. Slip tool over pin and squeeze. Bottom. Pin will break off when its sleeve has expanded on the blind side of the rivet.

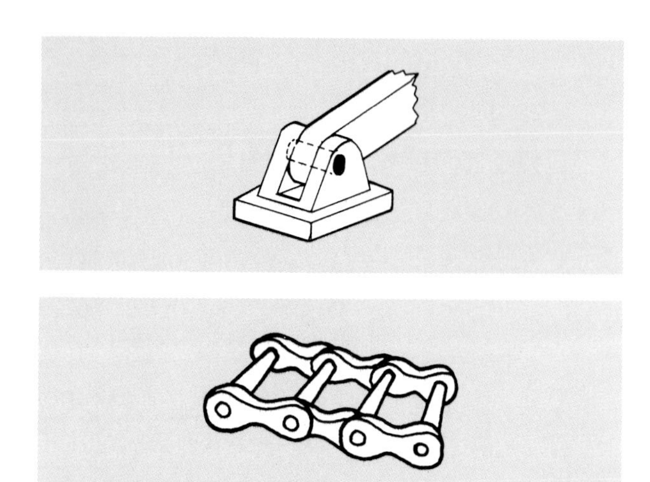

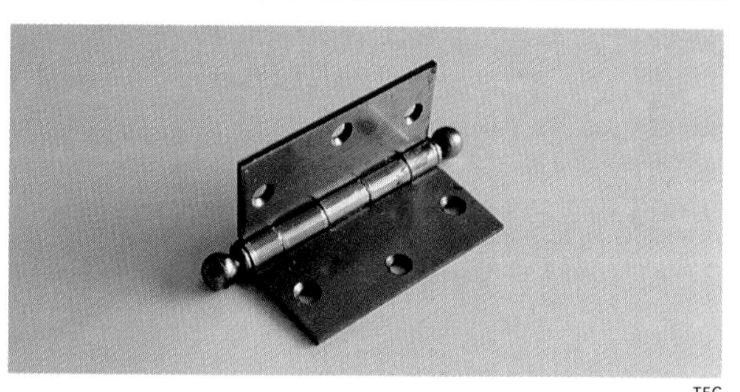

TEC

Figure 4.32. Common examples of pin hinges. Top. Knife switch. Middle. Bicycle chain. Bottom. Door hinge.

Knockdown (KD) joints. Some furniture is designed for "do-it-yourself" assembly. This requires special fasteners. They are strong and easy to assemble. The fasteners require no special tools or skills. Known as KD or "knockdown" joints, they can be taken apart and reassembled as needed. There are three common types, as shown in figure 4.31.

Movable mechanical joints. Some joints are made so that the joined parts can move. Think of how a door is joined to its frame. A hinge is used. A knife switch in figure 4.32 is another example of a movable joint. Both are pin hinges. They can move only back and forth. We say that they "move through one plane only."

A second type of movable joint allows movement in more than one plane, figure 4.33. The "joy stick" of a video game uses a ball and socket joint. There are other common examples. A camera tripod uses a lockable ball and socket. It can move in three different directions. The drive shaft of a car uses a "universal" joint. It permits the joint to move up and down or left and right as the shaft spins.

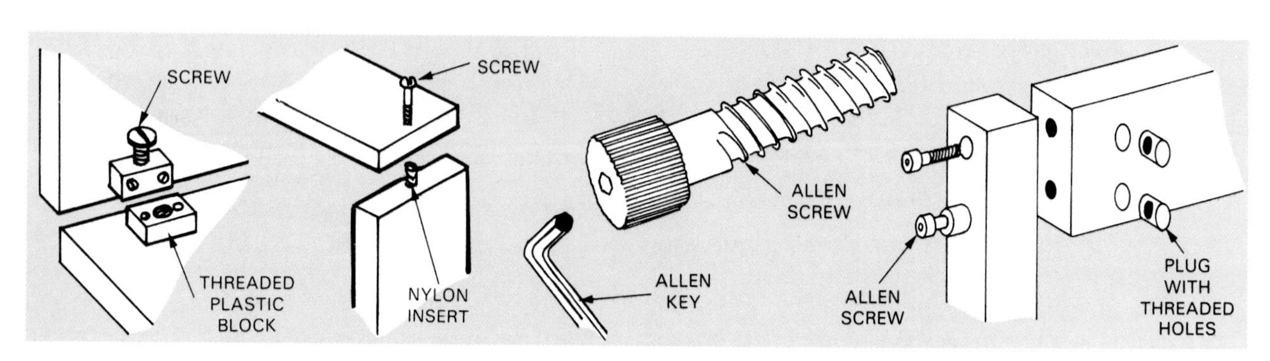

Figure 4.31. Knockdown furniture uses three basic fasteners.

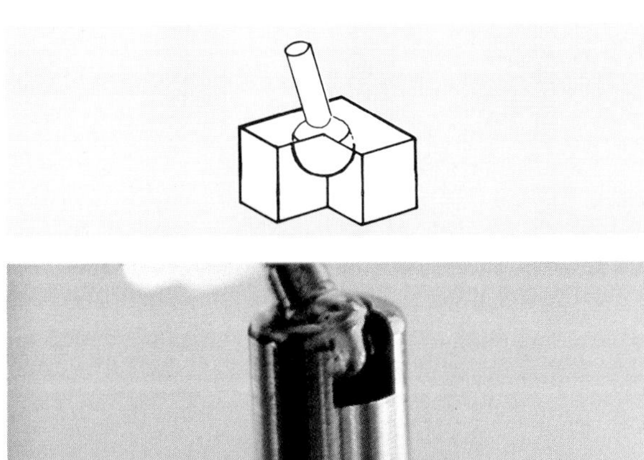

Figure 4.33. Top. Ball and socket joint is like the toggle on a video game. Middle. Camera tripod joint. Bottom. A universal joint.

A third type of movable joint uses a flexible material, figure 4.34. Polypropylene is used to make a variety of boxes and cases. The material, itself, acts as a hinge wherever it is folded. These are called "integral" or "living" hinges.

Wood joints. Joints are designed to hold wood pieces together. A wood joint's strength depends on two things:

- The amount of mechanical interlocking. This means that the wood parts are cut so that they cannot be pulled apart, figure 4.35.
- The surface area of the joint to be glued.

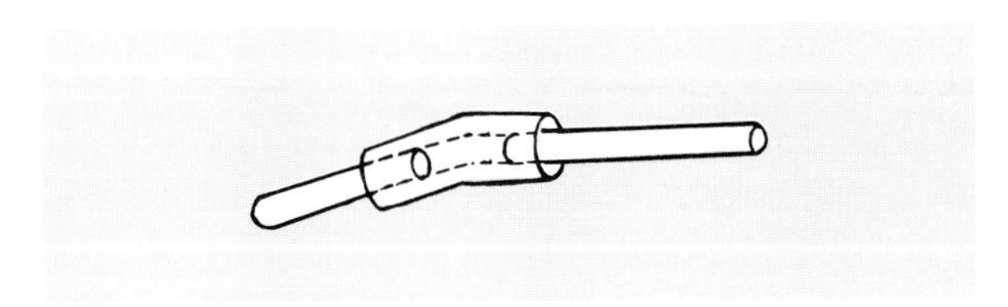

Figure 4.34. This is an example of a flexible joint. It can move in any direction.

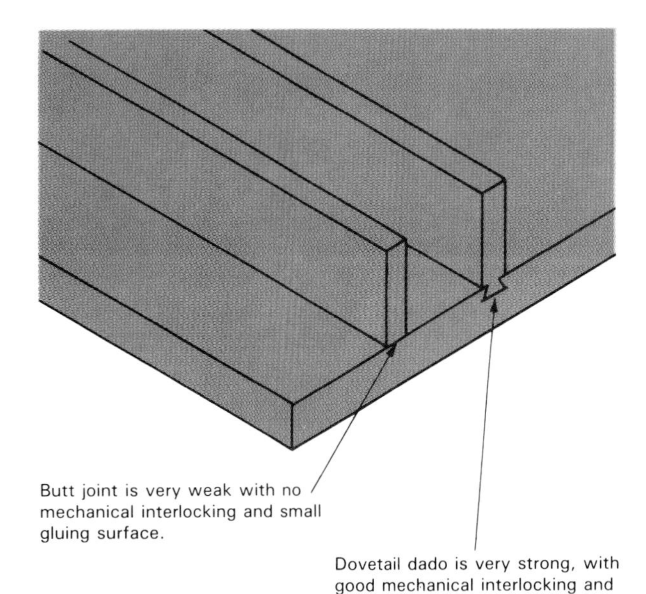

Butt joint is very weak with no mechanical interlocking and small gluing surface.
Dovetail dado is very strong, with good mechanical interlocking and increased gluing surface.

Figure 4.35. Joints are stronger when one piece is shaped to fit into the other. This is called an interlocking joint.

Wood joints can be grouped by type, figure 4.36. One group is used on frames. The other group is used to make boxes. Frame joints are found on chairs, windows, doors, and similar products. Box joints are used on items like cabinets, drawers,and storage boxes. Figure 4.37 shows eight different joints for constructing frames or boxes.

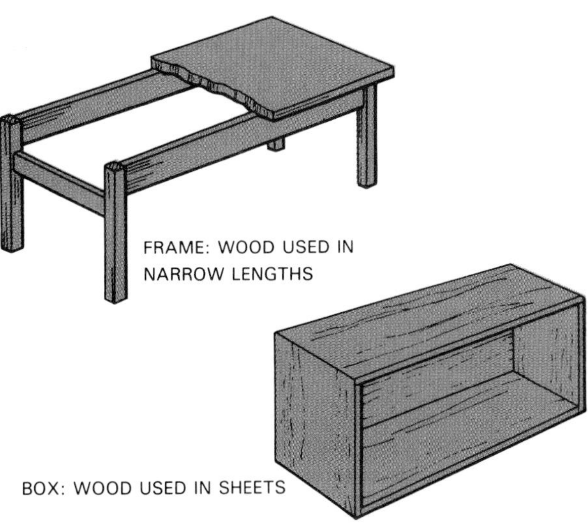

Figure 4.36. Two basic kinds of wood construction. Which do you think is easier to make?

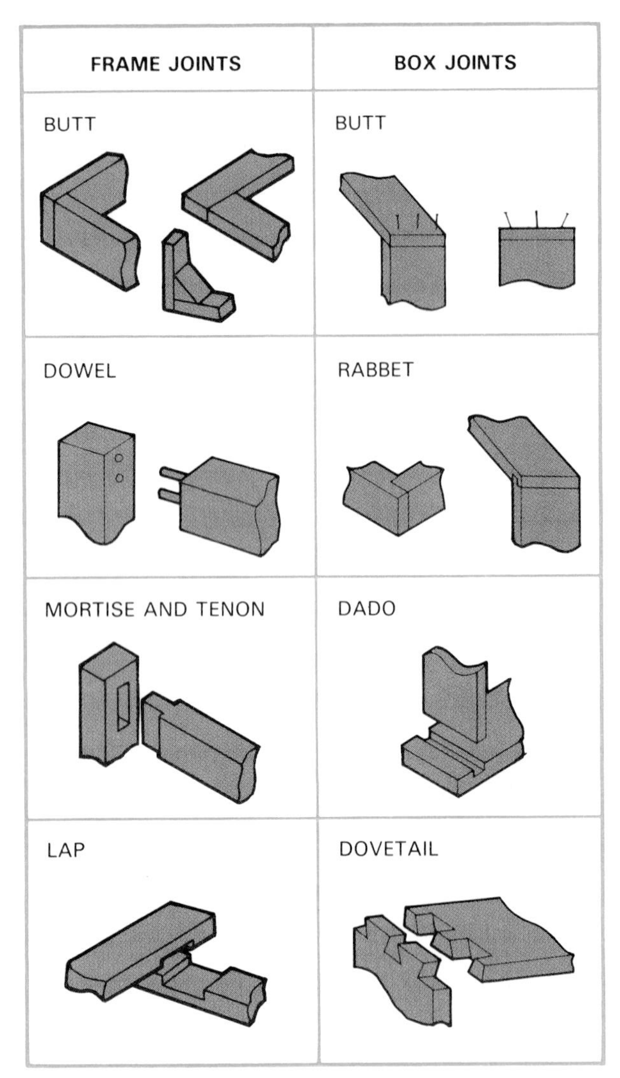

Figure 4.37. Which of these joints has the greatest mechanical strength? Which has the largest gluing surface?

Chemical joining

Mechanical joints are often strengthened by **chemical joining.** The chemicals include glues, adhesives, solvents, and cements.

Glues and adhesives are used for joining woods and metals. Glues are made from natural materials. These include animal bones, hides, and milk. They are rarely used today. Still, we use the term, ''glue.'' It is more correct to say ''adhesives.''

Adhesives come from petroleum products. These adhesives are of two types: thermoplastic and thermosetting. One common thermoplastic adhesive is liquid white glue. Also known as polyvinyl acetate, it is commonly used in wood joints.

Thermoplastic adhesives harden by loss of water or solvent. They may be softened by heat and are not waterproof.

Thermosetting adhesives include various types of resins. Heat will not soften them. They are waterproof.

How glues and adhesives work. Glues and adhesives work by adhesion. Adhesion makes use of a film. This sticks to each of the materials being joined. The adhesive material always remains in the joint. It holds the parts together.

How solvents and cements work. Solvents and cements are used to join plastics. In joining plastics, a pure solvent softens the areas to be joined. Cements dissolve a small amount of the plastic. They penetrate deeper into the two surfaces because the solvent evaporates much more slowly. A cement provides a stronger joint than a pure solvent.

Solvents and cements work on the principle of cohesion. (In cohesion the materials being joined become fluid. Then the molecules of each piece mix together. There is no foreign material in the joint.) Fluid edges flow together and fuse (become one). Solvent cementing of thermoplastics is an example of cohesion fastening.

Figure 4.38 shows uses for different solvents, adhesives, and glues. For safety and good results, follow these general rules:

- Make sure the surfaces are clean and dry. Remove grease, paint, varnish, or other coatings.
- Carefully read the instructions on the packaging.
- Carefully read any safety instructions or cautions.
- Secure a good fit between the two surfaces.
- Work in a well-ventilated area especially when using solvents and cements.
- Clamp the joint until the adhesive or solvent dries.

Solvent joining acrylic sheet. In solvent joining the low-viscosity (flows easily) solvent travels through a joint area by capillary action. (This is a force that causes a liquid to rise through a solid.)

Properly done, solvent joining yields strong, perfectly transparent joints. It will not work at all if the parts do not fit together perfectly.

CLASS	TYPE	USES	COMMENTS
Glues	Animal	Interior woodwork	Difficult to use Must be used hot Not waterproof or heat proof
	Casein	Interior woodwork	White powder mixed with water Sets in six hours Heat and water resistant
Adhesives	Polyvinyl Acetate (PVA—white glue)	Wood, leather, paper	White liquid ready to use Hardens in under one hour Not waterproof
	Plastic resins (urea and phenol)	Wood	Urea: powder mixed with water Phenol: ready to use Hardens in approximately two—six hours Urea is water resistant: phenol is waterproof Good strength
	Epoxy resin	Wood and metal	Two parts are mixed together Hardens in 12—24 hours Waterproof Very high strength
	Contact	Plastic laminates	Ready to use liquid Apply to both surfaces and let dry to touch Used in situation where clamps cannot be applied
	Instant or super glues (cyano acrylate)	Nonporous materials such as glass and ceramics	Ready to use liquid Hardens almost immediately Water resistant
Solvents	Pure Solvent (methylene chloride and ethylene dichloride)	Acrylics	Colorless liquid, ready to use Bonds almost immediately Waterproof
	Solvent cement	Acrylics	Colorless, viscous liquid Sets in 12—24 hours Waterproof
SAFETY NOTE: Use solvents and cements in well-ventilated areas			

Figure 4.38. Glues, adhesives, and solvents are designed for specific applications.

To solvent join plastic parts:

1. Hold the two pieces of acrylic in a jig, figure 4.39.
2. Apply solvent along the entire joint.
3. Work from the inside of the joint where possible.
4. Use a hypodermic syringe or needle or a nozzled applicator bottle to apply the solvent.
5. Allow the joint to dry thoroughly (from 24 to 48 hours).
6. Remove the part from the jig.

Dipping is a second method of joining acrylic sheet. As shown in figure 4.40:

1. Set up a tray of solvent. The tray must be larger than the plastic pieces.
2. Ensure that the tray is sitting level.
3. Dip only the very edge of the plastic part

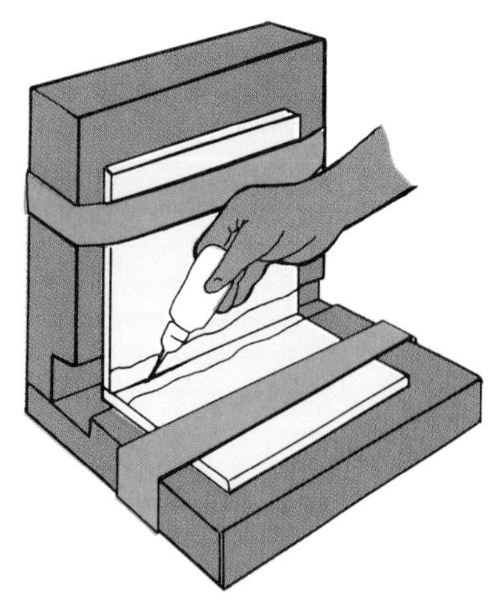

Figure 4.39. Solvent joining. Apply solvent to the inside edges of the parts where possible.

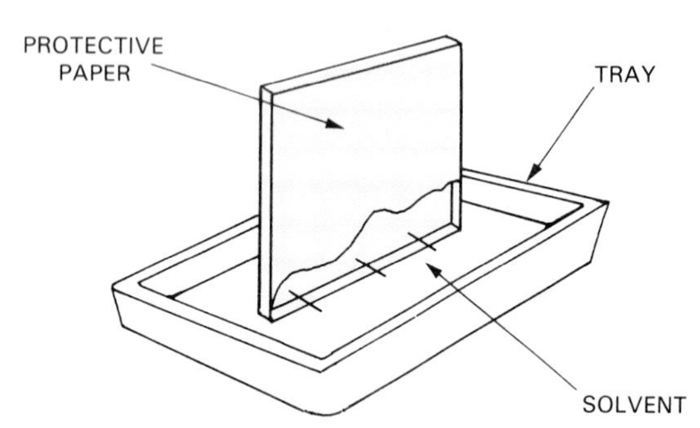

Figure 4.40. Solvent joining can be done by dipping, too.

into the solvent.
4. Use finishing nails in the bottom of the tray to keep the acrylic off the bottom.

SAFETY: Work with solvents only in a well-ventilated area!

Heat joining

Heat joining is used mostly on metals. It is also used to some extent on plastics. Two types of heat joining are used on metals:
- Welding.
- Brazing and soldering.

Welding brings metals to their melting point. When they melt, the metals flow together. When cooled, they solidify, becoming one piece. The joint is as strong as the original metal.

Brazing and soldering work differently than welding. The heat melts the metal being used to join the parts. It does not melt metal in the parts themselves.

Brazing uses a brass alloy to join the parts. The alloy melts at 1650 °F (900 °C). Lead alloy is used in soldering. It melts at only 420 °F (216 °C). At the melting temperature, the metal runs through the joint. As the metal cools it bonds the pieces.

The strength of the joining metal controls the joint's strength. The mild steel used for welding is the strongest. Brass, used for brazing, is much weaker. Still, it is stronger than the lead alloy used for soldering.

Plastics may also be joined by welding. This is possible with some thermoplastics such as PVC. A hot air torch heats the two parts of the joint. Heat fuses them.

To solder tin plate, copper, brass, and, sometimes, mild steel:
1. Heat an electric soldering iron.
2. Clean the tip with an old file.
3. Apply flux to the tip and add solder. This is called tinning. It helps the solder to flow.
4. Hold the tip of the soldering iron against the joint. When it becomes hot enough, touch the solder to the metal. As the solder melts, allow it to run along the joint, figure 4.41.
5. Allow the joint to cool slowly.

■ FINISHING MATERIALS

When a product is completed, its surface is usually finished. **Finishing** changes the surface by treating it or placing a coating on it. Finishing is done to:
- Protect the surfaces from damage caused by the environment.
- Prevent corrosion, including rust.

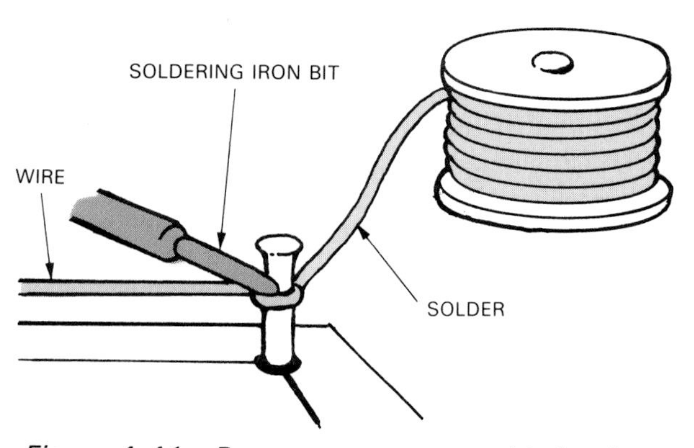

Figure 4.41. Proper way to use soldering iron.

Material	Abrasive	Comments
WOOD	Sandpaper was once the general name given to all abrasive papers used for smoothing wood. Today, the industry calls them coated abrasives.	
	Flint paper	Crushed flint or quartz used as the abrasive Wears out quickly Cuts slowly Normally used in grades coarse to extra fine (50-320 grit)
	Garnet paper	Uses garnet as the abrasive More durable than flint paper Normally used in grades coarse to extra fine (50-320 grit)
METAL	Emery cloth	Uses the natural abrasive, emery Dull black in color Normally used in grades coarse, medium, and fine (3-3/0) Oil may be added to the fine grade to give a mirror finish
WOOD AND METAL	Aluminum oxide	An artificial abrasive Gray-brown in color Tough, durable, and resistant to wear Normally used in grades coarse, medium, and fine (40-180 grit) Used on steel and other hard materials
WOOD, METAL AND PLASTIC	Silicon carbide paper (wet-and-dry paper)	An artificial abrasive Available in three common grades: coarse (50), medium (100), very fine (400) Paper is best used wet Creates a smooth, matte finish

Figure 4.42. Coated abrasives are designed for different materials. All prepare the surface of the material for finishing.

- Improve the appearance by covering the surface or treating it to bring out the natural beauty of the material.

Converted surface finishes

When the surface is treated to beautify or protect, it is called a converted surface. The material is chemically altered to change the way it reacts to elements in the environment. The protective coating is provided by the reaction of the chemical and the atoms of metal on the product's surface.

Some converted coatings are natural. Aluminum develops an oxide covering if exposed to the open air. The covering will resist the natural elements.

Surface coatings

Materials applied to a surface are called coatings. The most common coatings are paints, enamels, shellac, varnish, lacquer, vinyl, silicone, and epoxy.

The first step in finishing is to prepare the surfaces. They should be clean and smooth. Surfaces can be made smooth using abrasive papers or abrasive cloths. These are made in a wide range of grades and coarseness. Abrasive materials and their uses are described in figure 4.42.

You should follow these three general rules when using an abrasive:

- Clean inside surfaces before assembling the project.

- Begin with a coarse abrasive. Then gradually work up to a fine grade.
- Support the abrasive whenever possible, figure 4.43. A wood or cork block can be used for wood and plastic. Files can be used for holding abrasive papers while finishing metals, figure 4.44.

Figure 4.45 lists 10 different finishes. Some are for wood. Others are best used on plastics or metal. The finishes can be applied by:

- Wiping.
- Brushing.
- Rolling.
- Dipping.
- Spraying.

You can apply stain and oil to wood by wiping with a cloth. Brushes work well with most finishes. They are best with liquid plastic and paint. A roller works well for painting large surfaces. Items with many curves and parts can be dipped.

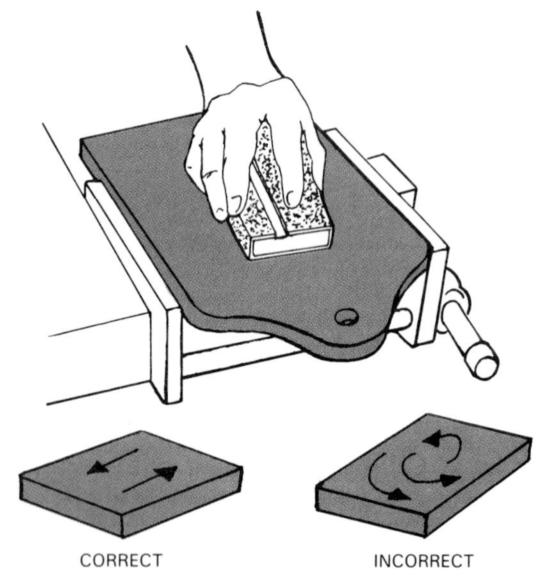

Figure 4.43. Always sand wood along the grain or you will cause scratches.

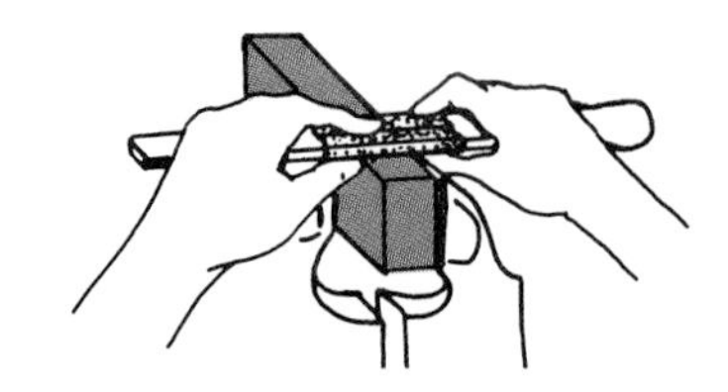

Figure 4.44. One way to use emery cloth on metals is to wrap it around a file.

Paint can be sprayed onto most shapes and materials. Aerosol spray cans are fast and easy to use but expensive. Spray guns use compressed air. They produce a high quality finish.

SUMMARY

The technologist must know how to select and use hand and machine tools safely. Marking out tools are used to measure and mark dimensions. Saws remove material quickly. Small amounts of material may be removed by filing and planing. Materials may be shaped by shearing and chiseling. Shears are used to cut thin metal and chisels to cut wood. Drilling is a process used to make holes in a material. Some sheet materials may be bent to shape. Others are formed or laminated. Casting and molding are methods of making shapes that are very difficult to produce by sawing, drilling, or filing.

Materials are joined together using mechanical devices, chemicals, or heat. Mechanical devices include nails, screws, nuts and bolts, rivets, and joints. Joints are frequently strengthened using chemicals such as glues, adhesives, solvents, or cements. Heat joining is used on metals and sometimes plastics. It includes welding, brazing, and soldering.

When the construction of an object is completed, a finish may be applied for purposes of:

- Protection of the object from damage caused by the environment.
- Corrosion prevention.
- Improvement of appearance by surface covering or surface treatment.

Finishes may be sprayed, brushed, rolled. or dipped onto the surfaces.

KEY TERMS

Bending
Casting
Chemical joining
Chiseling
Drilling
Filing
Finishing
Forming
Heat joining
Laminating
Marking out
Mechanical joining
Molding
Planing
Safety
Sawing
Shearing

Material	Type	Comments
WOOD	Liquid plastic (urethane)	Provides a clear coating Apply with a brush Gives a hard, water resistant, and long-lasting coating
	Stain	Changes the color of the wood Cheaper wood can be stained to resemble the color of more expensive woods Applied with brush or cloth Another clear, protective finish must be applied later
	Paint	Two types: latex (water based) and oil based The surface must be primed with a primer coat Read and follow manufacturers' suggestions
	Plastic laminate	Provides a decorative, durable surface The laminate is glued to a flat surface using contact cement
	Creosote and pressure treatment	Wood is immersed in creosote or a preservative is forced into the wood under pressure Exterior use only
	Oil	Teak oil is preferred, as linseed oil requires preparation Used on hand-made furniture
METAL	Paint	Surface must be completely free of oil and grease First apply primer coat, next an undercoat, and finally a top coat
	Plastic coating	Metal is heated and dipped into fine particles of PVC that soften under the heat to form a smooth coating Useful for tool handles
	Enameling	A thin layer of glass is fused onto a metal surface For decorative work, copper is the metal used
PLASTIC	Polish	Surfaces are polished using very fine silicone-carbide paper followed by using a buffing attachment on a power drill Buff with a light pressure to prevent melting the plastic
	Dye	Dip transparent plastic in a strong dye for a few minutes to give a tinted effect

Figure 4.45. Finishes are made for many different uses.

TEST YOUR KNOWLEDGE

Write your answers to these review questions on a separate sheet of paper.

1. All loose clothing should be fastened or removed ______.
   a. Before operating any machine.
   b. During the operation of the machine.
   c. After operating a machine.
   d. Only when you are assisting the teacher.
2. When passing hand tools to a friend you should ______.
   a. Slide them across the bench.
   b. Pass them with the sharp edge towards you.
   c. Pass them with the sharp edge away from you.
   d. Explain how they are to be used.
3. Safety glasses or a face shield must be worn ______.
   a. On the drill press only.
   b. On the band saw only.
   c. Every time you use a machine.
   d. Only by those people who wear prescription glasses.
4. Before cleaning or adjusting any machine you should ______.
   a. Allow the machine to come to a complete stop.
   b. Run the machine at full speed.
   c. Allow the machine to coast slowly.
   d. Make sure the power is switched on.
5. Before plugging in a portable electric drill you should ______.
   a. Place the chuck key into the chuck.
   b. Lock the drill switch in the ON position.
   c. Remove the drill bit.
   d. Make sure the drill bit is secure in the chuck.
6. To mark a piece of wood to length, use a ________ and ________.
7. A ________ is used to make straight cuts through metal and plastic.
8. The edge of a piece of metal is made smooth and flat using a ________.
9. The surface of a piece of wood is made flat using a ________.
10. Tin plate, aluminum, and copper sheet are cut using ________.
11. The process used to cut a round hole in wood, metal, and plastics is called ________.
12. Laminating means ____.
13. A strip heater is used to ____.
14. Pouring liquid metal or plastic into a mold is called ________.
15. Describe the difference between mechanical, chemical, and heat joining. Give examples of each.
16. Why is a finish applied to the surface of a material?
17. Which of the following materials does NOT require a protective or decorative surface finish? ______
   a. Pine.
   b. Acrylic.
   c. Maple.
   d. Steel.
18. A finish may be applied to a surface of a material by ________, ________, ________, ________, or ________.
19. What are the general rules to follow when using an abrasive to prepare surfaces for finishing?
20. If you were buying a basic kit of tools, which would you buy?

APPLY YOUR KNOWLEDGE

1. Design and make a simple game for a young child.
2. List what you consider to be the five most dangerous activities in the technology room. What would you do to reduce the dangers?
3. Name five objects in your home. Describe how the parts of each object are joined.
4. Choose five objects each with a different type of finish. Make a chart to show (a) the material, (b) the finish, and (c) the reason why that finish has been used.
5. Make a list of the tools you have in your home. State the material and process for which each is designed.

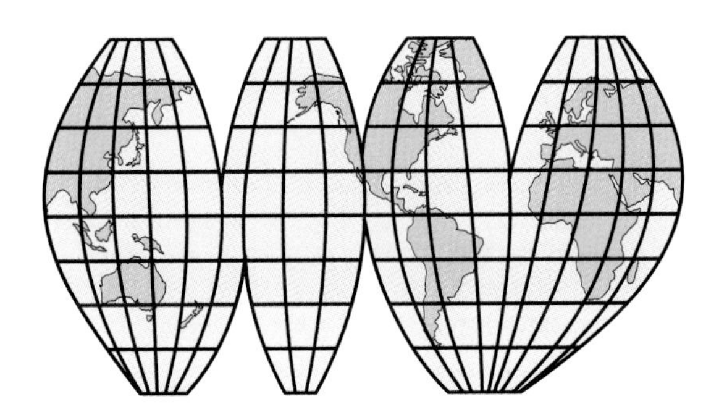

Chapter 5

Structures

OBJECTIVES

After reading this chapter you will be able to:

- Recognize many different types of structures, both natural ones and those made by humans.
- Understand that structures made by humans include bridges, buildings, dams, harbors, roads, towers, and tunnels.
- Identify the loads acting on structures.
- Analyze the forces acting on a structure.
- Demonstrate how structures can be designed to withstand loads.

Structures are all around us. We build them to live in or to cross a river. We build them to carry wires, to receive radio waves, and to transport people. Not only are houses, bridges, and towers, structures; so are airplanes, boats, and cars.

The main purpose of a structure is to enclose and define a space. At times, however, a structure is built to connect two points. This is the case with bridges and elevators. Still other structures are meant to hold back natural forces, as in the case of dams and retaining walls.

Everyone has built some kind of structure. Did you ever use a cardboard box to build a playhouse large enough to crawl inside? Have you ever constructed a ramp for a skateboard? Perhaps you built a treehouse from a variety of scrap materials. Maybe you have made a model crane, a dollhouse, a tunnel for a model railroad, or a sand castle on the beach. Figure 5.1 shows a simple structure. Do you recognize it?

Not all structures are made by humans. Living organisms, such as trees and our bodies are natural structures. A giant redwood tree must be rigid enough to carry its own weight. Yet it is able to sway in high winds. Grass is flexible. It must spring back after it is stepped on. The bones of a skeleton have movable joints. They permit activities such as running and lifting. Figure 5.2 shows both natural and human-made structures.

■ WHAT STRUCTURES HAVE IN COMMON

What do all structures have in common? They all have a number of parts. The parts are

CHRISTOPHARO

Figure 5.1. A scaffold supports workers while they are building a structure. It is a structure, too! It has connected parts. It carries the worker's mass without collapsing.

WASPS BUILD DWELLINGS. ECRITEK

THE TALLEST STRUCTURE. ECRITEK

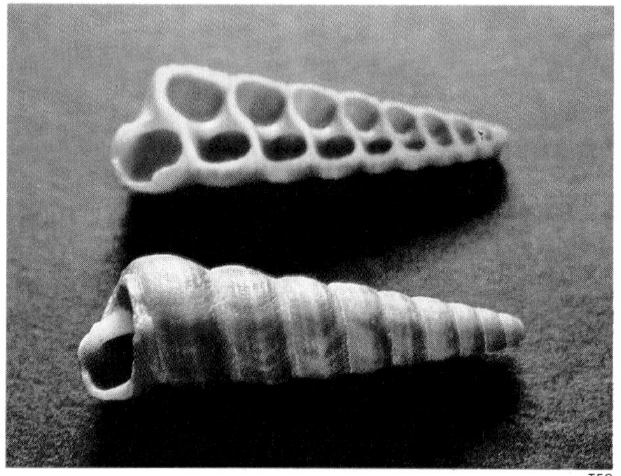

A "HOME" FOR A TYPE OF SHELLFISH. TEC

CHILDREN MAKE PLAY HOUSES IN THE SNOW. ECRITEK

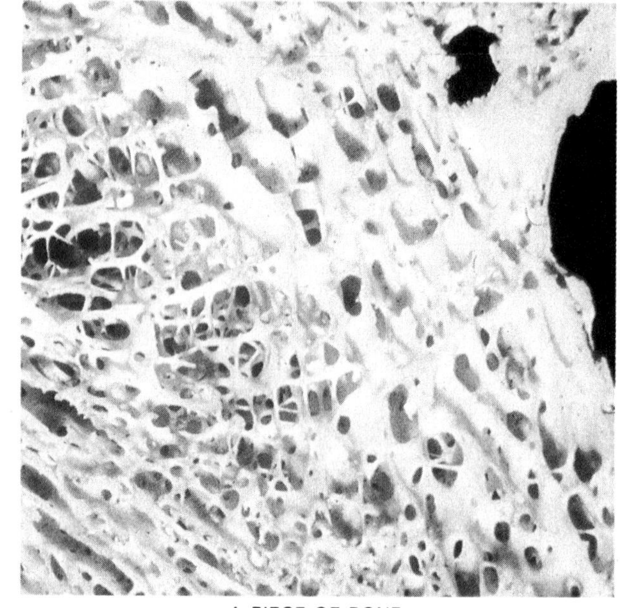

A PIECE OF BONE.

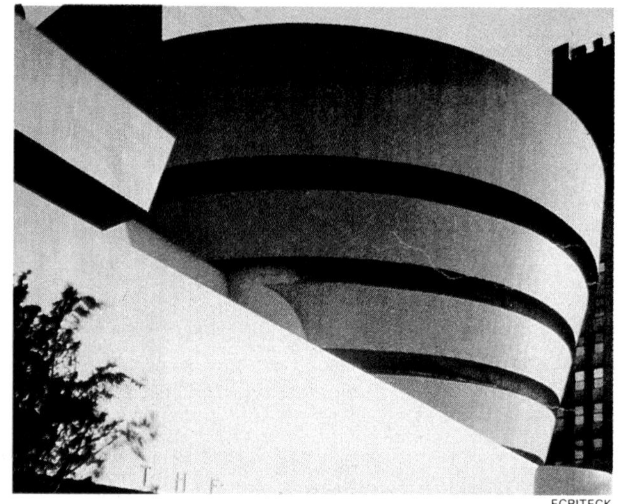

THE GUGGENHEIM MUSEUM. ECRITECK

Figure 5.2. Structures are found all around us. Some (above, left) are found in nature. Others (above, right) are planned and built by humans.

connected. The parts provide support so the structures can serve their purpose. One important job of all structures is to support a **load.** A load is the weight, mass, or force placed on a structure. For example, the "load" on a bridge would be the heavy vehicles crossing it. The load for a dam is the force of the water behind it. Both must also support the materials from which they are built. This is part of the load.

■ TYPES OF STRUCTURES

Structures vary greatly in size and type. Look at the photographs in figures 5.3 through 5.6. As you look, think about the loads that each of the structures must withstand. Think of the materials used in their construction. Think how the parts are connected together.

■ STATIC AND DYNAMIC LOADS

All structures must be able to support a load without collapsing. A roof must not only support its own mass but also a heavy blanket of snow. A dining chair must carry the load of a person sitting or fidgeting, figure 5.7.

There are two types of load: **static** and **dynamic.** Static loads are unchanging or change slowly. Dynamic loads are always changing.

Static loads may be caused by the mass (weight) of the structure itself. Columns, beams, floors, and roof are part of this load. They are also caused by objects placed in or on the structure. When you sit at a desk reading a book, both you and the furniture create a static load. Figure 5.8 is an example of such a load.

A ECRITEK

C ECRITEK

B TEC

D ECRITEK

Figure 5.3 Buildings. How is the framework of each building like a skeleton? A—Frame of a highrise. B—Geodesic dome. C—Modern frame tent. D—Frame houses under construction.

Figure 5.4. Roads, sidewalks, and bridges are important structures. They help us travel from place to place. A—Pipes that supply water, electricity, and fuel are often built under sidewalks or roads. B—Tunnel may provide a walkway under obstructions. C—Bridges span other obstacles. D—This expressway cloverleaf looks like spaghetti.

Figure 5.5 Cranes, towers, and platforms. What other similar structures have you seen? What do they lift or support?

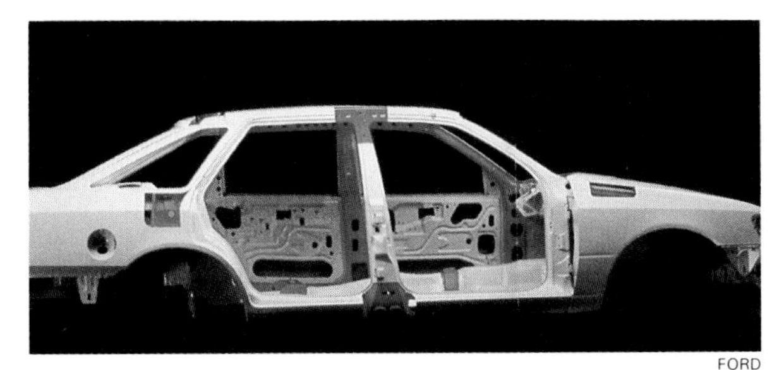

FORD

ECRITEK

Figure 5.6. Structures for transportation vehicles. Both structures must carry people and support other parts of the vehicle.

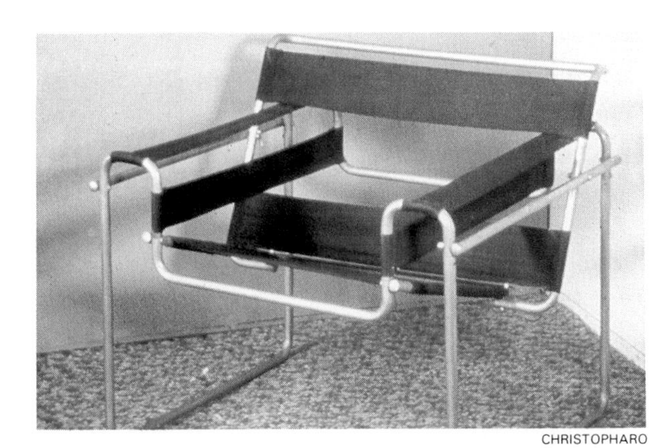

CHRISTOPHARO

ECRITEK

Figure 5.7. The structure of a chair must be such that it can carry the load of a person sitting on it.

Figure 5.8. Objects at rest create static loads.

Figure 5.9. Moving objects create dynamic loads.

Dynamic loads are loads that move. For example, the mass of a person bouncing on a chair creates a dynamic load. Other dynamic loads include the force of a gust of wind pushing against a tall building. A heavy truck crossing a bridge would also be a dynamic load, figure 5.9.

■ FORCES ACTING ON STRUCTURES

Both static and dynamic loads create forces. These act on structures. To understand these forces and what they do, imagine a plank placed across a stream, figure 5.10. When you (the load) walk across the plank (the structure) what would you expect to happen? The plank will bend in the middle. The forces acting on the bridge may be shown by the foam rubber in figure 5.11. Notice that parallel lines have been marked on it. Support the foam at each end. A vertical load applied to the center of the foam will bend it, figure 5.12.

Notice what has happened to the parallel lines. At the top edge, the lines have moved closer together. The lines at the bottom edge have moved farther apart. The top edge of the plank is in compression (being squeezed) and the bottom edge is in tension (being stretched). Along the center is a line that is neither in compression nor in tension. It has no force acting along it. This line is called the neutral axis.

Figure 5.10. A person standing on a plank is a static load. Bending will cause compression on its top surface and tension on its bottom surface.

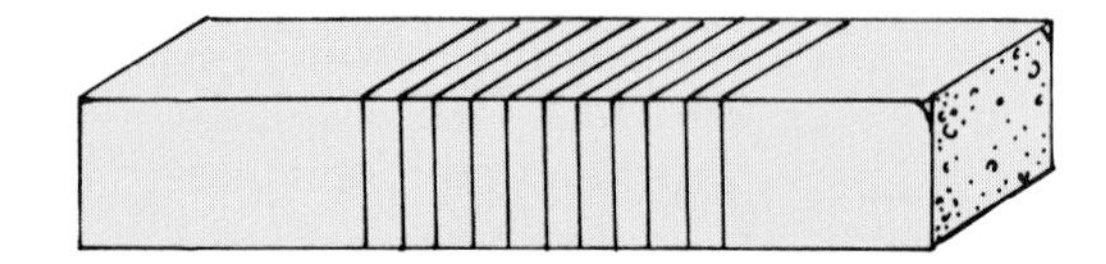

Figure 5.11. Foam rubber with parallel lines drawn on it will show what happens when a load is placed on a beam.

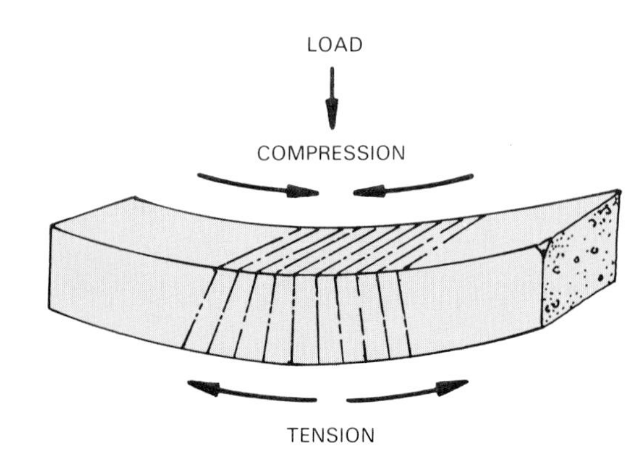

Figure 5.12. Stresses of compression and tension are caused by bending.

The design and construction of structures must minimize (reduce to least amount) the effects of bending. Parts must be shaped and connected so there is little chance to bend.

■ DESIGNING STRUCTURES TO WITHSTAND LOADS

As was shown by the foam rubber in figure 5.12 and the shelf in figure 5.13, the top and bottom surfaces of a beam are subject to the greatest compression and tension. These surfaces are where the greatest strength is needed. Strengthening a beam along these surfaces gives the shapes shown in figure 5.14. After members have been shaped to resist compression and tension, they must be

Figure 5.13. What happens when you load a long shelf with books?

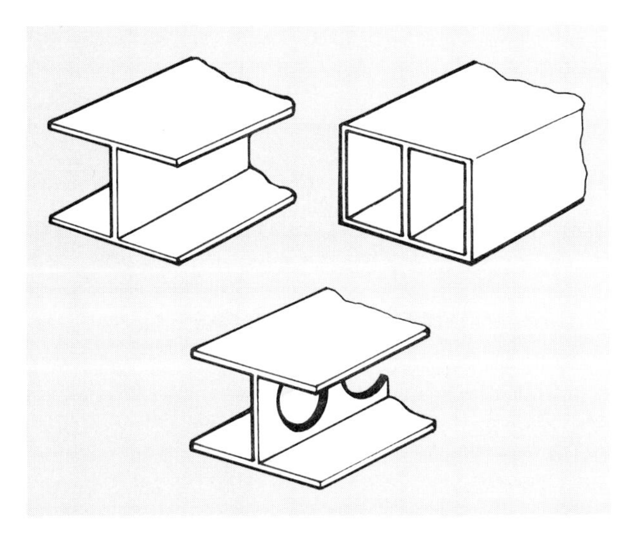

Figure 5.14. The shapes shown here will support heavy loads.

connected in a way that minimizes bending. Look at the structures in figure 5.15. What shape appears most often?

You can see that the triangle appears most often. To understand why the triangle is important in structures, look at figure 5.16. The frame is made of four connected members. If a load is applied at A, the frame retains its shape. However, if a load is applied at corners B or C, the frame will collapse. Now compare this frame to the one in figure 5.17.

Figure 5.15. Left. Pylon. Right. Geodesic dome. What shape appears most often?

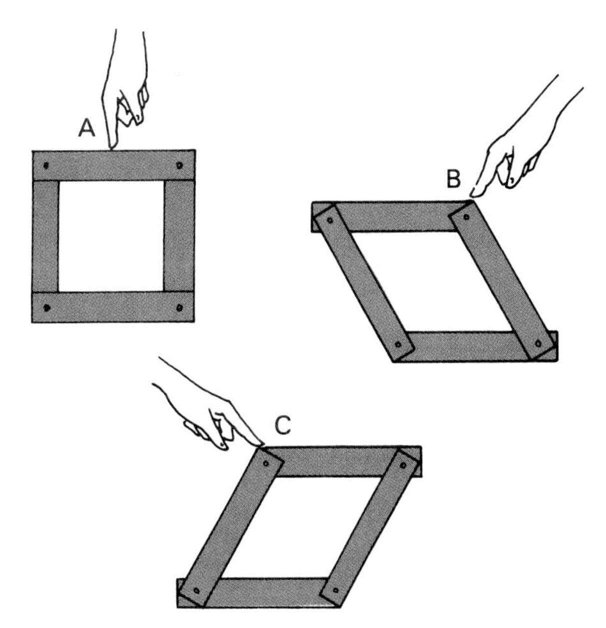

Figure 5.16. Loads placed on a frame. Two will collapse from the load. Why?

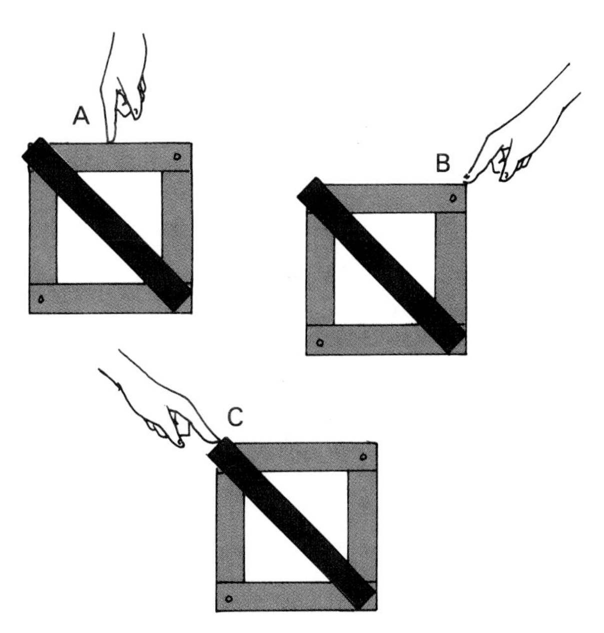

Figure 5.17. The frames retain their shape from loads at A, B, and C.

A rigid diagonal member (running from corner to corner) has been added. Once again when a load is applied at A the frame retains its shape. This time, however, it also retains its shape when a load is applied at corners B or C. At corner B, the load causes the diagonal to be in tension. A rigid member in tension is

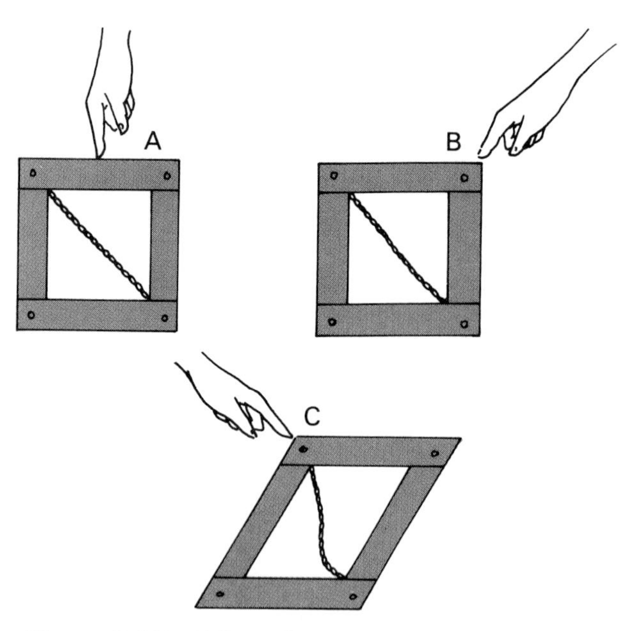

Figure 5.18. Why will a rope or chain work as a tie but not as a strut?

called a **tie.** When the load is applied at corner C, the diagonal is in compression. A rigid member in compression is called a **strut.**

What would be the effect of replacing the rigid diagonal member with a nonrigid member such as a rope, chain, or cable? Would the frame retain its shape when loaded in each of the three positions? When is the rope in compression and when is it in tension?

In addition to compression and tension there is a third force acting on structures. This force is called shear. To understand how shear takes place, imagine you are pulling the wagon in figure 5.19. Suddenly the wheels hit a rock. The effect is a sharp jolt on the pin. This force causes the material to shear. Let us see how bending and the forces of compression, tension, and shear are resisted in the design of structures. Then we will see why bridges are built the way they are.

A major problem with bridges is that they bend, figure 5.20. One common way to prevent a beam bridge from bending is to support the center with a **pier** (like a strut) as in figure 5.21. However, it is not always possible to build piers under a bridge. Piers may not allow the passage of ships. Sometimes the river is too deep. It may run too swiftly. It may have a soft bed with no firm foundation. Other ways have to be found to strengthen the beam bridge.

One solution is to make the beam much thicker. This, however, would make the beam very heavy. Its own mass would make it sag in the middle.

The beam could also be strengthened at the center where it is most likely to bend or break, figure 5.22. Once again, notice that the strongest shape is the triangle. As we saw in figure 5.17, a triangle does not have to be solid; it can be a frame and still be very rigid.

Figure 5.19. Shear is a force that causes materials to break.

Figure 5.20. A simple beam bridge bends easily.

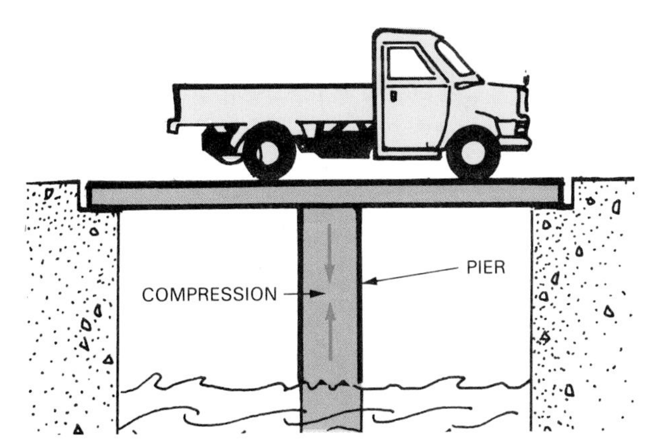

Figure 5.21. The pier of a beam bridge is compressed by the load on it.

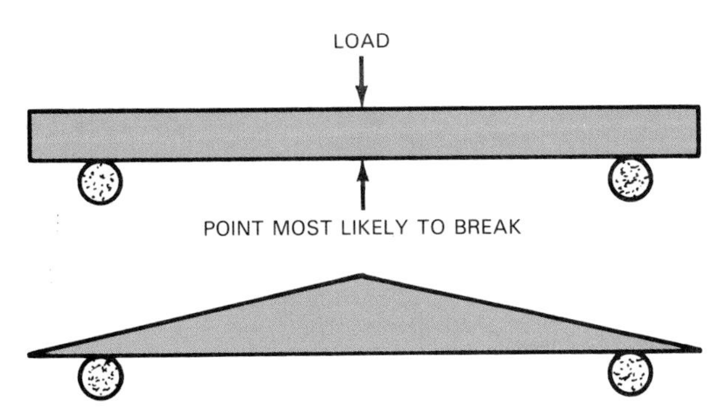

Figure 5.22. One way to strengthen a beam bridge is to make it thicker in the middle.

Truss bridges make use of the triangle in their design, figure 5.23. As the truck crosses the bridge its mass causes the bridge deck (roadway) to bend. Member "A" moves down. This in turn pulls down on members "B" and "C," pulling them towards the end of the bridge and carrying the forces out to the bridge supports. This simple truss bridge can be used for bridges up to 60 ft. (18 m) long.

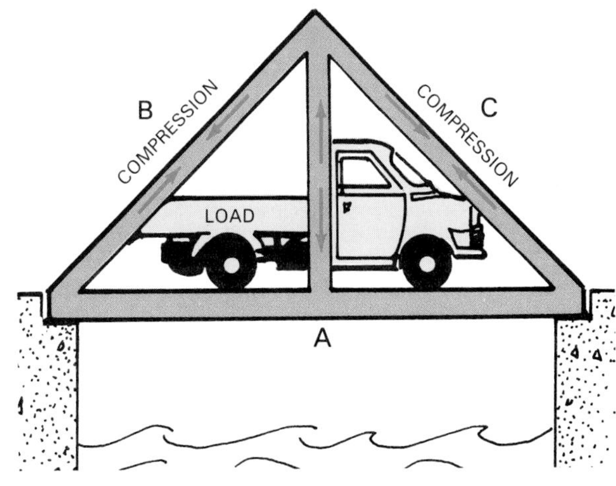

Figure 5.23. A simple truss bridge. Is the center (vertical) beam under tension or compression?

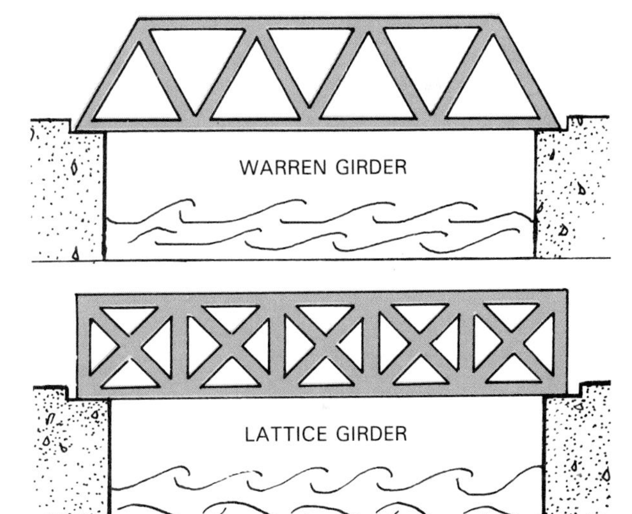

Figure 5.24. Truss bridges. A truss is a long beam made up of shorter beams or girders that give strength to one another.

Most truss bridges are longer than that. Therefore, a series of triangular frames are used to construct them, figure 5.24. A bridge deck can also be supported from above. Cables, called **stays,** provide the support, figure 5.25. Notice that the pylons are in compression and the stays are in tension.

This principle is used for **suspension bridges.** Suspension bridges are the longest, figure 5.26. The bridge deck is suspended (hung) from hangers attached to a continuous cable. The cable is securely anchored into the ground

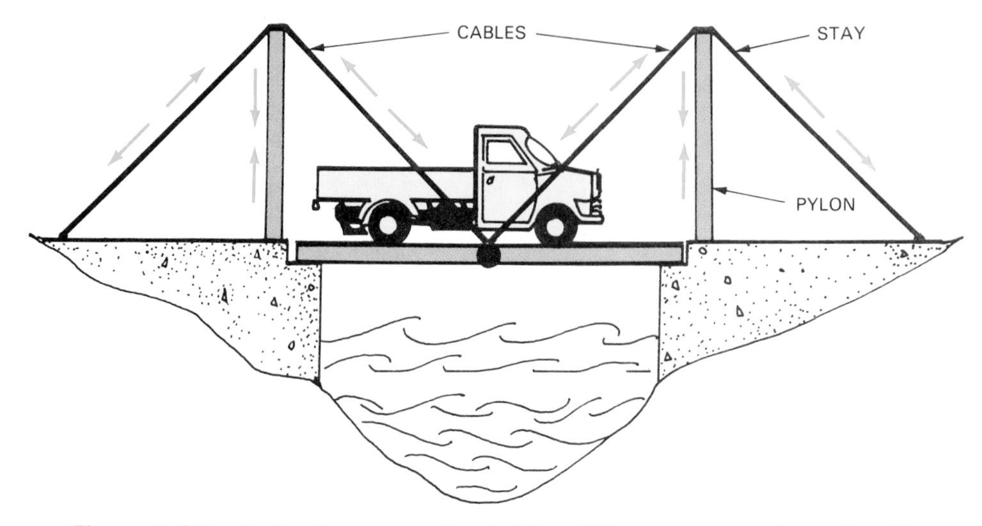

Figure 5.25. A simple suspension bridge. What supports the cables?

Figure 5.26. Examples of suspension bridges. Top. Note the steel cables supported by the tower. Bottom. The Humber Bridge in England, longest single-span suspension bridge in the world, stretches across the Humber Estuary.

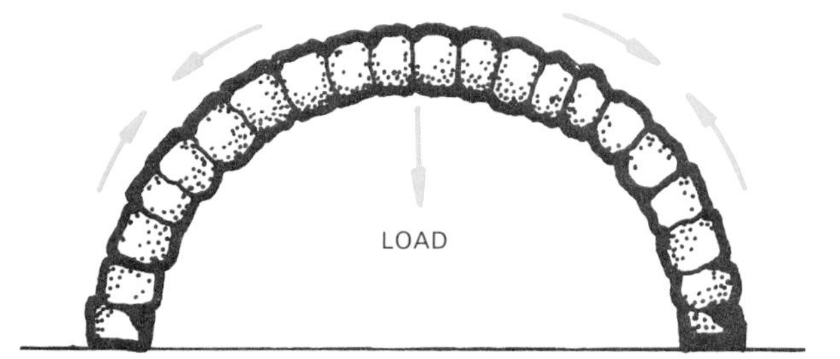

ECRITEK

Figure 5.27. Top. A simple arch bridge. Arch transfers load back to its ground supports. Bottom. World's longest arch bridge spans the New River Gorge in West Virginia. It is 3030 ft. (923.5 m) long.

at both ends. The cables transfer the mass of the deck to the top of the towers. From there, compression transfers the mass to the ground.

There are many other types of bridges. Their design follows the same general principle: try to reduce bending. Two of the most common types are **arch** and **cantilever** bridges.

In an arch bridge, the compressive stress created by the load is spread over the arch as a whole. The mass is transferred outward along two curving paths. The supports where the arch meets the ground are called **abutments.** They resist the outward thrust (push) and keep the bridge up, figure 5.27.

A beam can support a load at one end provided that the opposite end is anchored or fixed. This is known as a cantilever beam. The principle of a cantilever is seen in figure 5.28. A cantilever bridge has two cantilevers with a short beam to complete the span, figure 5.29.

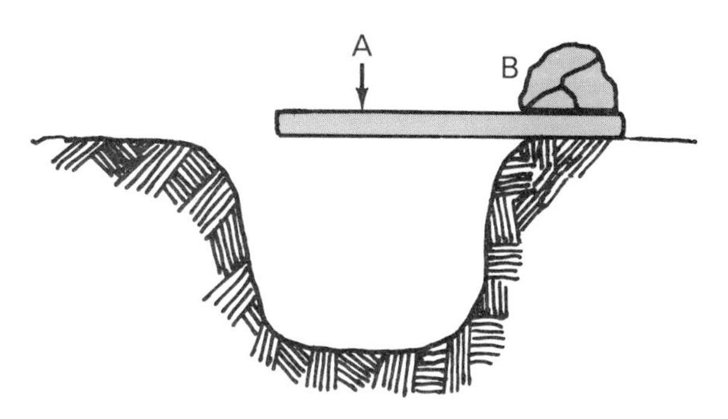

Figure 5.28. This shows the principle of the cantilever. Load at A is tranferred to B.

Bridges are made from many materials. The most common are stone, steel, and concrete. Stone is cheap and resists weather, fire, and corrosion. It is strong under compression but weaker under tension. Steel is fairly cheap, strong under compression and tension but needs maintenance to prevent corrosion (rust). Concrete is cheap and resists fire and corrosion. It is strong under compression but weak under tension. However, it can be strengthened with steel rods.

Reinforced concrete

Most modern bridges use steel and concrete. Steel cables made of wire rope are used to support the mass of the roadway and the traffic load on it. The towers of many bridges are made of steel. Steel trusses, like enormous beams, give rigidity (stiffness) to the bridge deck. They also resist bending.

Many bridges use concrete even though it is weak in tension. To overcome this weakness, the concrete is reinforced with steel rods wherever it is in tension. The embedding of steel rods to increase the resistance to tension is the basic principle of **reinforced concrete,** figure 5.30.

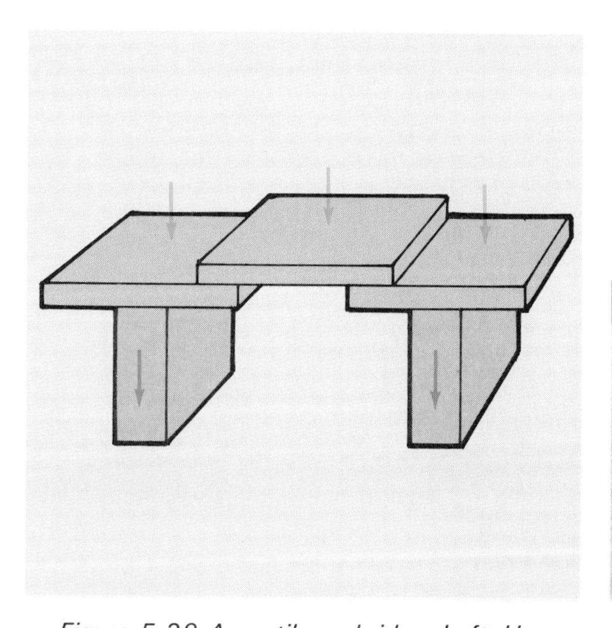

ECRITEK

Figure 5.29.A cantilever bridge. Left. How two cantilevers work. Right. An actual cantilever bridge.

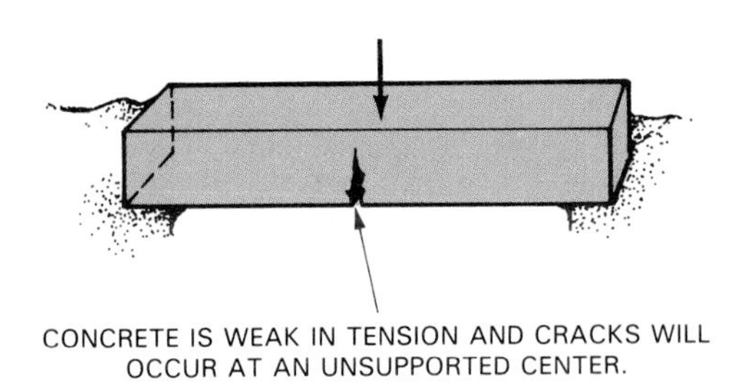

CONCRETE IS WEAK IN TENSION AND CRACKS WILL OCCUR AT AN UNSUPPORTED CENTER.

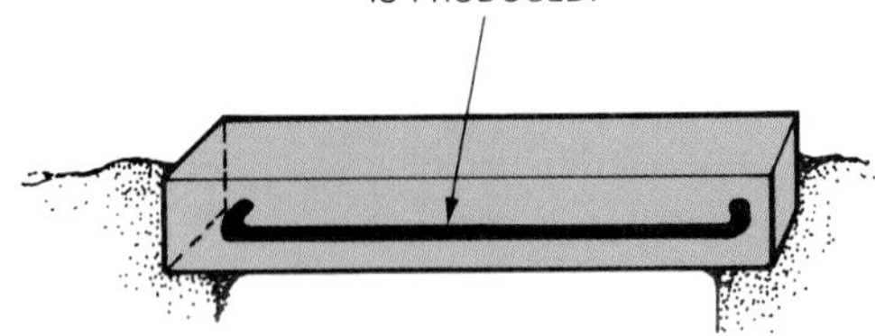

REINFORCED CONCRETE USES STEEL RODS TO RESIST TENSION. IF THESE RODS ARE STRETCHED WHILE THE CONCRETE IS HARDENING, PRE-STRESSED CONCRETE IS PRODUCED.

Figure 5.30. Concrete is made stronger with steel reinforcing rods.

SUMMARY

All structures comprise a number of connected parts. These parts provide support and withstand a load without collapsing. There are two types of load: static and dynamic. These loads create the forces of compression, tension, and shear. Individual members of a structure must be designed to minimize the effects of these forces. The members are then connected together in such a way as to minimize bending.

Bridges provide an example of how structures are designed to resist forces. A truss bridge uses the rigidity of the triangle to resist the forces of compression and tension. These same forces are resisted by cables and pylons in a suspension bridge. There are many other types of bridges. As with all structures, they are designed to withstand loads and minimize bending.

KEY TERMS

Abutment	Static load
Arch bridge	Stays
Cantilever	Structure
Dynamic load	Strut
Load	Suspension bridge
Pier	Tie
Reinforced concrete	Truss

TEST YOUR KNOWLEDGE

Write your answers to these review questions on a separate sheet of paper.

1. Name three natural structures and three structures made by humans.
2. Which of the following is NOT a natural structure?
   a. Spider's web.
   b. Bridge.
   c. Tree.
   d. Beaver's dam.
3. All structures are:
   a. Built to withstand heat.
   b. Made in factories.
   c. Built to withstand a load.
   d. Designed to house people.
4. Name the two types of load acting on structures. Give one example of each.
5. What forces are acting on the top and bottom surfaces of a beam loaded from above?
6. To strengthen a beam loaded on the top surface, it must be reinforced at the:
   a. Top surface only.
   b. Bottom surface only.
   c. Center.
   d. Top and bottom surfaces.
7. Which geometric shape gives the greatest rigidity to a structure?
   a. Square.
   b. Circle.
   c. Rectangle.
   d. Triangle.
8. A beam in compression is called a:
   a. Strut.
   b. Tie.
   c. Post.
   d. Stay.
9. A beam in tension is called a:
   a. Strut.
   b. Tie.
   c. Post.
   d. Stay.

10. A bridge that uses a series of triangular frames is called a __________ bridge.
11. The world's longest bridges are __________ bridges.
12. Using notes and diagrams, explain how an arch bridge resists loads.
13. Using notes and diagrams, explain the principle of a cantilever bridge.
14. What are the most common materials from which bridges are built?
15. Concrete is weak in tension. How is this problem overcome?

APPLY YOUR KNOWLEDGE

1. Look at the natural structures in the illustrations. Next look at the structures made by humans. For each of the structures made by humans, name the natural structure it most closely resembles.
2. Look at the structures in figures 5.4 to 5.6. Write the location or address of a structure in your town that most closely resembles each one.
3. Name five different structures. For each structure list the loads to which it is subjected. State whether each load is static or dynamic.
4. Draw a diagram of a plank bridge with a load on it. Label your diagram to show the forces of tension and compression.
5. Using only one sheet of newspaper and 4 in. (10 cm) of clear tape, construct the tallest freestanding tower possible.
6. Using drinking straws and pins, construct a bridge to span a gap of 20 in. (508 mm) and support the largest mass possible at midpoint.

TEC

Habitat '67. Each concrete module supports its own load and those of the apartments above it.

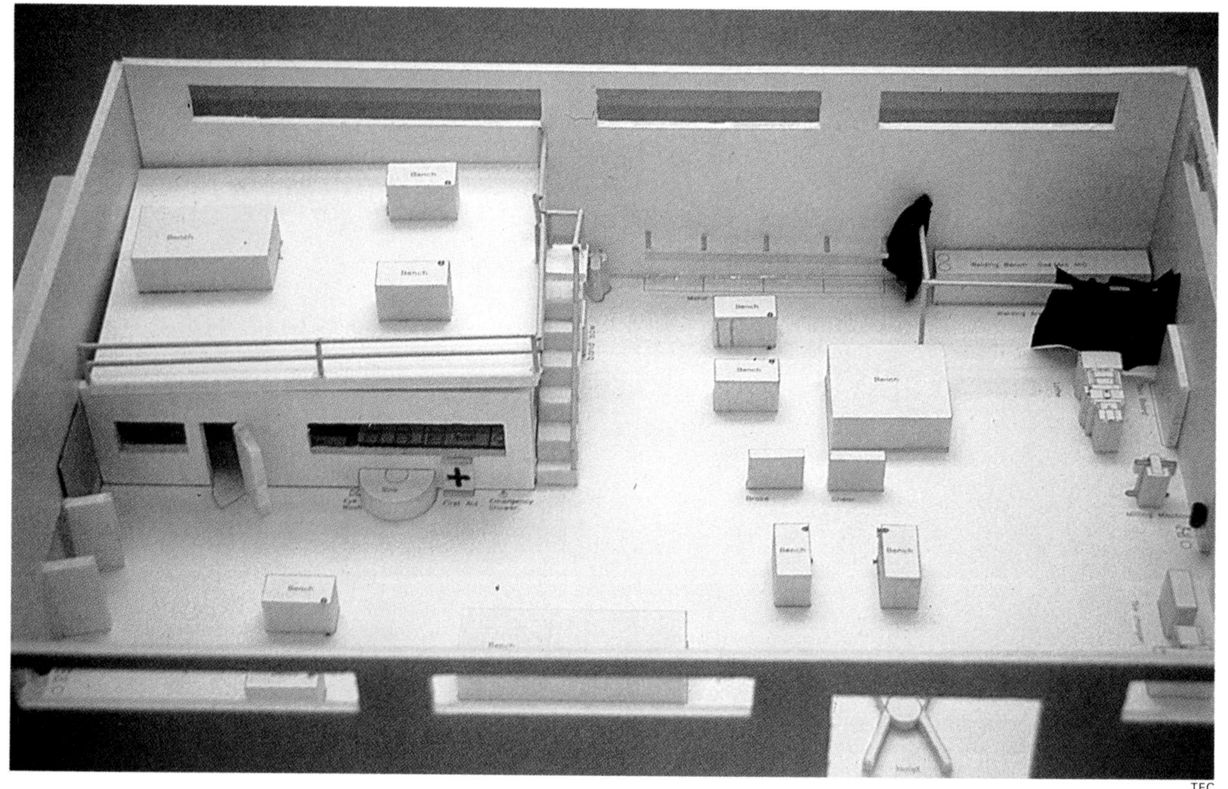

TEC

When designing a school workshop, an architect will build a scale model.

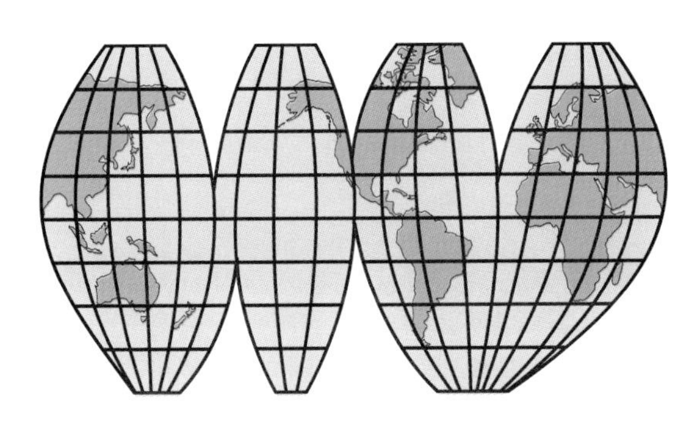

Chapter 6
Construction

OBJECTIVES

After reading this chapter you will be able to:

- O Identify the principal types of residential buildings found in modern communities.
- O Discuss the steps involved in selecting and buying a building site.
- O Identify the component parts of a typical house and describe the function of each part.
- O Identify the principal materials used in the construction of a house.
- O Recognize that all systems have inputs, a process, and outputs.
- O Use a systems model to explain an example of technology.
- O Differentiate between an open- and a closed-loop system.
- O Design, build, and evaluate a scale model of a home.

■ THE STRUCTURE OF A HOUSE

The structure most familiar to you is your home. Many materials have been used for different kinds of homes, figure 6.1. Traditionally, people have built houses from the handiest material. They used what was most commonly available nearby. In North America, clay and straw, called **adobe,** were used in the Southwest. Thick walls kept the home warm in winter and cool in summer. Pioneers who homesteaded the Great Plains used sod cut by plow. The sod served as walls and sometimes roofs. In wooded regions of the north where lumber was plentiful, log cabins were built. Wood is still a popular building material in North America. Treatment of the wood to retard fire and decay has made the frame

MEXICAN HOUSE MADE OF BRANCHES, CLAY, AND STRAW. IT HAS A FLAT ROOF AND FEW WINDOWS FOR LIFE IN A HOT DRY, LAND.
AN INUIT IGLOO BUILT OF BLOCKS OF SNOW.
NORTH AMERICAN INDIAN TENT MADE OF SKINS STRETCHED OVER A POLE FRAME.
LIBERIAN HUT MADE OF MUD, STICKS, AND FOLIAGE.

Figure 6.1. The shape of homes is often the result of materials found locally.

house more durable than ever.

Building a home entirely on its site often takes a number of months. A new method is often used to speed up construction. It is called **prefabrication.** The term means building parts of the house in a factory. This is much faster because the parts can be made on an assembly line with power tools and heavy equipment. Workers are not affected by bad weather.

Prefabricated parts are moved to the building site for final assembly. Time is saved in the factory because of mass production methods. Time is saved on the site because much of the assembly has already been done.

A popular type of prefabrication is known as **modular construction.** Modules are basic units such as rooms. Modules of different sizes and shapes can be combined on site.

The house frame

The frame of a house provides a supporting structure. The simplest framed structure is a **post-and-lintel** or post-and-beam, figure 6.2. The lintel is a beam simply supported on the posts. It carries the roof load. The posts are vertical struts compressed by the lintel. Post-and-lintel structures may be built one on top of another to frame multistory buildings.

Like the bridges you have read about, houses must also support loads. The structure of a typical house must support the static loads. These are the mass (weight) of the materials from which it is built and its contents. The house must also withstand dynamic loads created by weather conditions outside and the movement of people inside. To understand how these loads are supported in a house, look at figure 6.3.

Loads are usually applied to horizontal members such as **joists** or **beams.** The total load then moves downward. It transfers to (travels down) columns or bearing walls to the foundation. Then the load transfers to the soil. Therefore, all loads and forces tend to push downward. Horizontal forces are sometimes

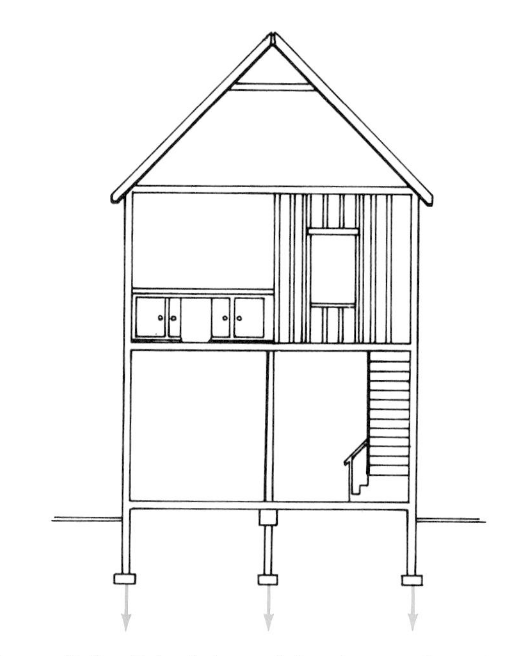

Figure 6.3. Principles of load transfer. Arrows show how load is carried through posts down to the foundation.

Figure 6.2. Post-and-lintel structures. A—Heavy posts support the horizontal lintel or beam. B—A post-and-beam building under construction.

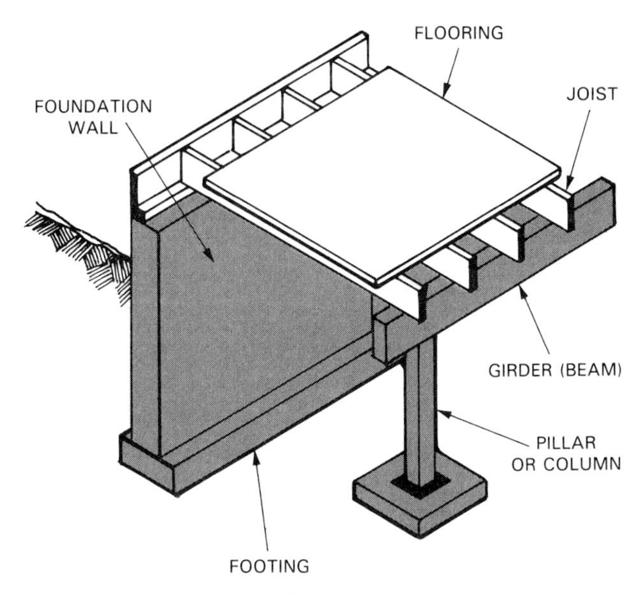

Figure 6.4. Loads on horizontal surfaces move to the soil through vertical members of a building.

produced by sloped roofs and stairs. However, these are also translated into vertical forces.

Think about the loads placed on the floor of your home if you and your friends are dancing. The mass of the people acts as a load. This load is transferred to a beam, then to a column. The load on the column is, in turn, transferred to the footing and to the underlying soil, figure 6.4.

■ PLANNING FOR A HOME

Before construction of a house can begin, two major decisions must be made. One is to decide on the basic type of house. The other is to plan the house site. (A site is the land where a house is built.)

Four types of single family houses are shown in figures 6.5 to 6.8. The type one chooses must be based on a number of considerations. How much room is needed? How much land is available for a home? Which type has the most appeal? Will any family member have difficulty with steps? How much money do you wish to spend? Is the money available? Often banks or other lending institutions make loans for building.

Planning inside space

The second major task in the design and building of a home is to plan the interior

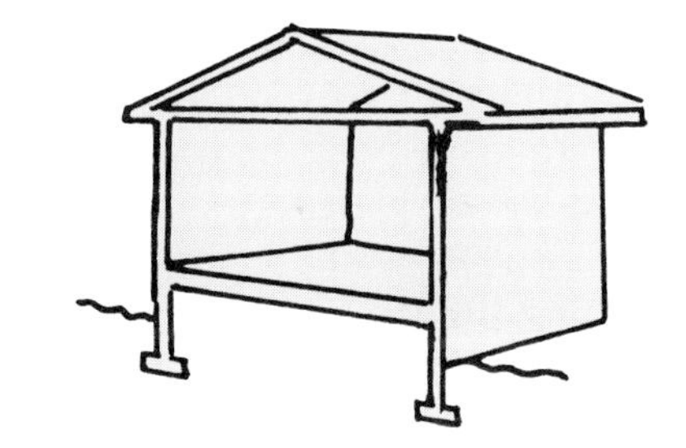

ECRITEK

Figure 6.5. A bungalow has only one story.

ECRITEK

Figure 6.6. A one-and-one-half-story home. The second-level rooms extend into the roof.

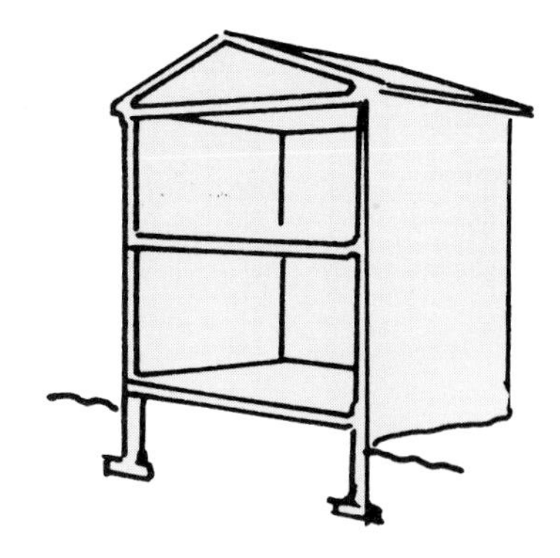

ECRITEK

Figure 6.7. A two-story home. Roof extends above the second story.

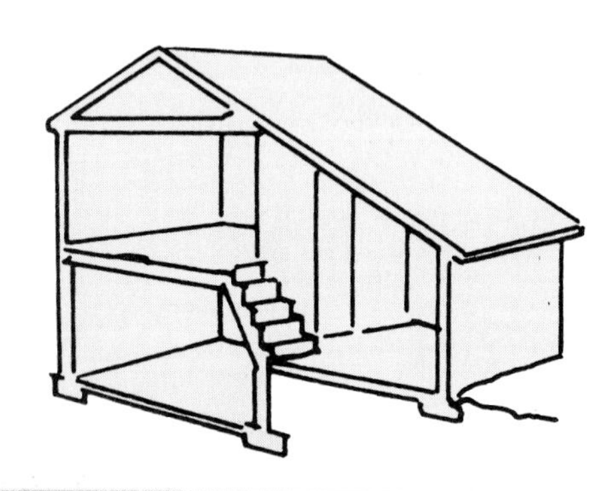

ECRITEK

Figure 6.8. A split-level home is divided vertically. Floors of one part are located midway between the floors of the other part.

spaces. This task is often performed by an architect.

What if there were no interior walls in your home and you had to divide the space into a number of rooms? How would you do it?

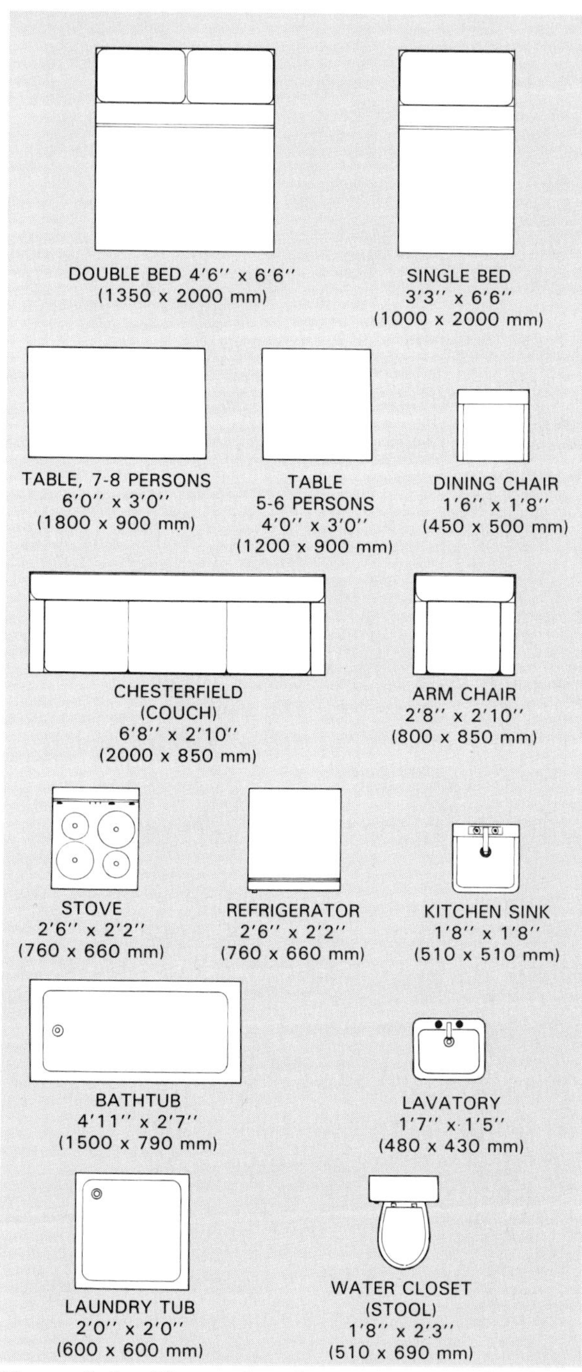

Figure 6.9. These drawings represent furniture and fixtures found in your home. They are used on floor plans so you can visualize (imagine) how they will fit in a home.

First of all, think about the spaces needed for the things in a home. You must be able to fit in all of your furniture and household equipment. You should also leave sufficient space to move around. The sizes of most items of furniture and equipment are fairly standard. Figure 6.9 shows a plan view and floor space required for common items.

Finding space needs. To determine the overall size and shape of any individual room, you can use small pieces of card stock. Cut the card stock to represent the size and shape of each piece of furniture. Position the shapes in different ways. Study each arrangement. Continue until you find a suitable arrangement. Leave space for people to circulate around the room. There must also be room to open doors and pass through doorways. The result would provide the size and shape of the room, figure 6.10. Next, fit the rooms into the overall shape of the house. Generally, rooms are grouped according to their functions: living, sleeping, and working. There are a number of reasons for grouping rooms:

- To separate noisy and quiet areas. Family members can work or play without disturbing others who are resting or studying.
- To place bedrooms and bathroom close to each other for convenience in washing, bathing, and dressing.

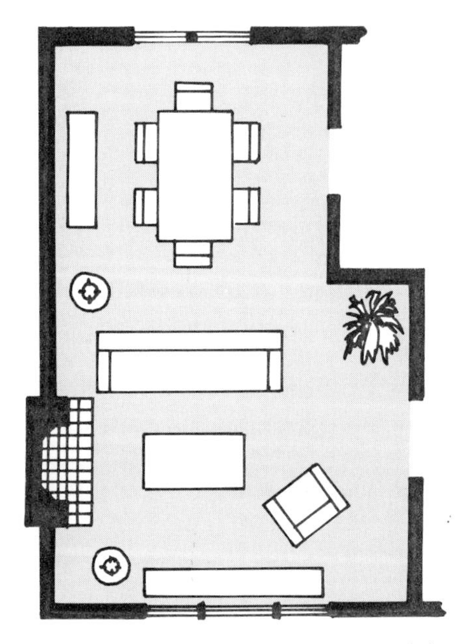

Figure 6.10. A room planned around furniture. What furniture items are shown by symbols?

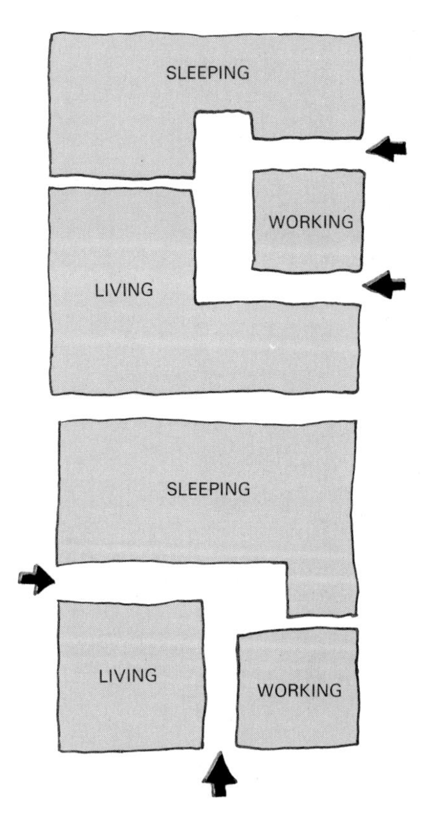

Figure 6.11. Hallways allow access to all rooms without passing through other rooms.

- To give direct access between kitchen and dining area for convenience in carrying hot foods from kitchen to table and for clearing the table.

Rooms must be connected by hallways, stairs, and doorways. The ideal is to have good traffic patterns. You should be able to move from one area to another without passing through a third area, figure 6.11. For example, it should be possible to walk from the kitchen to the front door without going through several rooms.

When you feel you have a good arrangement of rooms you can make a drawing. An architect calls this a **floor plan**. The floor plan in figure 6.12 shows a two-bedroom apartment. Figure 6.13 shows a three-bedroom bungalow.

Finding and preparing a site

The land on which you will build is called a site or lot. It can be any size. City lots are usually small. Those outside the city may be as large as several acres.

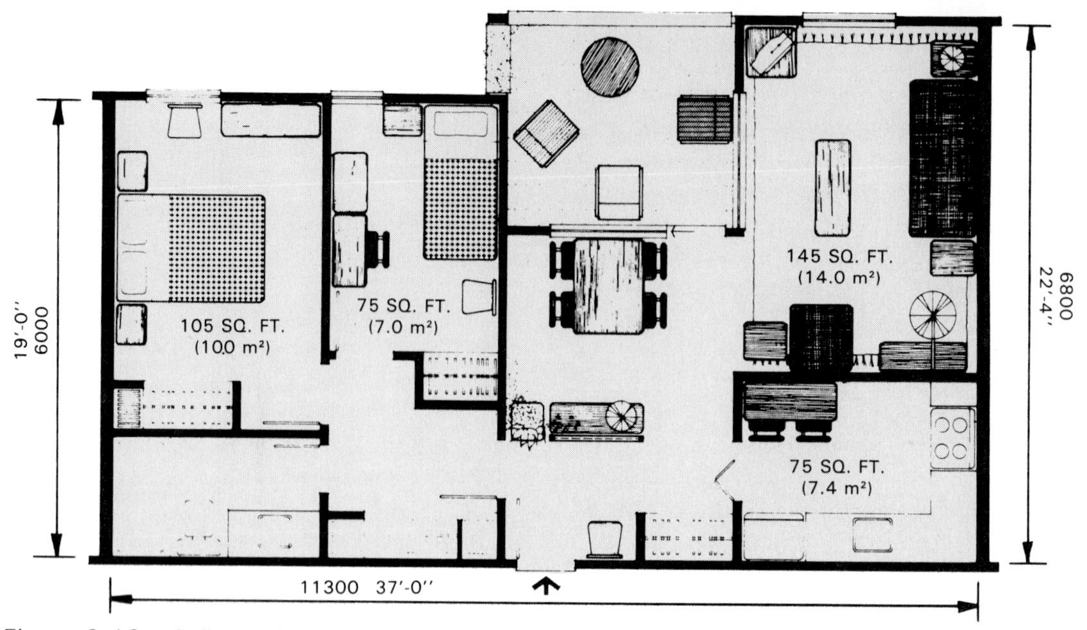

Figure 6.12. A floor plan imagines that you are looking down on the apartment from above.

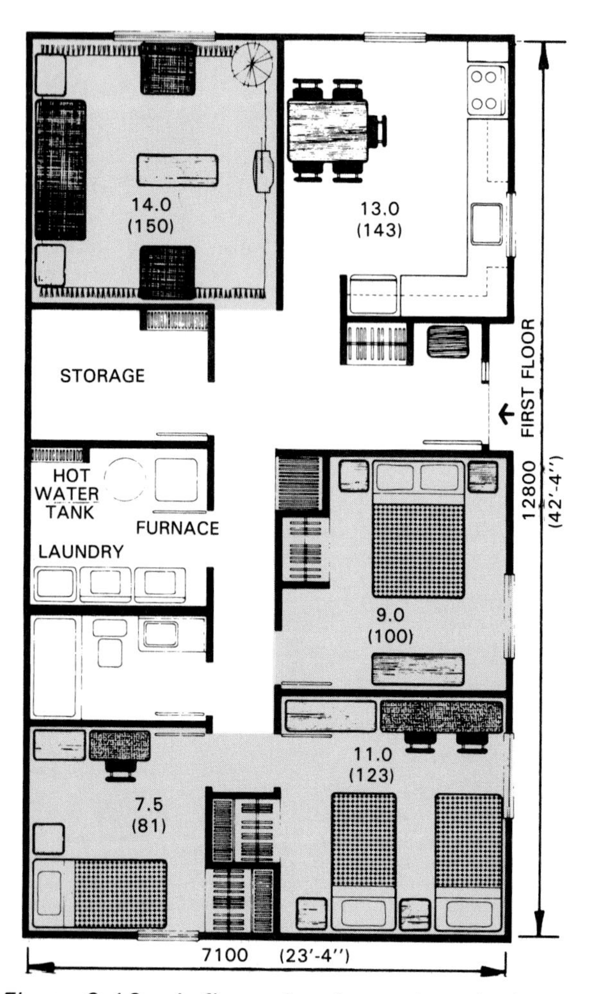

Figure 6.13. A floor plan for a three-bedroom bungalow. See how the plan separates the sleeping and living areas from the dining and kitchen areas.

Planning the site is just as important as designing the home itself. It involves several important steps.

Selecting the site. Where you locate a new home is important. You may want it to be in a certain community. Perhaps it should be close to your job, shopping, and schools. A quiet wooded area may be preferred.

Acquiring the site. To purchase (buy) a building site, both the owner and you must agree on a price. If you are not willing to pay the asking price, you must negotiate. This means finding a price at which the owner will agree to sell.

Going into contract. When a price is agreed upon, you sign a contract. This is a legal document (paper) which sets down all the conditions of the purchase. Among other things, it lists the selling price agreed upon. When it is signed by both parties, it is binding. This means that the seller must sell for the agreed price. The buyer is obliged to purchase at that price.

Site preparation. Site preparation means getting the site ready for the home. One of the first steps is to do a soil test. You need to know how well the subsoil will carry the mass (weight) of your home. For example, it is important to find out if there is hard rock underground. It may be expensive to remove. There may be ground water too close to the surface. This could cause flooding in the house. Other soils may be too light to carry loads well.

Once a soil engineer has determined that the site is suitable, a contractor will clear the site of boulders and excess soil. Grading may be needed to level a spot for the foundation. Lines and grades must be established to keep the work true and level. Figure 6.14 shows how **batter boards** are used for this purpose. Small stakes are located at what will be the corners of the house. Nails driven into the tops of these stakes mark the four corners of the house. Straight lines between these nails indicate the outside edge of the foundation walls.

Once the four corners have been located, larger stakes are driven into the ground at least 4 ft. (1.3 m) beyond the lines of the foundation. The batter boards are nailed horizontally to these stakes. The boards must be level. Strong string is next held across the tops of opposite boards and adjusted exactly over the tacks in the small corner stakes. A plumb bob may be used to set the lines exactly over the nails.

Next, a saw kerf (cut) is made to mark where the string crosses the top of each batter board. Some carpenters drive a nail at this point. This is done so the strings can be removed during excavation. Later, the strings can be stretched from corner to corner, across the batter boards, to relocate the corners of the building.

Main parts of a home

Now let us look at the main structural components and materials used in house construction. There are five major components: foundation, floor, wall, ceiling, and roof, figure 6.15. These separate structural components fit together to form a house.

Foundation. Most structures rest on a foundation, figure 6.16. Normally a foundation lies below the surface of the ground. Notice that there are two major parts to this type of foundation: the **footing** and the foundation wall.

To understand the importance of the footing, think about the reason for wearing snowshoes. If you try to walk on deep, soft snow without them, you might sink down to your thighs. Snowshoes spread your body mass over a larger area of the snow's surface.

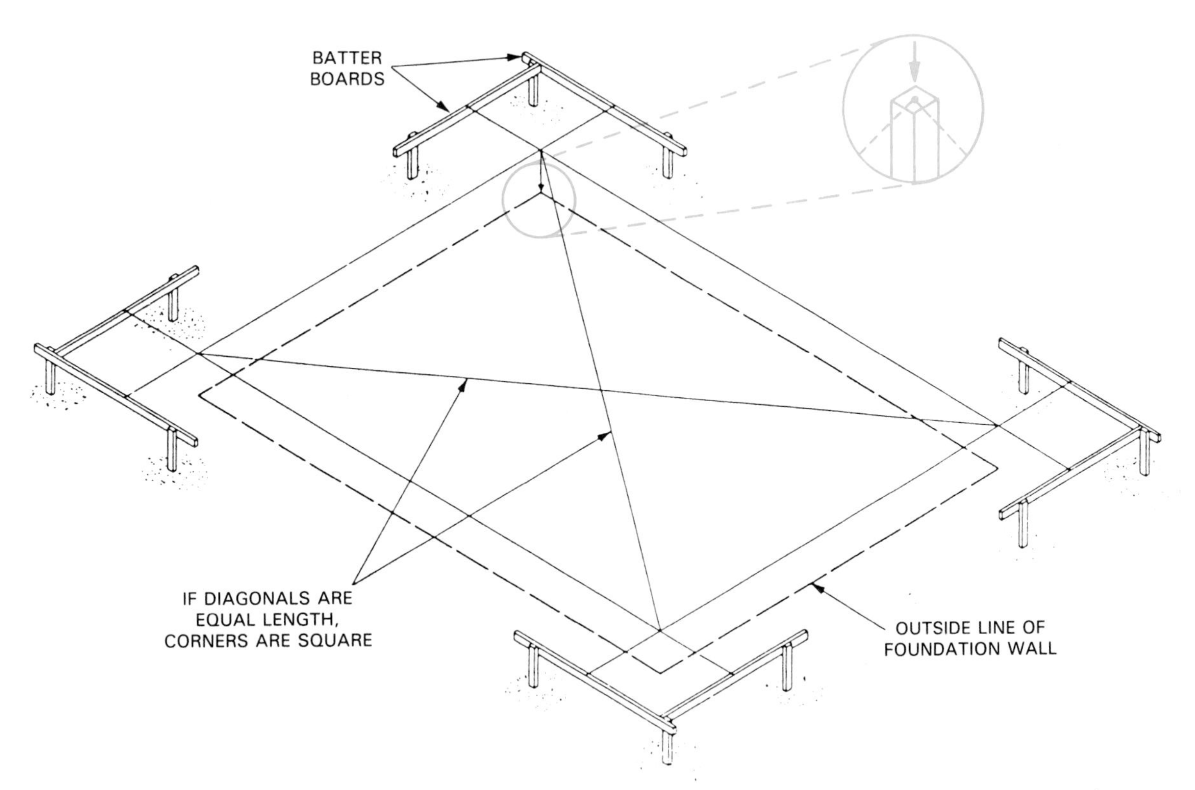

Figure 6.14. Batter boards support the lines set up to locate the building so excavating can begin for the foundation.

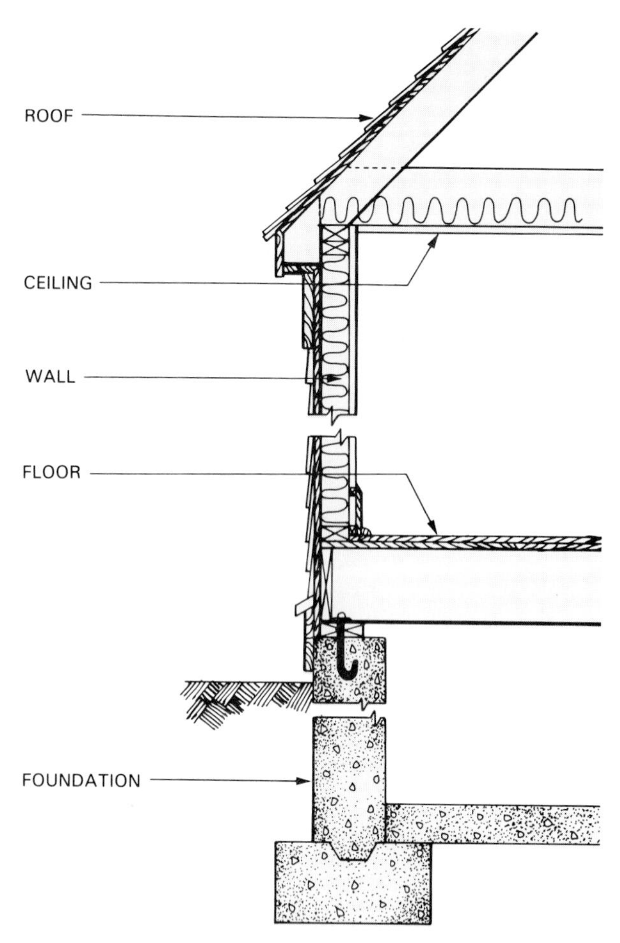

Figure 6.15. A typical section through a house shows its main structural parts.

This prevents you from sinking. The same principle is used to build a foundation. The load of the building is first transmitted to the foundation wall. Then it is spread over a larger area by the footing. Thus, the building is prevented from sinking into the ground.

In most locations, it is necessary to drain away any subsurface (underground) water to avoid damp basements and wet floors. Tile laid around the wall footings serves this purpose. These are known as drain tile, perimeter tile, or weeping tile.

Foundation materials. The two materials most commonly used for foundation walls are poured concrete and concrete blocks. Concrete is strong enough to support heavy loads. Its strength may be increased by embedding steel rods or wire mesh in it. This combination is known as reinforced concrete. The various parts of foundation walls, their functions, and the materials used in their construction are summarized in figure 6.18.

In warm climates, there is either no frost or the frost does not penetrate very far below the ground. Therefore, a combined slab and foundation is commonly used, figure 6.19.

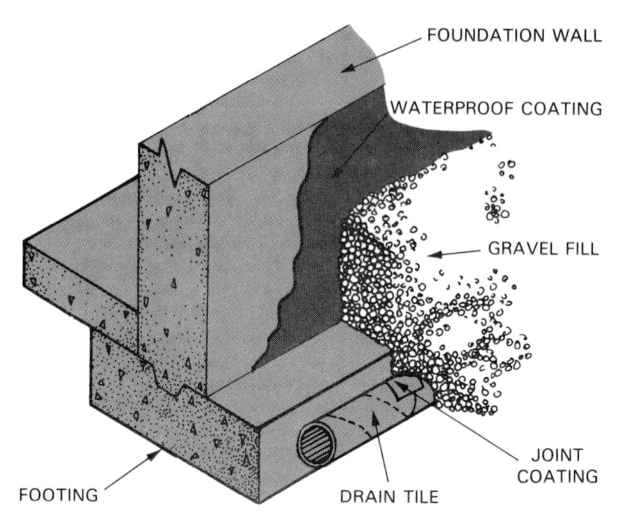

Figure 6.16. Many foundations are of this type. Explain why the footing is so wide.

ECRITEK

Figure 6.17. Why doesn't the boy sink into the snow?

PART	FUNCTION	MATERIAL
Footing	To transmit the superimposed load to the soil	concrete
Foundation wall	To form an enclosure for the basement and to support walls and other building loads	concrete concrete blocks
Weeping or perimeter tile	To provide drainage around footings	clay plastic
Gravel fill	To permit water to drain into the weeping tile	gravel

Figure 6.18. Parts of a foundation. Though not approved by all codes, treated wood has begun to be used as foundation wall material.

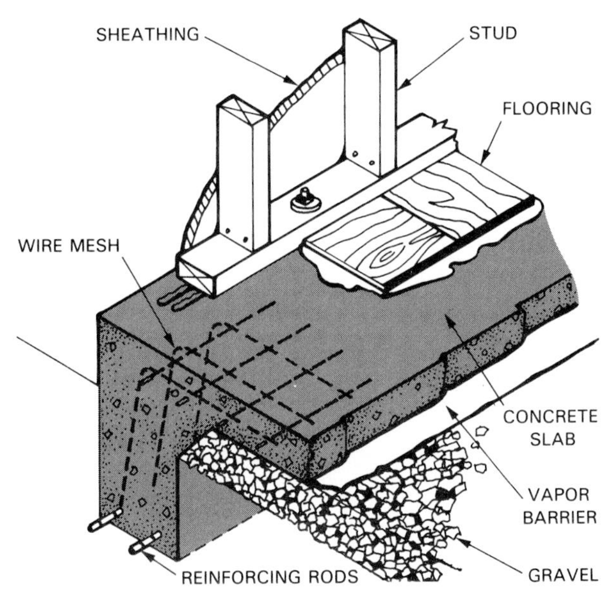

Figure 6.19. Where frost does not penetrate deeply, a combined slab and foundation can be used.

Floor. When the concrete for the foundation wall is poured, anchor bolts are set in the top. These bolts are used to fasten a sill to the foundation, figure 6.20. Joists are nailed to the sill on edge. This forms a framework. This framework is, in turn, supported by a beam. (Joists are usually made of wood nearly 2 in. thick and 10 or more inches wide.)

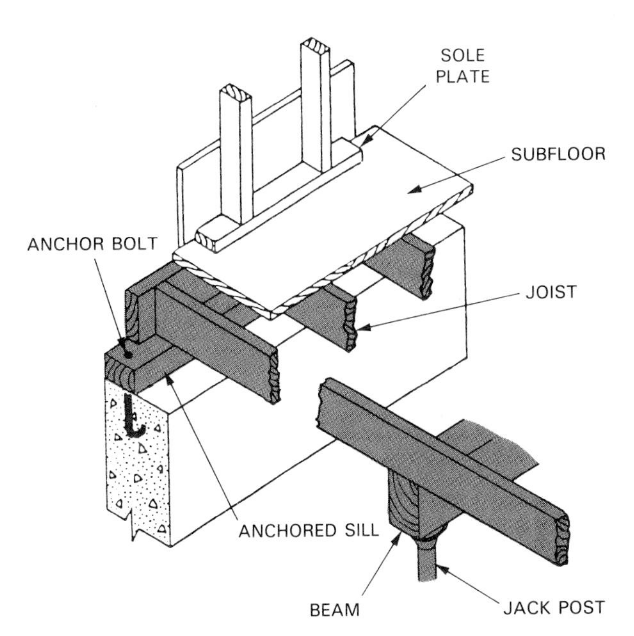

Figure 6.20. A floor is supported by joists.

When beams must span a long distance, they are supported in the middle by jack posts. The joists support a subfloor. A subfloor is a covering over joists. It supports other floor coverings. The various parts of floors, their functions, and the materials used in their construction are summarized in figure 6.21.

Walls and finish flooring. The subfloor is fastened to the joists. Then the walls for the

PART	FUNCTION	MATERIAL
Beam	To support joists when long distances are spanned	wood (pine or spruce) steel
Joist	To support a floor	wood (pine or spruce)
Subfloor	To support finish flooring	board or sheet material (e.g., tongue-and-groove pine, plywood)
Sill	To support joists where they meet the foundation	wood (pine or spruce)
Jack post	To support beams	wood (pine or spruce) steel

Figure 6.21. Parts of a floor frame.

first floor are laid out and built. The many parts of this type of wall are shown in figure 6.22. The various parts of walls, their functions, and the materials used in their construction are listed in figure 6.23.

Ceiling and roof. The construction of a ceiling often requires joists the same as for a floor. A roof is made up of sloping timber called rafters. Modern roofs are usually built of a series of prefabricated trusses. The truss is shaped like a triangle. Its base supports the ceiling materials. Braces on the inside create triangles to support and strengthen the rafters.

A typical ceiling and roof construction is shown in figure 6.24. The various parts of ceilings and roofs, their functions, and the materials used in their construction are given in figure 6.25.

Figure 6.26 shows how all the many parts of a house fit together. The next time you see a building going up, see how many of the parts you can identify.

Finishing the house. The final stages in building a house include trimming, painting, decorating, and landscaping. Trimming involves covering rough edges and openings with moldings. For example, a baseboard is the trim used to cover the small space between a wall and a floor. Painting protects and improves the appearance of interior and some exterior surfaces. Wallpaper is the most common interior decorating material. Paneling or tongue-and-groove boards are also used to decorate walls.

Landscaping, figure 6.27, is designing the exterior space that surrounds a home. It involves planning the location of lawns, hedges, trees, shrubs, and plants. The plan will also show the location of paths, such as driveways and walkways. It will also show special features such as patios, fences, walls, and plant boxes. After accesses (paths) and features have been built, topsoil is added to the site. Topsoil is a layer of rich earth. It is needed so that trees, shrubs, lawns, and plants can grow.

■ SYSTEMS IN STRUCTURES

What happens when you telephone a friend? After lifting the receiver, you dial a number. Signals travel to a central location where automatic switching equipment sends

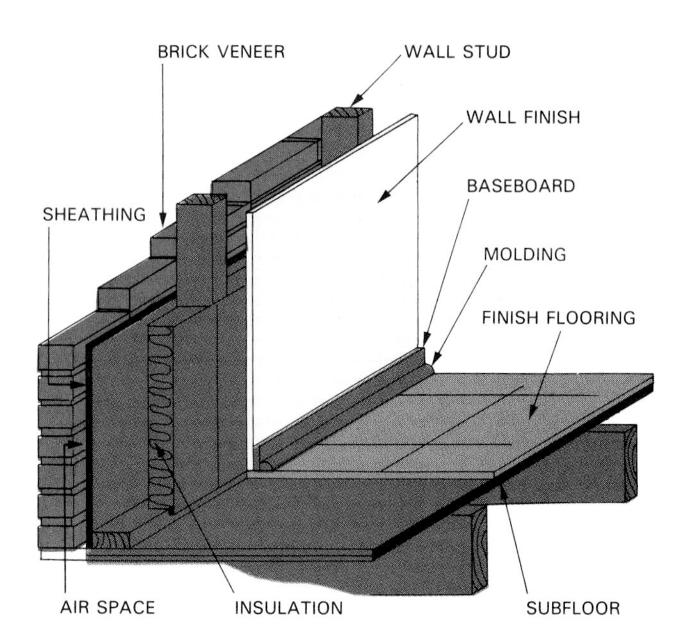

Figure 6.22. A section of a wall and floor. Refer to figure 6.23 for function of parts.

PART	FUNCTION	MATERIAL
Exterior surface	To provide protection and decoration to the outside of a building	brick aluminum siding wood
Air space	To provide a barrier against the passage of moisture	
Sheathing	Reinforces studs Provides insulation	wood fiberboard
Wall stud	To provide a framework for walls or partitions	wood (pine or spruce)
Insulation	To resist heat transmission	fiberglass polyurethane vermiculite
Vapor barrier	To retard the passage of water vapor or moisture	polyethylene sheet
Interior wall surface	To cover the interior wall framing	plasterboard wood paneling plaster
Finish flooring	To cover a subfloor and provide a decorative surface	parquet ceramic linoleum carpet

Figure 6.23. Parts of walls and finish flooring.

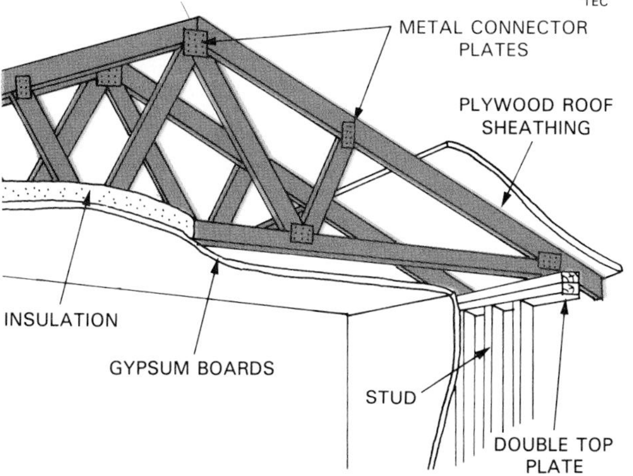

Figure 6.24. The construction of a truss roof. Gypsum board ceiling is attached to the underside.

PART	FUNCTION	MATERIAL
Joist	To support a ceiling	wood (pine or spruce)
Insulation	To resist heat transmission	fiberglass polyurethane vermiculite
Vapor barrier	To retard the passage of water vapor or moisture	polyethylene sheet
Interior surface	To form the ceiling	plasterboard plaster
Roof truss	To form a framework for the roof and to support loads applied to it	wood (pine or spruce)
Exterior finish	To provide protection from rain, snow, and other weather conditions	asphalt wood shingles tar and gravel

Figure 6.25. Parts of a ceiling and roof. Study their purpose.

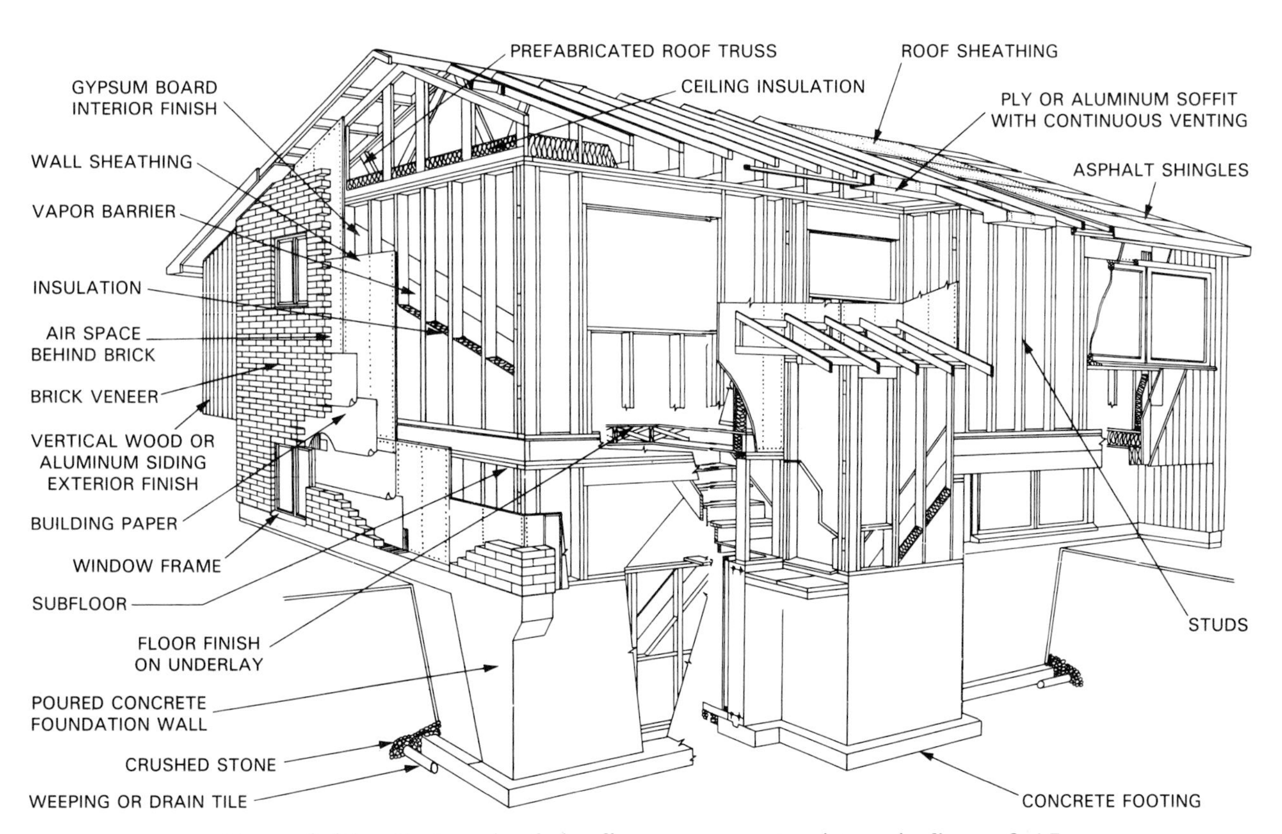

Figure 6.26. Find each of the five components shown in figure 6.15.

Figure 6.27. A well-landscaped house. Flowers, shrubs, trees, grass, and walkways are pleasing.

your call to your friend's house. Your friend answers and your voices are carried over the lines. At the end of your conversation, replacing the receivers disconnects the lines. The telephones, cables, and automatic switching equipment are part of a system.

Some systems are very large. Others are quite small. The sun and the planets that revolve around it form the solar system. The skeletal system of your body is made up of more than 200 bones. Together, they support the body's mass. They also give the body shape and protect important organs. The fuel system of a car pumps gasoline from a fuel tank, through fuel lines, to the carburetor or injectors into the cylinders of the engine.

What is a system?

What do all systems have in common? How is a system defined? A system is a series of parts or objects connected together for a particular purpose. There are two types of systems: open-loop and closed-loop.

Open-loop system. A portable space heater without a thermostat is an example of an **open-loop system.** When plugged in and switched on, the heating element warms the air passing over it. It continues to heat the room until switched off. There is no method of controlling whether there is too much or too little heat.

Closed-loop system. In a **closed-loop system,** the same heater would be connected to a control mechanism such as a thermostat. When the room air reaches the temperature you set on the thermostat, the heater shuts off. It will switch itself on again when the temperature falls below the set limit.

In both cases, the systems contain: input (cool air and fuel), process (burning the fuel), and output (warmed air). Input, process and output are characteristics of all systems. However, in a closed-loop system there is also a feedback device. It provides control. Control of our environment is a major reason why technological systems have been created.

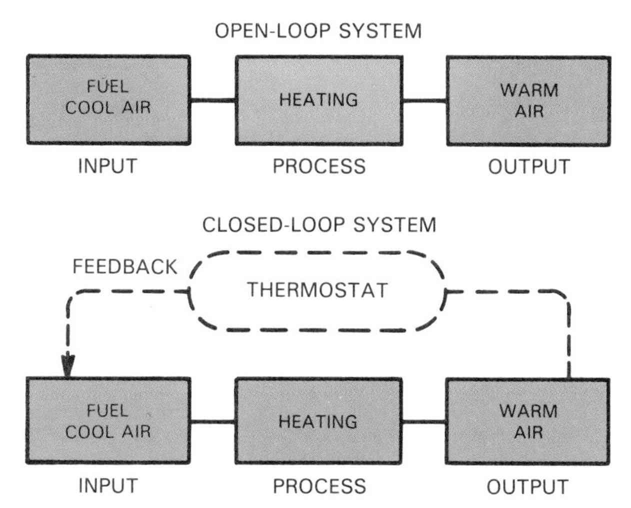

Figure 6.28. These diagrams compare open- and closed-loop systems.

In our homes there are four major systems: heating, electrical, plumbing, and communications. Each helps to control the environment in the home.

Heating system

Figure 6.29 illustrates the major parts of a forced air heating system. (Some heating systems use water to carry heat to rooms. However, forced air is a popular way to move heat from a furnace to various rooms.)

Cool air enters the bottom of the furnace. Here the filter traps dirt. A blower forces the filtered air up into a compartment, called a heat exchanger. The exchanger has passageways. These are heated by electricity or the combustion gases from burning oil or gas. The blower forces the warmed air through a network of ducts into each room. Cooler, heavier air sinks to the floor and flows through a return air duct leading back to the furnace. Control switches turn the blower, and also the supply of heat, on and off. Thus, the furnace controls the temperature of the circulating air.

This heating system is, in fact, composed of several subsystems. A subsystem is a smaller system that operates as a part of the

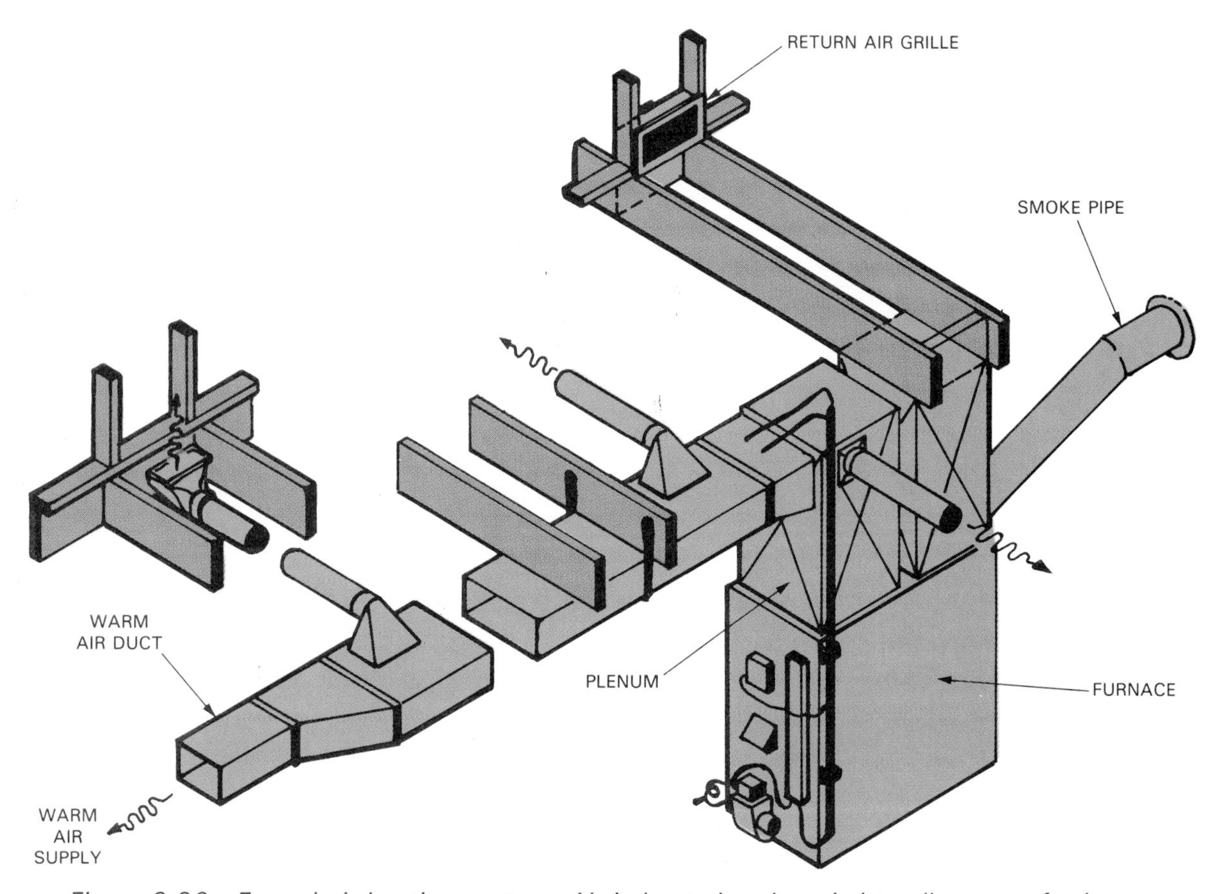

Figure 6.29. Forced air heating system. Air is heated and carried to all rooms of a house.

larger system. The subsystems within the home heating system are:

- Heater to produce heat.
- Blower unit for pushing the heat through the duct work.
- Network of ducts for carrying the heated air.
- Thermostat for providing continuous feedback.

Some heating systems may have other subsystems, including:

- A humidifier.
- Air conditioning.
- An electronic filter.
- A heat pump.

Electrical system

An electrical system supplies electricity for light, heat, and appliances. This electricity is carried throughout the home by a number of separate circuits. (A circuit is a pathway for electrical current.) Each circuit has three wires running inside the walls and ceilings. A

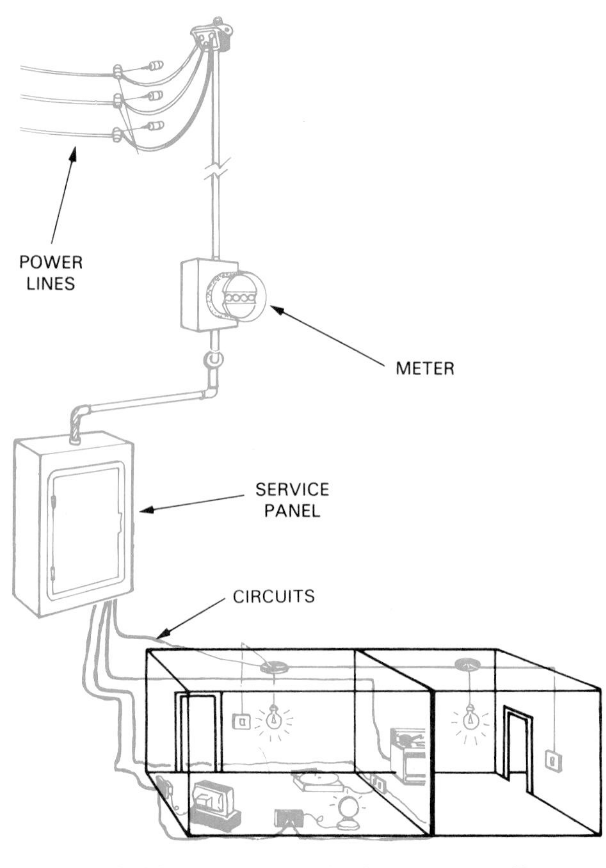

Figure 6.30. An electrical system. Try to remember names of all the parts.

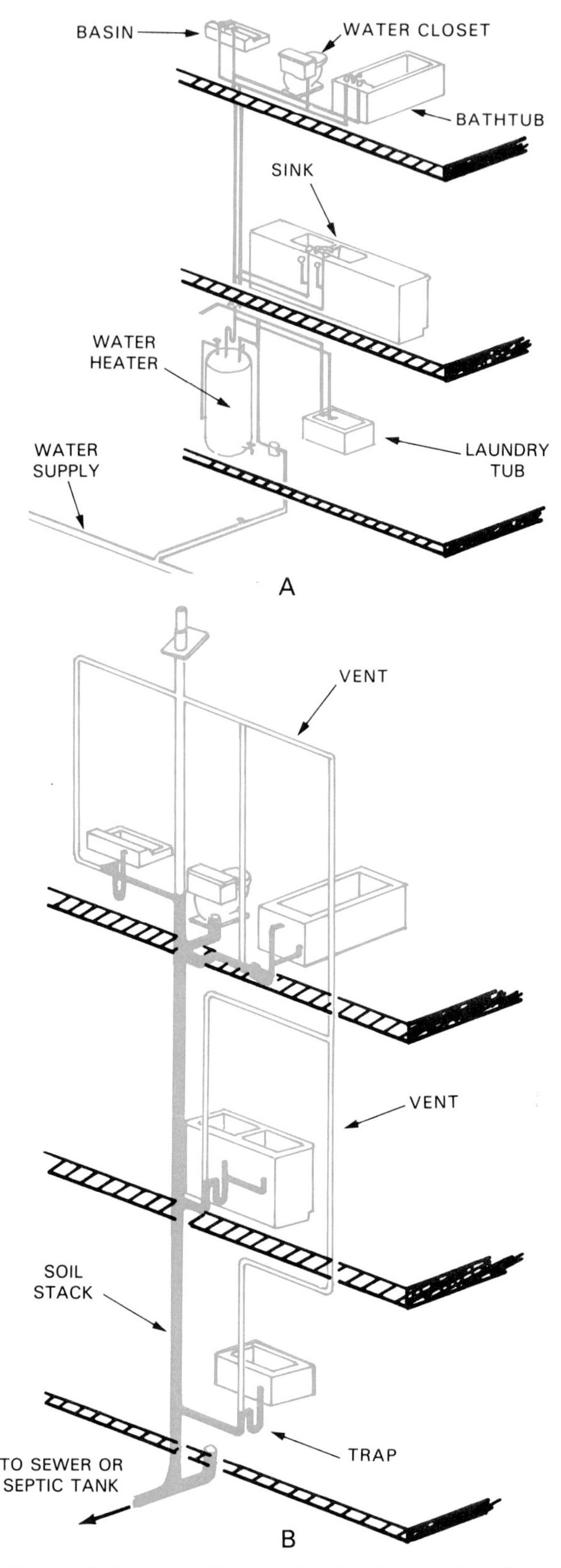

Figure 6.31. A—Water distribution system brings fresh water to different rooms of the home. B—Drain system carries away waste water.

circuit carries a current from a power source. Electric current travels to lights, motors, or heaters and back to the source. To supply these circuits, electricity from a utility company's wires must pass through a meter and a service panel. A service panel distributes the power among the separate circuits. Lamps, television sets, and small appliances are connected to 120 volt, 15 ampere circuits. Appliances, such as a refrigerator, toaster, and power tools, are connected to 120 volt, 20 ampere circuits. Separate 240 volt, 30 ampere circuits are provided for a clothes dryer or an electric range.

Plumbing system

The plumbing system in a home is basically very simple. Potable (clean) water is brought into the house. It is piped directly to all the faucets and outlets such as sinks, toilets, baths, and washing machine. It is also piped to the hot water heater. From this heater, heated water goes to all hot-water faucets, washing machine, and dishwasher. The used water is drained from the house and disposed of by a separate system.

Communication system

The communication systems in your home include the telephone system, the radio and television broadcasting system, and the cable television system.

Telephone service is provided to most homes by copper wires or fiber optic cables. A nationwide switching system enables your telephone to be connected to any other telephone. The same copper or fiber cables can be used for data transmission so that any home can have access to a national or international computer network.

Radio and television signals are received in each home. The programs may be broadcast from local stations. Satellites and antennas allow you to receive, live and instantaneously, radio and television coverage events from around the world, figure 6.32.

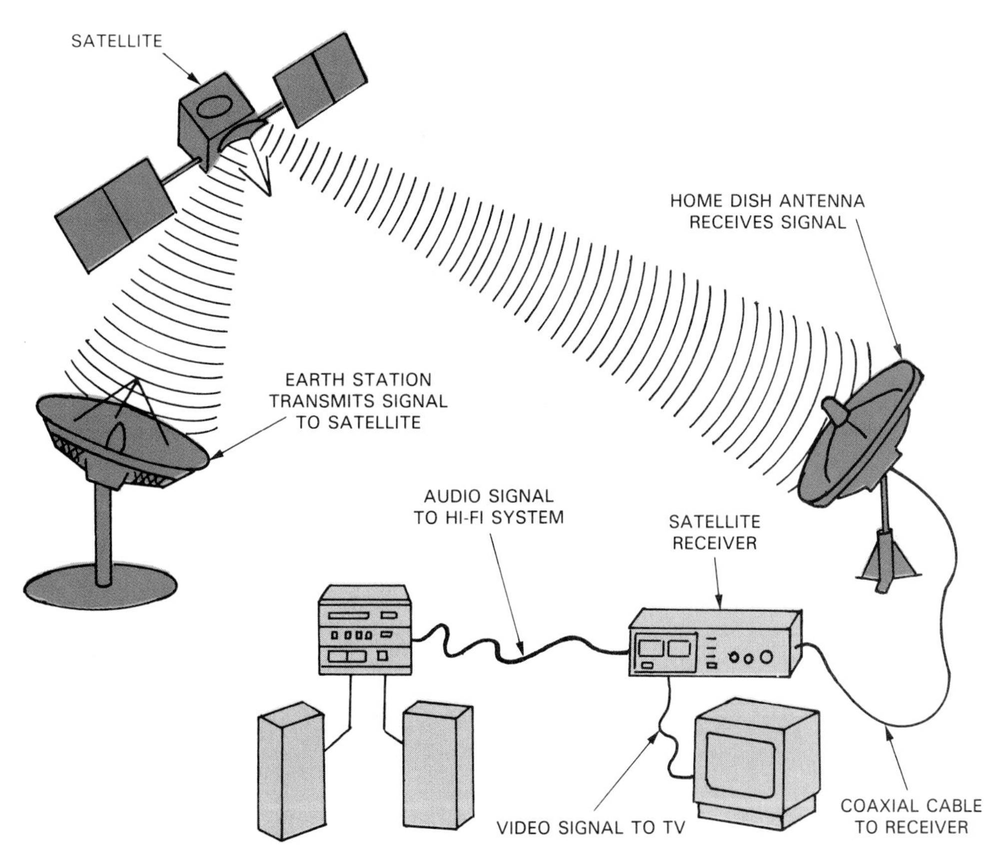

Figure 6.32. A telecommunication system linked to a satellite in space. Signals can be received in an instant in spite of great distances.

SUMMARY

The structure of a house must support the loads of the materials from which it is built, plus its contents. It must also support the dynamic loads created by the weather conditions outside and the movement of people inside.

There are four types of single family houses. These are: bungalow, one-and-a-half-story, two-story, and split-level. Space within a house must accommodate all furniture and equipment and still leave space to move around. Rooms are grouped according to their function: living, sleeping, and working.

The first step in building a house is to check the subsoil conditions. Next, a contractor clears the site and establishes lines and grades to ensure that the construction will be level. There are five major components of a house. These are the foundation, floor, walls, ceiling, and roof.

In a typical home there are four major systems: heating, plumbing, electrical, and communication. These all work as either open- or closed-loop systems. In a closed-loop system there is a feedback mechanism to provide control.

KEY TERMS

Adobe	Joist
Batter boards	Landscaping
Beam	Modular construction
Closed-loop system	Open-loop system
Communication system	Plumbing system
	Post-and-lintel
Electrical system	Prefabrication
Floor plan	Site
Footing	Subfloor
Foundation	System
Heating system	Wall stud
Insulation	

TEST YOUR KNOWLEDGE

Write your answers to these review questions on a separate sheet of paper.

1. List the materials used in the construction of the following different types of homes:
   a. Mexican house.
   b. Indian tent.
   c. Inuit igloo.
   d. Liberian hut.
2. What factors determined the use of a particular material when early settlers built their homes?
3. Sketch and name the parts of a post-and-lintel structure.
4. List two static and two dynamic loads that a typical house must withstand.
5. List four types of single-family houses.
6. How could you plan alternative arrangements of furniture in a room?
7. In the floor plan of a house, rooms are grouped according to their function.
   a. What are the three functions?
   b. State three reasons why rooms are grouped by function.
8. The five major structural components of a house are walls, floors, ceilings, foundation, and roof. List the order in which these are built on a construction site.
9. Describe the similarity between the footing of a foundation wall and a pair of snowshoes.
10. Draw and label the parts of a house foundation.
11. Complete the following sentences:
    a. A foundation wall is connected to a ________ by means of ________ ________.
    b. A subfloor is supported by a number of ________.
    c. Wall studs provide a framework for ________________________.
    d. Heat transmission is resisted by the use of ________________________.
    e. Interior wall surfaces are usually made of ________________________.
    f. The function of a roof truss is to form the ________________________.
12. What do all systems have in common?
13. Why is a portable space heater without a built-in thermostat an example of an open-loop system?
14. Using notes and diagrams, describe the difference between an open-loop system and a closed-loop system.
15. List the four major systems in a home.

APPLY YOUR KNOWLEDGE

1. Collect five pictures to illustrate the different types of residential buildlings. Name each type of building.
2. Make a simplified copy of figure 6.15. Label the five major components and state the function of each.
3. Take a close look at the inside and outside of your home. List all the materials used in its construction.
4. Use the systems model in figure 6.28 to describe how a spaghetti dinner is prepared.
5. Give one example of a device in your home that uses an open-loop system and one example that uses a closed-loop system. Draw a system diagram of each.
6. Construct a scale model of the floor and walls of one room in your home. Think of the furniture you would like to choose if the room were empty. Cut blocks of styrofoam or cardboard to represent the furniture. Position them in the room.

A

B

C

D

The construction process. A—The construction crew consults the plans. B—The building takes shape floor by floor. C—The new hockey arena is nearly complete. D—Together, these buildings shape our cities.

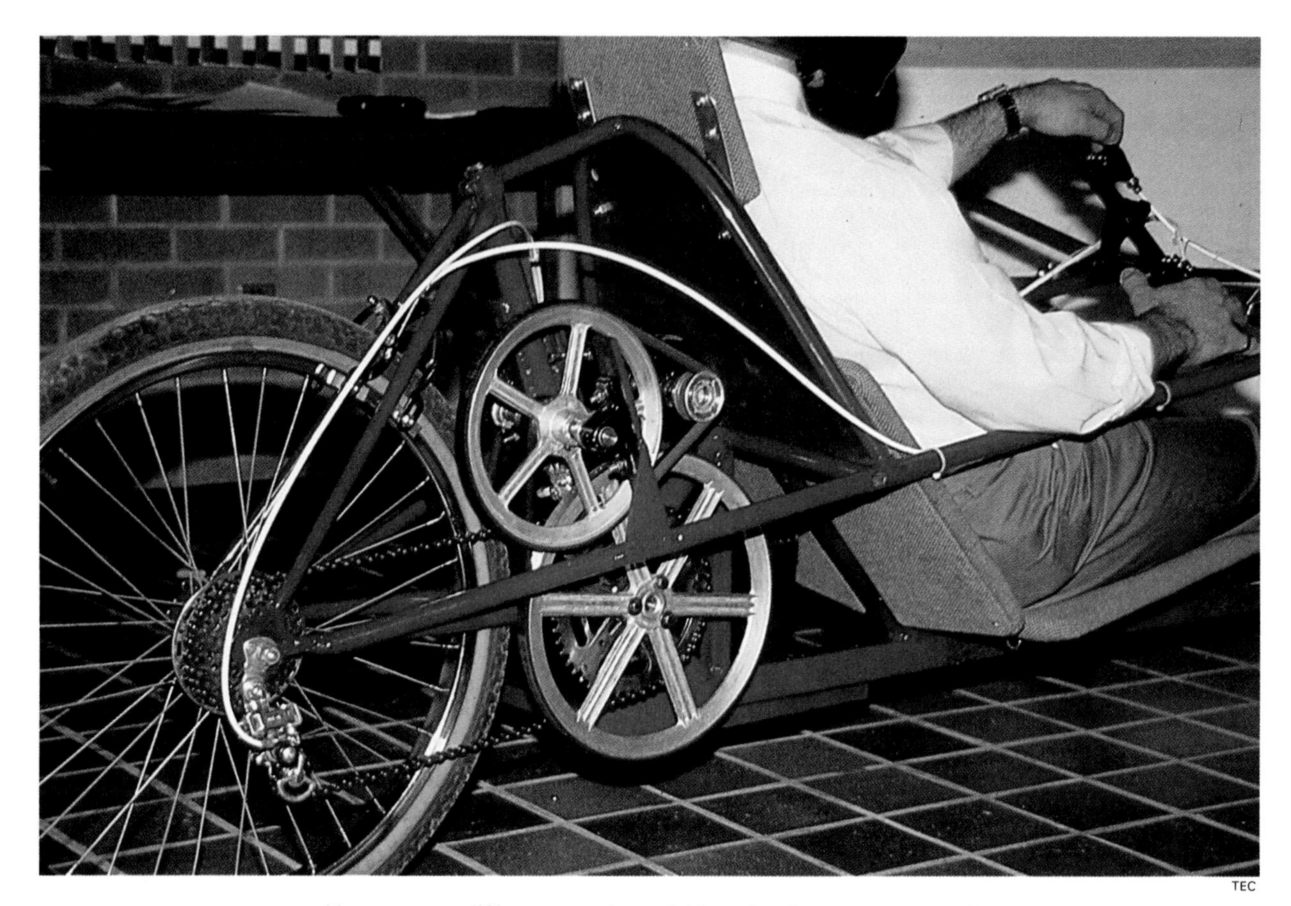

TEC

How many different styles of bicycles have you seen?

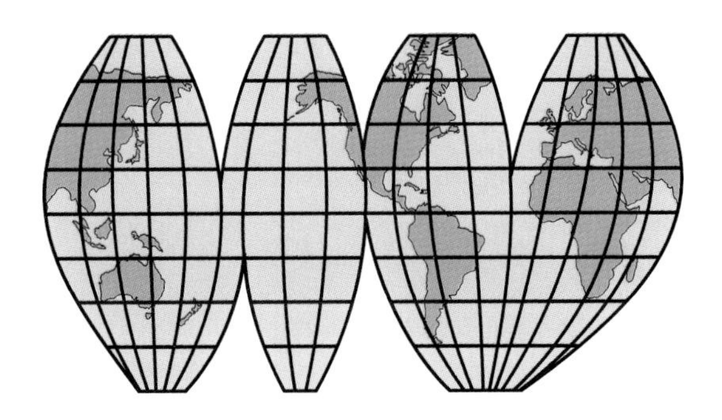

Chapter 7

Machines

OBJECTIVES

After reading this chapter you will be able to:

- Identify different types of simple machines.
- State where these simple machines exist in everyday objects.
- List the advantages of simple machines.
- Identify systems having mechanical operations and explain their function.
- Describe the principle of a hydraulic or pneumatic system.
- Design and construct a simple object with a mechanism.

Our society relies heavily on **machines.** Machines help us to produce televisions in factories, and food on the land. They help us to dig trenches and bore tunnels through mountains. Around the home, the automatic washing machine, the lawnmower, and the vacuum cleaner lighten our work load. The dentist's drill removes decay from our teeth. The jet plane speeds us to our vacation site.

Machines do not need to be large or complicated. A knife, a bottle opener, and a claw hammer are also machines. A machine is a device that does some kind of work by changing or transmitting energy.

Machines have been used for thousands of years to make work easier. In ancient Egypt, a shadoof used a lever to raise a leather bucket filled with water, figure 7.1.

Early civilizations also used a **pulley,** figure 7.2. The first pulley may have been developed by throwing a rope over a tree branch in order to pull in a downward direction. A further development of the pulley was the windlass. This was often used to raise water from wells, figure 7.3. People in ancient times used sections of tree trunks to move heavy loads. They found that it took less effort than sliding them, figure 7.4. Later, this idea led to the development of the **wheel and axle,** figure 7.5.

Figure 7.1. The shadoof from ancient times used a lever and leather bucket to lift water to a trough. The trough carried the water to fields.

Figure 7.2. A pulley is designed to change the direction of applied force. As the woman pulls down on the rope the pail moves upward.

Figure 7.3. The windlass is a wheel and axle machine. It is also somewhat like a pulley.

Blocks of stone used to build the pyramids were raised to great heights by means of a ramp. Called an **inclined plane,** figure 7.6, it allowed lifting of heavy loads with a minimum of effort.

Two further developments of the inclined plane were the **wedge** and **screw.** One of the most important uses of the wedge was the plow, figure 7.7.

Archimedes, who lived in the third century B.C., designed a type of screw. It was used to raise water for irrigation of the Nile Valley.

■ SIMPLE MACHINES

Today we refer to the lever, pulley, wheel and axle, inclined plane, wedge, and screw as simple machines. These six simple machines can be divided into two groups. One is based on the lever and the other on the inclined plane, figure 7.9.

Figure 7.4. What machines are used today to move a heavy load?

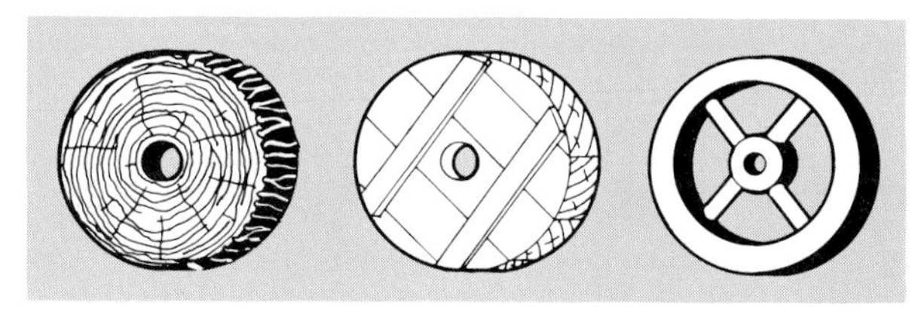

Figure 7.5. Over time, wheels became better made. Roman chariots used wheels with wooden spokes. How has the wheel changed since then?

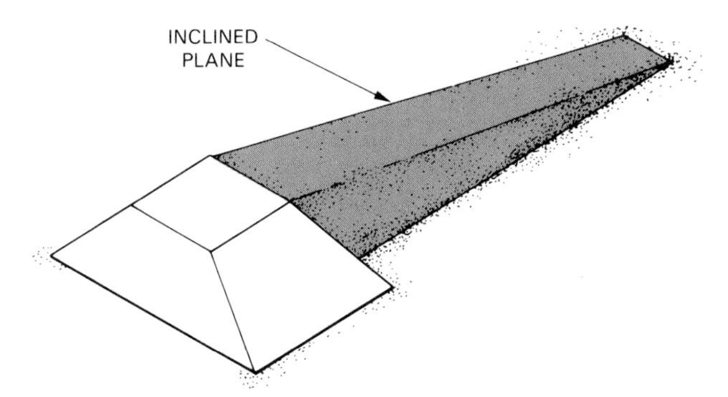

Figure 7.6. The Pyramids could not have been built without the inclined plane.

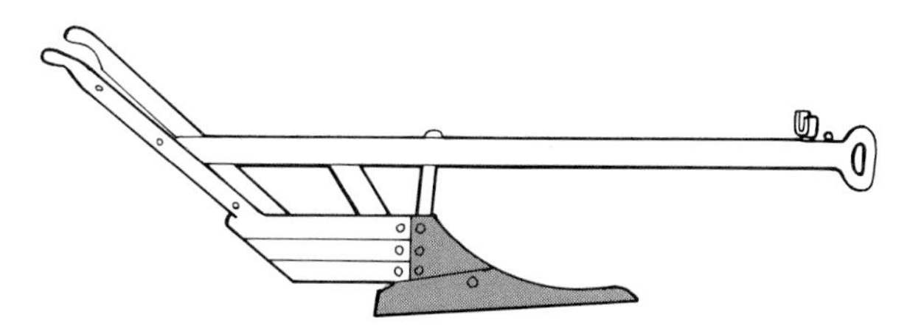

Figure 7.7. A plow is a type of inclined plane called a wedge.

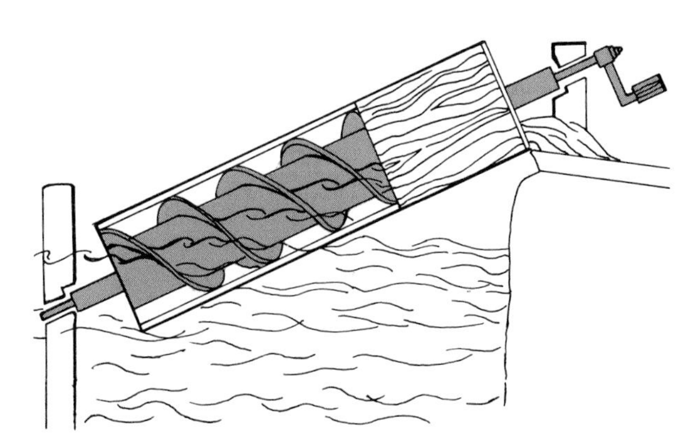

Figure 7.8. Archimedes was a Greek inventor and mathematician. His invention could lift more water faster than any other method.

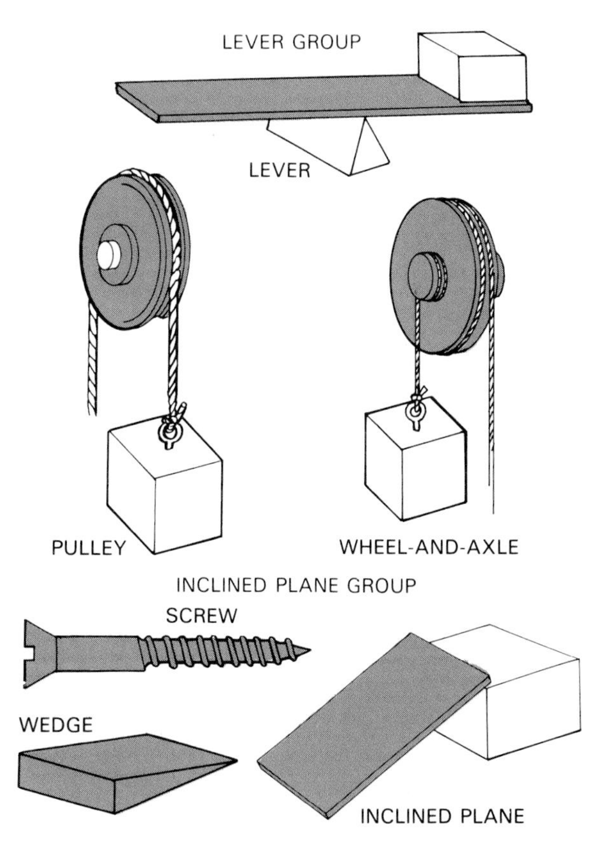

Figure 7.9. The six simple machines were important inventions.

Levers

You have very likely played or seen a game of baseball, figure 7.10. A baseball bat is a lever. It has a fulcrum, effort, and resistance. The fulcrum is the point where the bat is held. The effort is supplied by the batter's muscles and the resistance is the ball, figure 7.11.

To understand the principle of the lever, look at the boy in figure 7.12. He is using a branch to move a heavy rock. The branch is the lever. The mass of the rock is the resistance (R). The boy's muscle power pushing down on the lever provides the effort (E). The rock on which the lever is pivoting is the fulcrum (F). These three elements—resistance, effort, and fulcrum—are always present in a lever. However, they can be arranged in different ways to create three different classes of levers.

In a Class 1 lever, the fulcrum is placed between the effort and the resistance, figure 7.13. Some applications of Class 1 levers are illustrated in figure 7.14.

TEC

Figure 7.10. Baseball players use a bat as a lever to strike the baseball with greater speed.

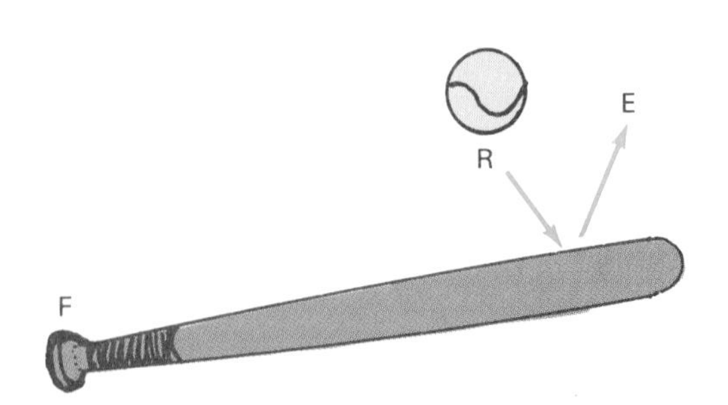

Figure 7.11. Baseball bat as a lever. F—Fulcrum. R—Resistance. E—Effort.

In Class 2 levers, the resistance is placed between the effort and the fulcrum, figure 7.15. Some applications of class 2 levers are illustrated in figure 7.16.

In Class 3 levers, the effort is applied between the resistance and the fulcrum, figure 7.17. Some applications of Class 3 levers are illustrated in figure 7.18.

From the many examples shown, you can see that some levers are designed to increase the force available. Examples are a wheelbarrow, a bar used to move a crate, and a garden spade. Other levers are designed to increase the distance a force moves or the speed at which it moves. Examples are a fishing rod and a human arm.

Figure 7.12. An example of a lever in use to move a heavy load.

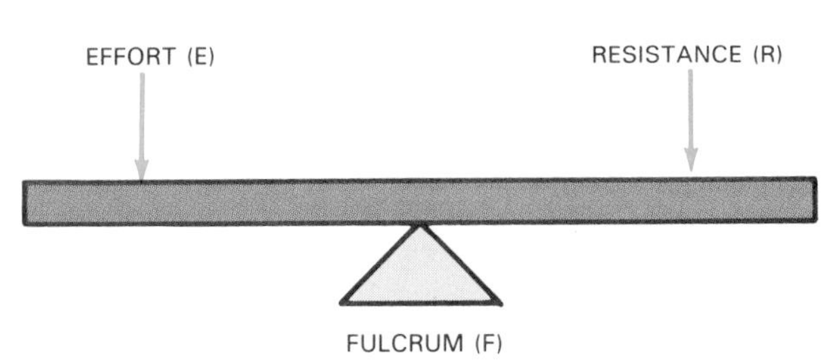

Figure 7.13. Class 1 lever. Fulcrum is between effort and resistance.

Mechanical advantage. When a small effort is applied to a lever to move a large resistance there is obviously an advantage. This is called the **mechanical advantage** of the lever.

Mechanical advantage is equal to the resistance divided by the effort. The greater the resistance that can be moved for a given effort, the greater the mechanical advantage. The formula is:

$$\text{Mechanical advantage (M.A.)} = \frac{\text{Resistance}}{\text{Effort}}$$

For example, if a lever can make it possible to overcome a resistance of 90 newtons (N) when an effort of 30 N is applied, the mechanical advantage will be 3. (The newton (N) is the metric unit of force or effort.) The formula is:

$$\text{M.A.} = \frac{90\text{ N}}{30\text{ N}} = 3$$

Figure 7.14. These are familiar examples of the use of Class 1 levers.

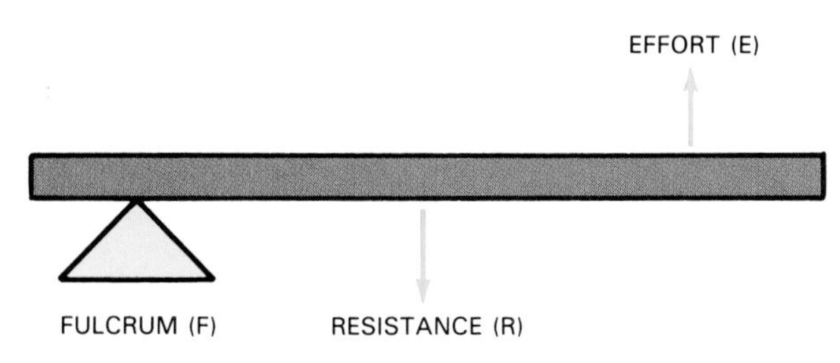

Figure 7.15. An illustration of a Class 2 lever. Do you see how it is different from a Class 1 lever?

In other words, the human effort applied is being multiplied by the machine. (In this case, it is the lever.) The effort required becomes less as the fulcrum and the resistance are brought closer together. However, the disadvantage is that as the fulcrum and resistance are moved closer, the load moves a shorter distance, figure 7.19.

Moments and levers

Imagine a lever with a fulcrum in the middle. On one side is an effort and on the other a resistance. When at rest this lever is said to be balanced. If the effort is increased, the lever will turn in a counterclockwise direction. If the resistance is increased, the lever will turn in a clockwise direction, figure 7.20. The turning force is called a **moment.**

The moment depends on the effort. It also depends on the distance of the effort from the fulcrum.

$$\text{Moment} = \text{Effort} \times \text{Distance of effort from fulcrum}$$

If a beam is in balance, the clockwise moments are equal to the counterclockwise moments.

$$4 \times 50 = 8 \times 25$$

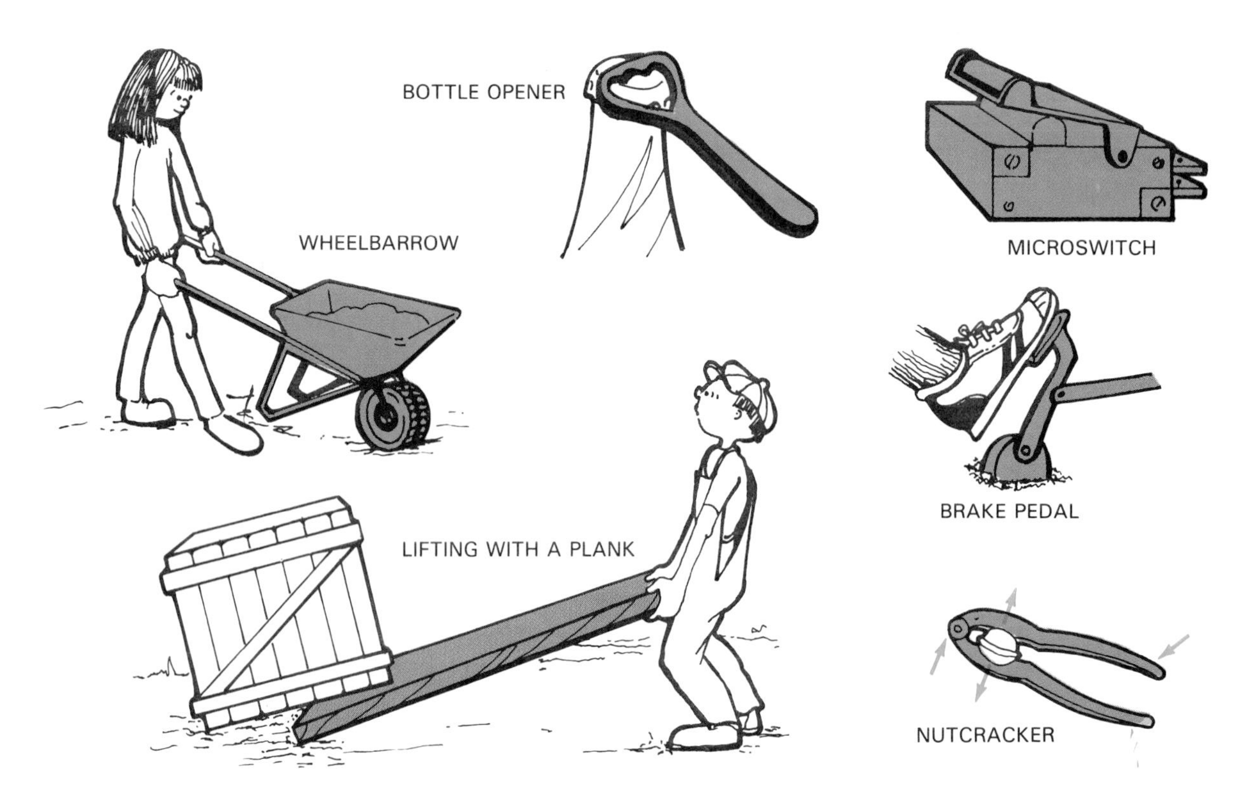

Figure 7.16. Do you recognize these examples of Class 2 levers?

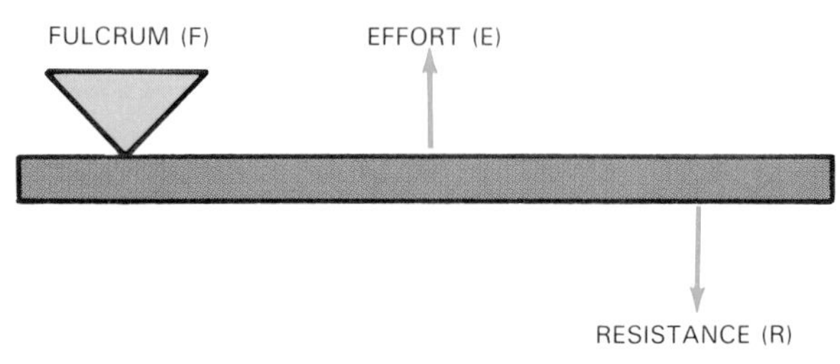

Figure 7.17. How is a Class 3 lever different from a Class 1 lever?

Levers discussed so far have been used to increase force, distance moved, or speed. Levers can also be used to reverse the direction of motion.

Think of a lever with a fulcrum in the center. If it pivots about its fulcrum, the ends move in opposite directions. One end moves down and the other end moves up, figure 7.21. A single lever with a pivot in the center reverses an input motion.

This idea is used in linkages. A linkage is a system of levers used to transmit motion. Figure 7.22 illustrates a reverse motion linkage.

The input force and output force are equal. If the pivot is not at the center, the input force is increased or decreased at the output. This is shown in figure 7.23.

The pulley

The pulley is a special kind of Class 1 lever, figures 7.24 and 7.25. Its action is continuous. The resistance arm is the same length as the effort arm. The length of each arm is the radius of the pulley, figure 7.26.

Pulleys are used for lifting heavy objects. A bale of hay can be lifted into a hayloft using a single pulley suspended from a beam. Car engine hoists enable one person to lift a car engine having a mass of over 450 lb. (200 kg). Cranes use pulleys to lift enormous loads.

Two types of pulleys are used to lift heavy objects. One is called fixed; the other movable.

In a single fixed pulley system, figure 7.2, the effort is equal to the resistance. There is no mechanical advantage. It is easier, however, for the operator to pull down instead of up. There has been a change in direction of force. The distance moved by the effort (ef-

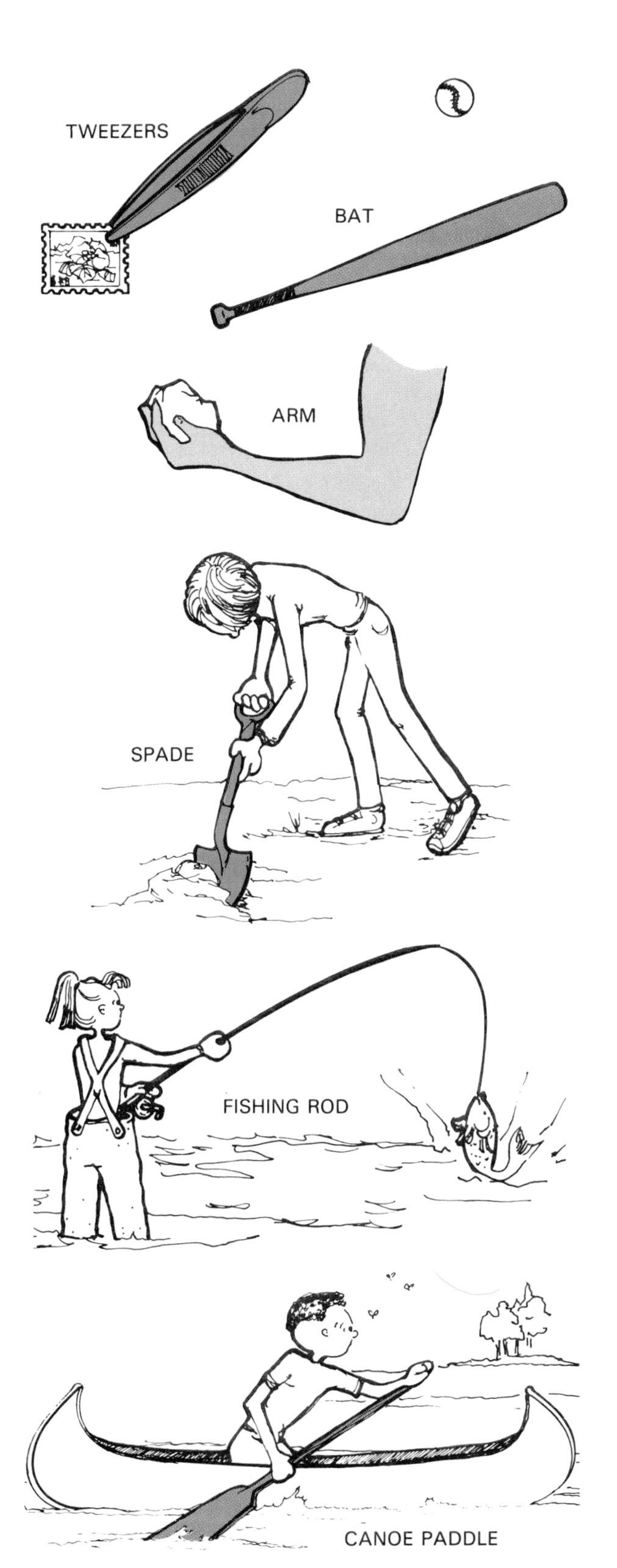

Figure 7.18. No doubt you have used some, if not all, of these types of Class 3 levers.

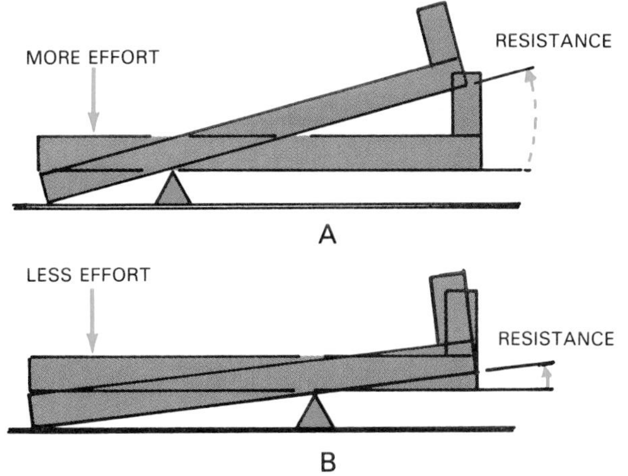

Figure 7.19. What is the effect of moving the fulcrum on a lever?

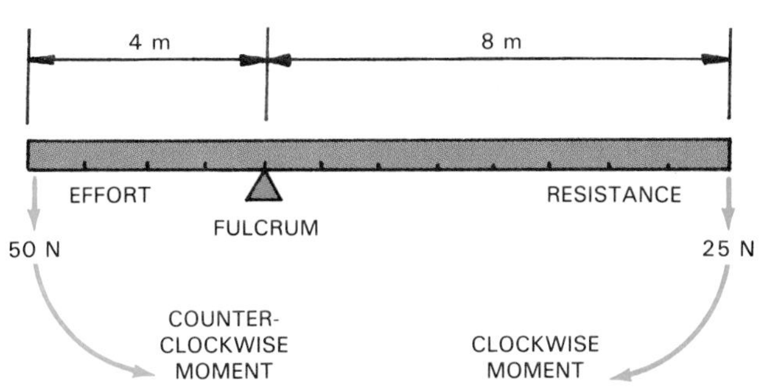

Figure 7.20. Forces acting on a lever are called moments.

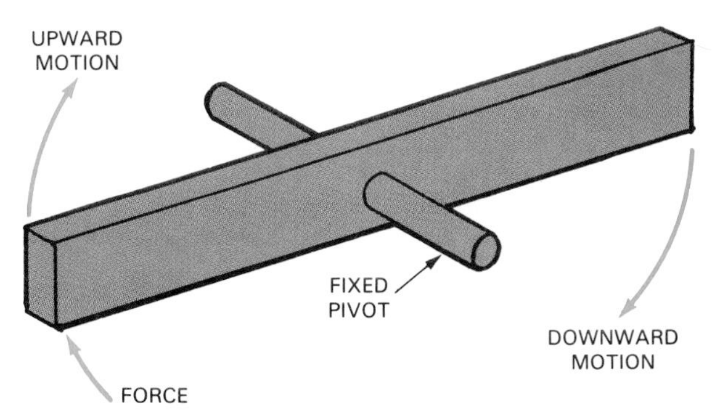

Figure 7.21. Some levers are designed to change motion.

fort distance) is equal to the distance moved by the resistance (resistance distance).

A single, movable pulley system has a mechanical advantage of two. The resistance is supported by both ropes equally. The amount of effort required is half that of the

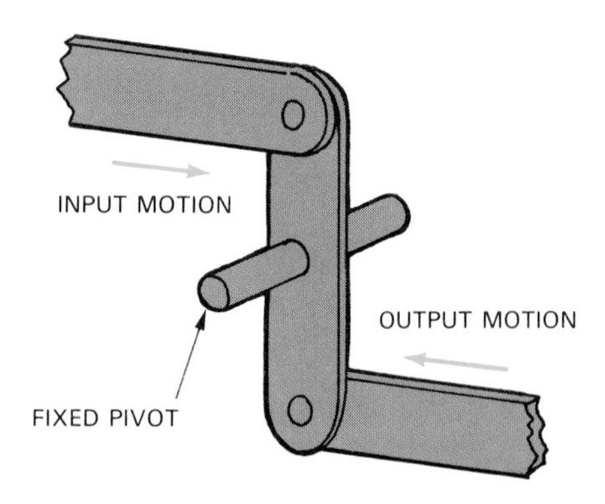

Figure 7.22. An example of a reverse motion linkage. The pivot is fixed at the center of one lever. Input force equals output force.

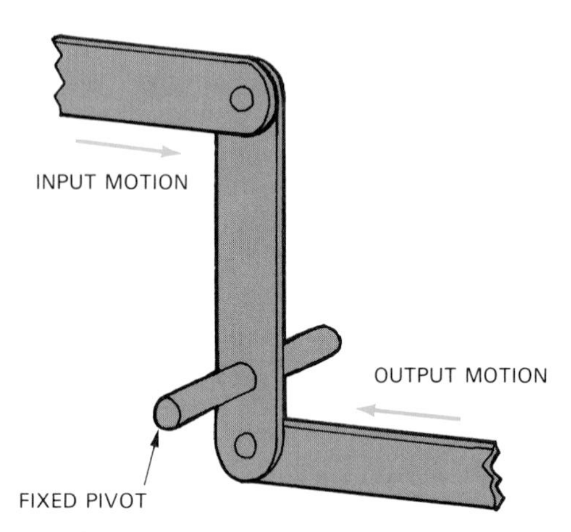

Figure 7.23. In this example, is the input force increased or decreased at the output?

resistance. The disadvantage is that the operator must pull upwards. Also the effort distance is two times the resistance distance. In all pulley systems, as the effort decreases, the effort distance increases, figure 7.28.

To have the advantages of change of direction and decreased effort, movable and fixed pulleys can be combined as shown in figure 7.29. In both examples, the mechanical advantage is two. However, the effort must be exerted over twice the distance.

Pulleys may also be used to transmit motion, increase or decrease speed, reverse the direction of motion, or change motion through 90 degrees, figure 7.30. These types of pulley systems may be used in cars (the fan belt), upright vacuum cleaners, washing machines, and electrical appliances.

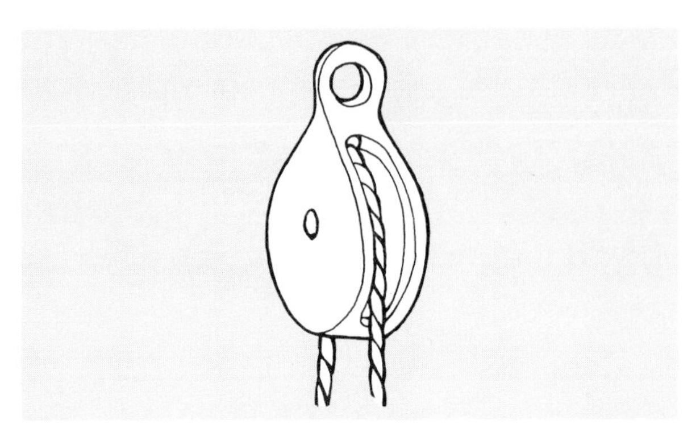

Figure 7.24. A simple pulley. The pulley rope changes direction once.

ECRITEK

Figure 7.25. A chain hoist is a pulley designed to lift heavy loads.

The wheel and axle

The wheel and axle is basically a large-diameter disk or wheel. This is attached rigidly to a smaller diameter bar (axle). Think of it as a special kind of Class 1 lever, figure 7.31.

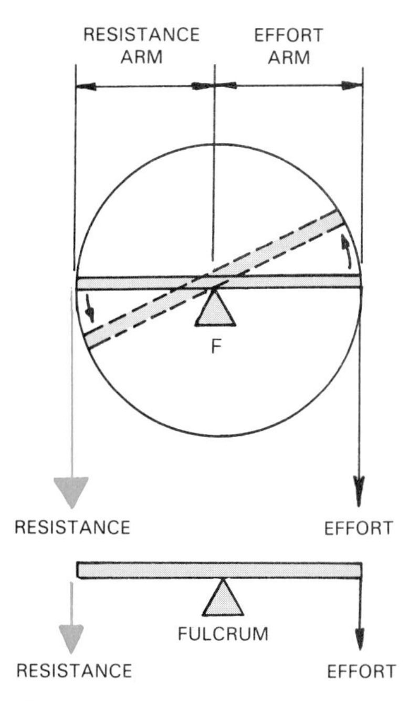

Figure 7.26. Do you see why a pulley is a special kind of Class 1 lever?

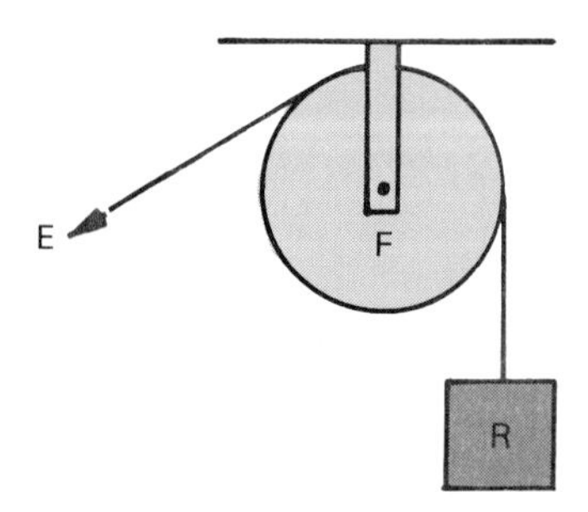

Figure 7.27. A single fixed pulley. The effort is equal to the resistance. There is no mechanical advantage.

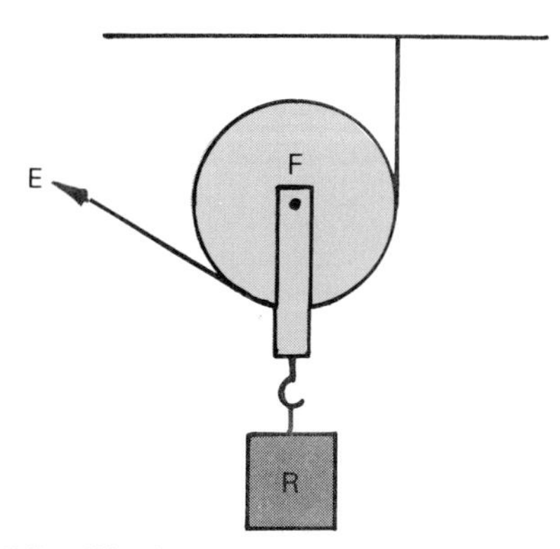

Figure 7.28. Single, movable pulley. Effort equals half the resistance and the effort moves twice the distance of the resistance.

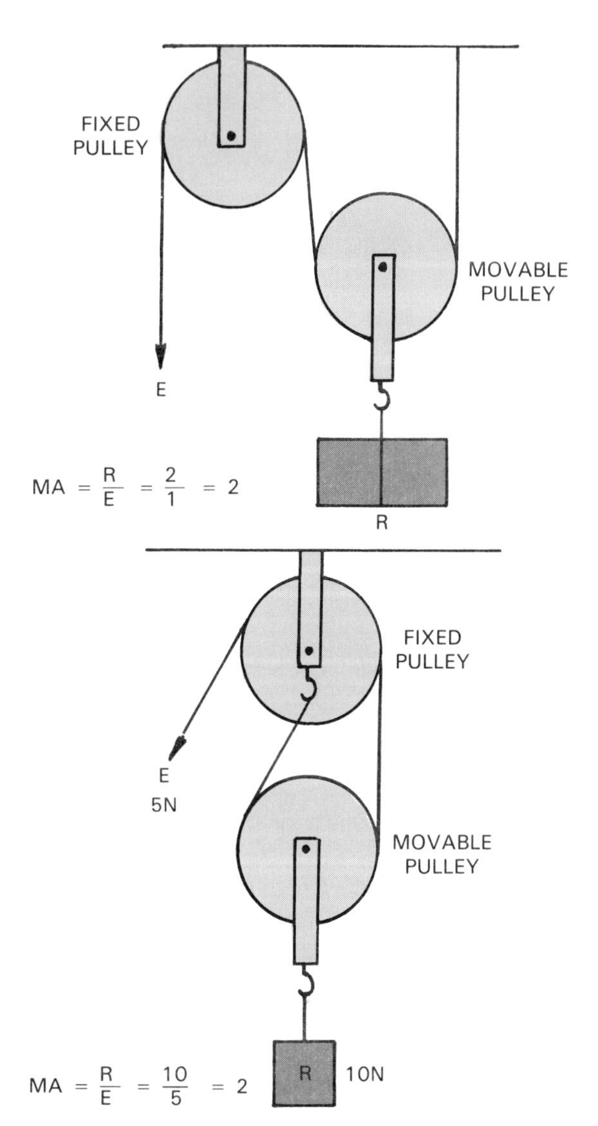

Figure 7.29. A fixed pulley changes direction of the effort. A movable pulley decreases effort.

Like levers, the wheel and axle contains three elements: effort, fulcrum, and resistance. In the case of the door knob, the effort is applied to the rim of the wheel (knob). The knob multiplies the effort and transmits it through the axle (bar). The resistance is the door latch.

A similar application is a car steering wheel. The driver's effort, applied to the wheel, is increased. The result is that the car can be steered with little effort, figure 7.32.

The drive wheel and axle of a car are another example, figure 7.33. The effort is supplied by the engine through the axle to the cir-

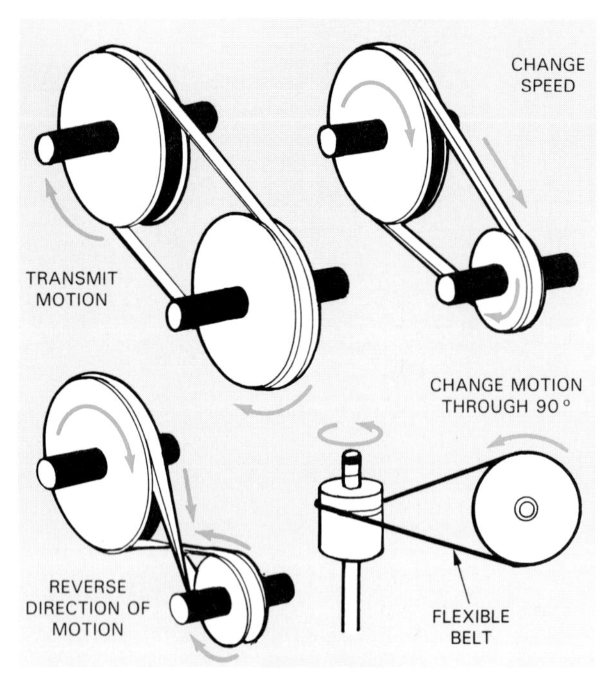

Figure 7.30. Pulleys can transmit motion from one point to another.

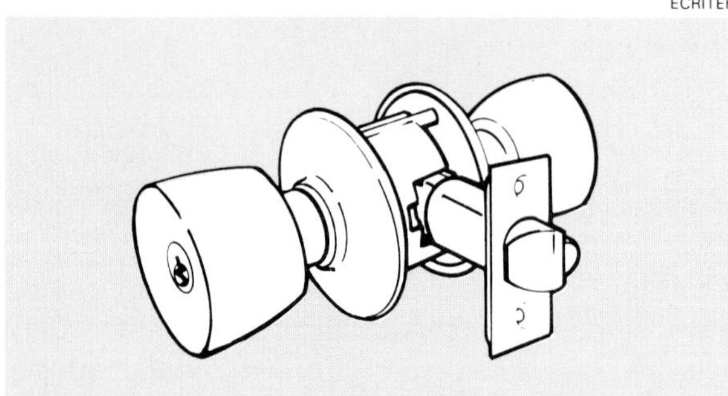

Figure 7.31. Top. A door knob is a common example of the wheel and axle. Bottom. Where are the effort, fulcrum, and resistance?

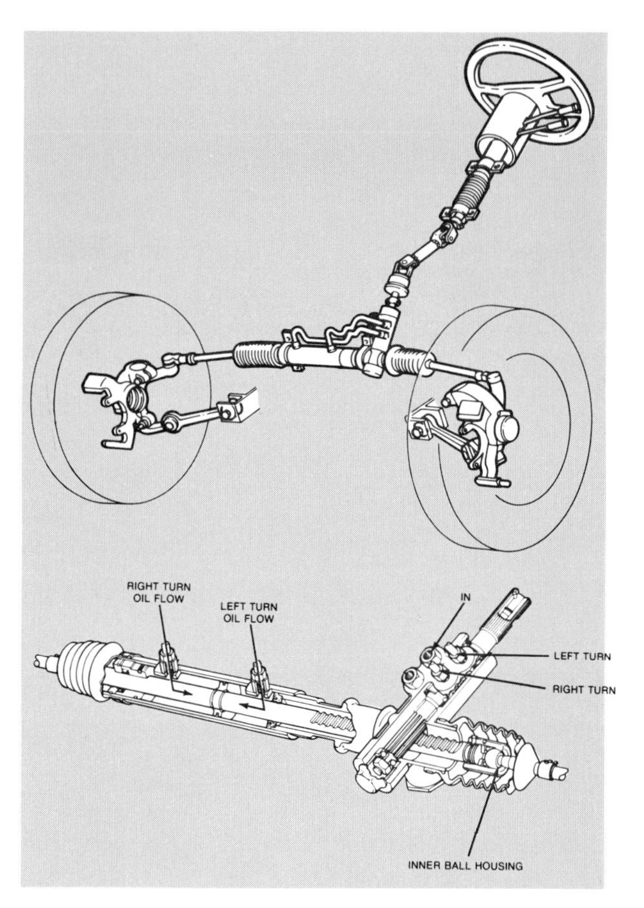

Figure 7.32. A car's steering wheel multiplies the effort applied by the driver to steer the vehicle.

Figure 7.33. Can you trace the path of the effort from the engine to the rim of the wheel?

cumference (rim) of the wheel. The resistance is the force of the road on the wheel.

The inclined plane

There are two ways of loading the motorcycle onto the truck, figure 7.34. One way is

ECRITEK

ECRITEK

Figure 7.34. Left. It would be impossible for one person to lift the motorcycle onto the truck. Right. One person could push it up a ramp onto the truck. Would a shorter ramp make the job harder or easier?

simply to lift it. However, it is much easier and safer to push the motorcycle up a sloping plank. The sloping surface formed by the plank is called an **inclined plane** or ramp.

There are many uses for an inclined plane. Many mountain roads zigzag up the mountainside rather than running straight up to the top.

Common uses are shown in figures 7.35 through 7.37:

- Loading and unloading a car transporter.
- Reaching different levels of a multistory parking lot.
- Replacing steps so handicapped people can enter a building.

TEC

ECRITEK

ECRITEK

Figure 7.35. We use inclined planes for many purposes. Top left. For loading cars onto a transporter. Bottom left. To park cars on different levels of buildings. Right. To help persons in strollers or wheelchairs move from one level to another.

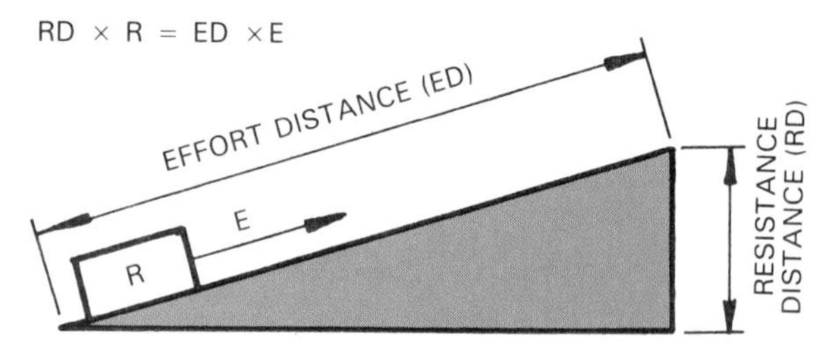

Figure 7.36. The mass of an object and the distance it is raised are related to the length of the inclined plane.

Figure 7.37. An axe is a wedge, which is a special kind of inclined plane. However, its uses are much different from the inclined plane.

Calculating inclined planes. How do people who build ramps decide about their length? Guessing about it is too costly. They might have to tear down the ramp and start over. Therefore, they use mathematics.

In general, the length of a ramp depends on the mass of the object to be moved. Heavier objects need longer ramps or more effort to move them upward. The more gradual the rise, the less effort required. We can say, then, that the length of the inclined plane is directly related to the mass of the object. This means that as mass increases, ramp length increases. Figure 7.36 shows this relationship. So does the following formula:

Effort (E) × Effort Distance (ED) = Resistance (R) × Resistance Distance (RD)

Now, let us see how we might use the formula to work out a problem. Suppose, for example, that a mass of 450 lb. (200 kg) needs to be raised 6 ft. (2 m). Then suppose that the most force that can be exerted is 100 lb. (50 kg). How long must the ramp be so the force is able to move the object?

$$ED = \frac{R \times RD}{E} = \frac{450 \times 6}{100} = \frac{2700}{100} = 27 \text{ ft.}$$

$$ED = \frac{R \times RD}{E} = \frac{200 \times 2}{50} = \frac{400}{50} = 8 \text{ m}$$

The wedge

The **wedge** is a special version of the inclined plane, figure 7.37. It is two inclined planes back-to-back, figure 7.38.

The shape is effective because the force exerted pushes out in two directions as it enters the object, figure 7.39. Do you see the difference between the wedge, figure 7.38, and the inclined plane in figure 7.36?

Figure 7.38. A wedge is like two inclined planes joined together.

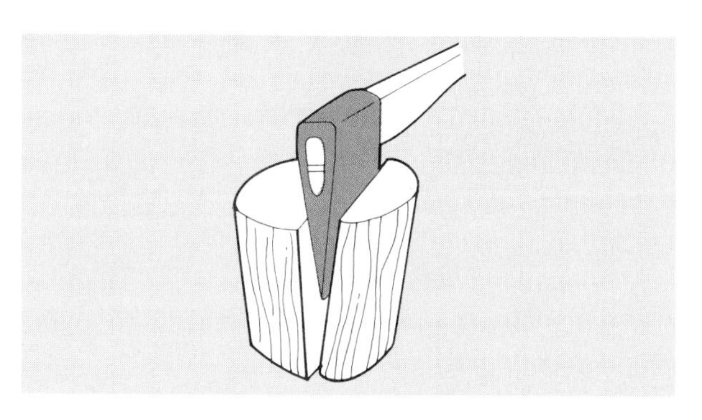

Figure 7.39. Because of its shape, an axe enters material easily and splits it apart.

Other applications of the wedge include the plow, a doorstop, the blade of a knife, and the prow of a boat, figure 7.40.

The screw

It is not hard to find examples of how we use the screw. When you see someone using a scissor jack to lift a car, you are seeing a screw at work, figure 7.41 and figure 7.42.

A screw is an inclined plane wrapped in the form of a cylinder. To illustrate how this works, take a rectangular piece of paper and cut it along a diagonal. This triangle will remind you of an inclined plane. Now wrap it around a pencil. Roll from the short edge of the triangle to shape it like a screw, figure 7.43.

LASER

CHRISTOPHARO

HEWITT

Figure 7.40. We use the wedge in many products. Top. Wedge-shaped prow (front) of a boat cuts through the water easily. Middle. A knife cuts through many materials. Bottom. Bucket of a backhoe is wedge-shaped to enter soil more easily.

CHRISTOPHARO

Figure 7.41. A scissor jack lifts heavy loads with less effort applied.

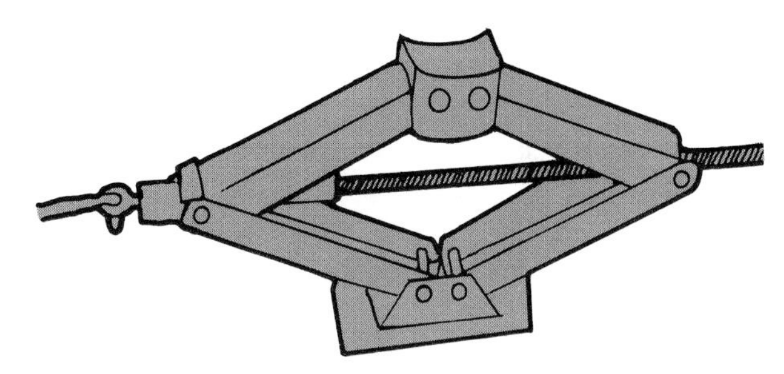

Figure 7.42. As the screw of the jack is turned, the two ends of the jack are forced together and the car is raised.

When we examine the inclined plane we find that the time taken for the load to be pushed up a longer inclined plane increases while the effort decreases. The same is true in the case of the screw threads on a nut and bolt. The greater the number of threads, that is, the shallower the slope, the longer it takes to move the nut to the head of the bolt. However, it will be easier to move the nut against a resistance.

The wedge-shaped section of a tapering wood screw reveals another application of the wedge, figure 7.44. It allows the screw to force its way into the wood.

Screws threads may be used in two quite different ways. They may be used to fasten as

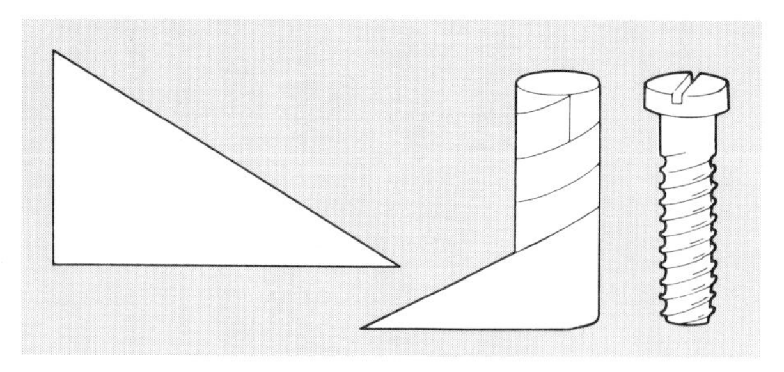

Figure 7.43. A screw is an inclined plane wrapped into a cylinder.

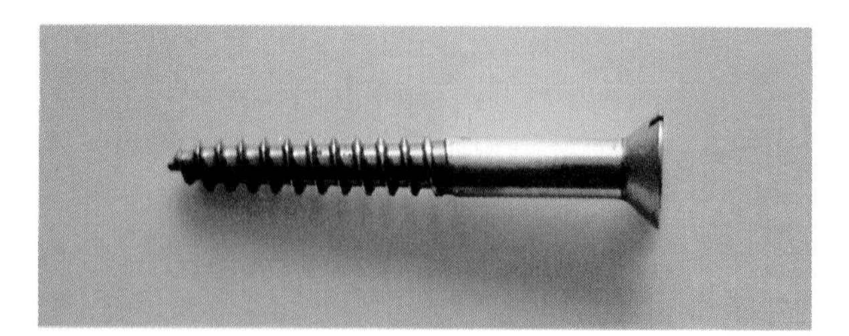

Figure 7.44. A wood screw also has a wedge shape to push aside the wood fibers as it enters the wood.

with wood screws, machine screws, and light bulbs. They may also transmit motion and force. Examples are C-clamps, vises, and car jacks. Note also that a screw converts rotary motion into straight-line motion, figure 7.45.

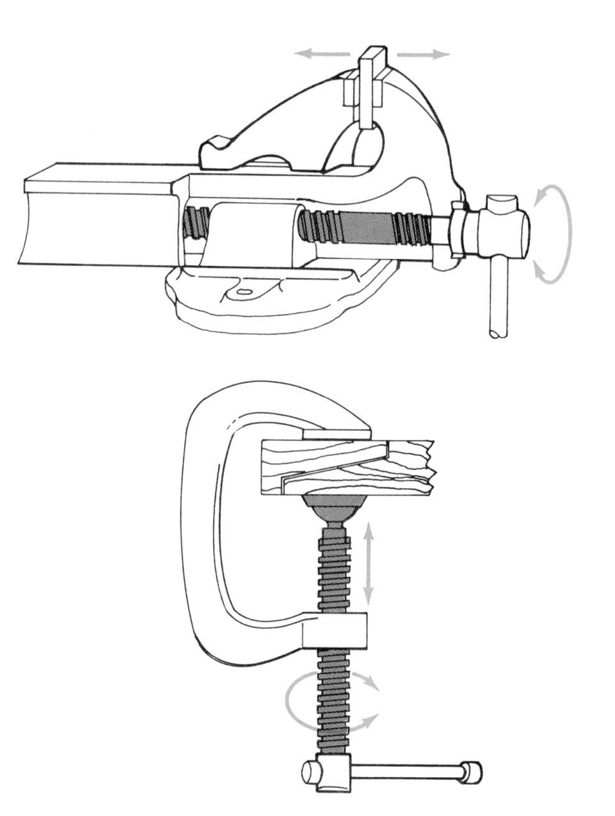

Figure 7.45. Screw threads are used on vises and C-clamps. They convert rotary motion to straight-line motion.

■ GEARS

Gears, figures 7.46 and 7.47, are not classified as one of the six simple machines. They are similar to pulleys in that their motion is usually circular and continuous. Gears have an advantage over pulleys. They cannot slip.

Many bicycles have gears to help make pedaling as effortless as possible. When climbing hills, the cyclist selects a low gear to make pedaling easier. When descending a hill, a higher gear provides a high speed in return for slower pedaling.

A mechanical clock contains many different sized gear wheels. They are arranged so that they rotate the clock hands at different speeds.

Gears, like pulleys, are modified levers. They transmit rotary motion. They increase or decrease speed, change the direction of motion, and transmit a force. This force, known as torque, acts at a distance from the center of rotation. To understand this concept more easily, think of torque as a measure of turning effort. It is similar to the use of a wrench to tighten a bolt, figure 7.48 and figure 7.49.

Figure 7.46. In what way are gears similar to pulleys?

Figure 7.47. Gears cannot slip because of the teeth.

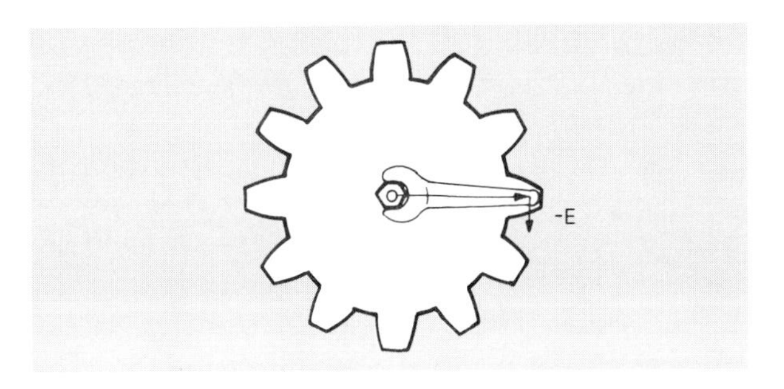

Figure 7.49. Compare a gear to a wrench turning a nut.

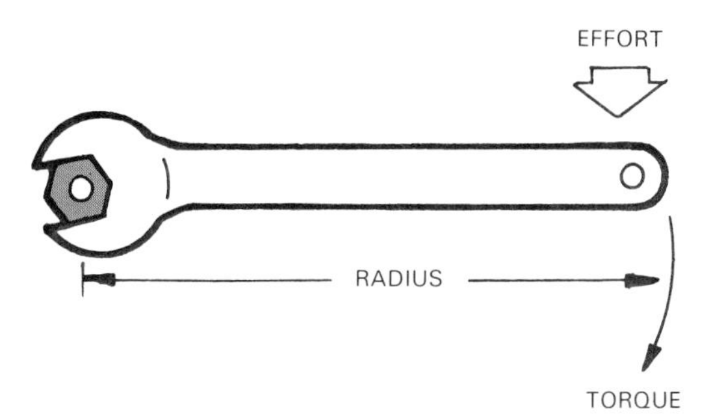

Figure 7.48. Torque is force applied to a radius.

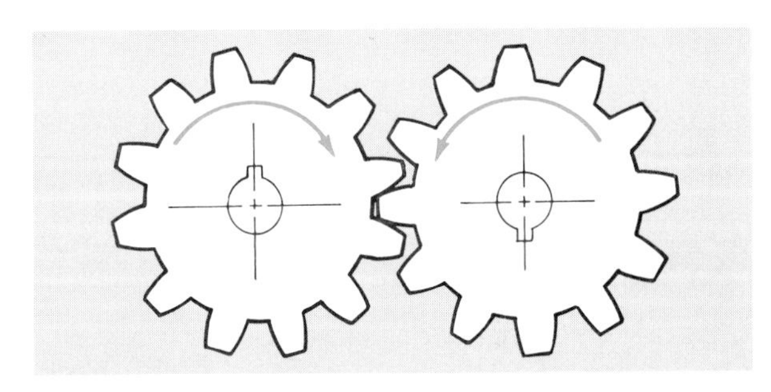

Figure 7.50 A gear train. As one gear moves, its torque is transmitted to another gear.

An effort (E) is applied at a distance (R) from the center of the nut. The torque on the nut is calculated by multiplying the effort (E) by the distance (R). The applied force is measured in pound (newtons) and the distance from the center of rotation is measured in feet (meters). Therefore, torque is measured in ft./lb. (N/m).

As you look at a gear, you can consider the center of the gear to be like the nut. Consider the end of the gear tooth to be like the end of the wrench, figure 7.49.

Gears are used in groups of two or more. A group of gears is called a gear train. The gears in a train are arranged so that their teeth closely interlock or mesh. The teeth on gears that mesh are the same size and of equal strength. The spacing of the teeth on each gear is the same.

When two gears of the same size mesh, they act as a simple torque transmitter. They both turn at the same speed but in opposite directions, figure 7.50. The input motion and force are applied to the driven gear. The output motion and forces are transmitted by the driven gear.

When two gears of different size mesh, they act as torque converters, figure 7.51. The

Figure 7.51. Two gears act like levers to convert torque.

larger gear is called a wheel. The smaller gear is called a pinion. The pinion gear revolves faster, but the wheel delivers more force.

The amount of torque delivered by a gear is described as a ratio. For example, suppose that a gear of 10 teeth meshes with a gear of 30 teeth. The small gear will make three revolutions for each revolution of the larger gear. As the small gear makes one revolution, its 10 teeth will have meshed with 10 teeth on the larger gear. The large gear will have turned through ten-thirtieths or one-third of a revolution. The small gear will have to make three revolutions to turn the large gear through one revolution. The gear ratio is, therefore, 3:1, figure 7.52.

Gear trains are either simple or compound. In a simple gear train there is only one meshed gear on each shaft. Figure 7.53 shows an idler gear placed between the driver gear and the driven gear. The driver gear and the driven gear now rotate in the same direction. The idler gear does not change the gear ratio between the driver and the driven gear.

A compound gear train has a driver gear and a driven gear, but the intermediate gears are fixed together on one common shaft, figure 7.53. The gear wheels on the intermediate shaft are not idlers, for one is a driven gear and the other is a driver gear. They do affect the ratio of the gear train.

Just like the six simple machines, gears provide mechanical advantage. This advantage is calculated in the following way:

$$\text{Mechanical advantage (MA)} = \frac{\text{Number of teeth on driven gear}}{\text{Number of teeth on driver gear}}$$

TEC

Figure 7.52. A gear train with a 3:1 ratio. If the smaller gear is driving the larger one, the input torque will be multiplied by three.

The velocity (speed) of the driven gear is calculated as follows:

$$\text{Velocity of driven gear} = \frac{\text{Number of teeth on driver gear} \times \text{velocity of driver gear}}{\text{Number of teeth on driven gear}}$$

An example of these calculations for the gears is shown in figure 7.54.

Gears are designed in a variety of types for a variety of purposes. The five most common gear types are spur, helical, worm, bevel, and rack-and-pinion.

The spur gear is the simplest and most fundamental gear design, figure 7.55. Its teeth are cut parallel to the center axis of the gear. The strength of spur gears is no greater than the strength of an individual tooth. Only one tooth is in mesh at any given time.

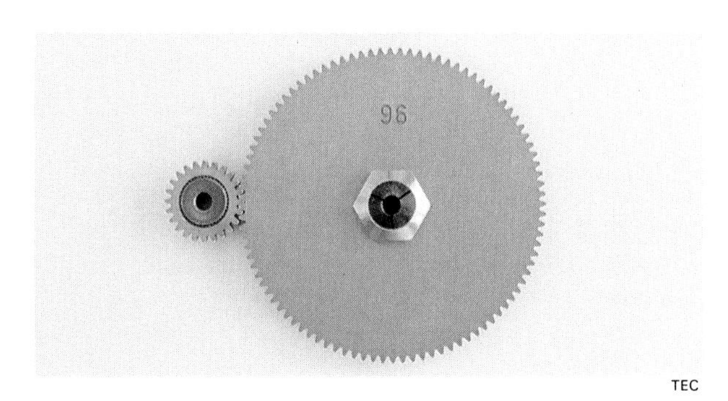

TEC

$$\text{MA} = \frac{30}{15} = \frac{2}{1} = 2$$

$$\text{Velocity} = \frac{15 \times 60}{30} = 30 \text{ rpm}$$

Figure 7.54. You can calculate the mechanical advantage and velocity of compound gears by using the formula above.

TEC

TEC

Figure 7.53. A compound gear train.

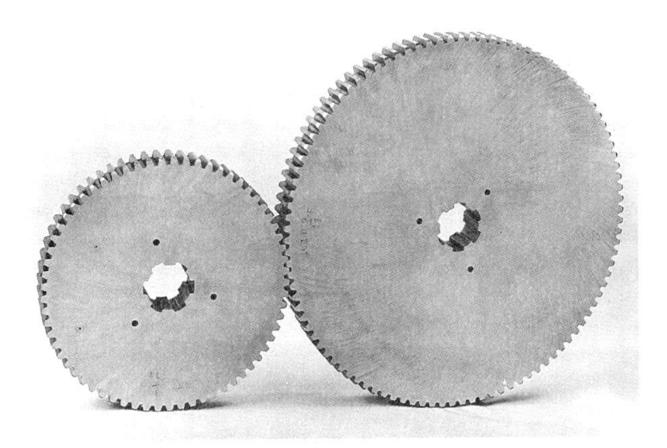

Figure 7.55. Spur gears. Teeth are cut straight across the width of the gears.

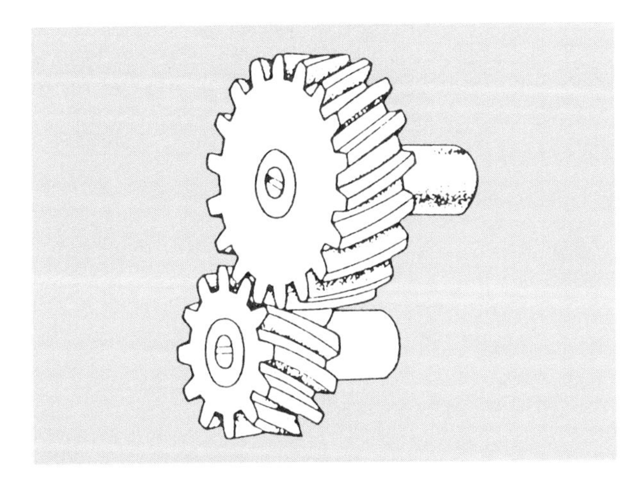

Figure 7.56. Helical gears, cut at an angle, allow several teeth to engage at one time.

To overcome this weakness, helical gears are sometimes used, figure 7.56. Since the teeth on helical gears are cut at an angle, more than one tooth is in contact at a time. The increased contact allows more force to be transmitted.

A worm is a gear with only one tooth. The tooth is shaped like a screw thread. A wormwheel meshes with the worm, figure 7.57. The wormwheel is a helical gear with teeth inclined so that they can engage with the threadlike worm. This system changes the direction of motion through 90 degrees.

It also has the ability to make major changes in mechanical advantage (M.A.) and speed. Input into the worm gear system is usually through the worm gear. A high M.A. is possible because the helical gear advances only one

FRANKE

Figure 7.57. A worm and wormwheel. The worm, right, has only one tooth. It spirals like a screw thread.

tooth for each complete revolution of the worm gear. The worm gear in figure 7.57 will rotate 40 times to turn the helical gear only once. This is an M.A. of 40:1. Worm gear mechanisms are very quiet running.

Bevel gears, figure 7.58, change the direction in which the force is applied. This type of gear can be straight cut in the same way as spur gears. Or, they may be cut at an angle, similar to helical gears.

Rack-and-pinion gears, figure 7.59, use a round spur gear (the pinion). It meshes with a spur gear that has teeth cut in a straight line (the rack). The rack-and-pinion transforms rotary motion into linear (straight-line) motion and vice versa. Figure 7.60 shows two uses for rack-and-pinion gears.

■ PRESSURE

You have seen how simple machines and gears are able to move a greater resistance with a smaller effort. Now you will see that pressure can increase the effort applied. Figure 7.61A shows a diagram of a simple system for multiplying force through use of pressure.

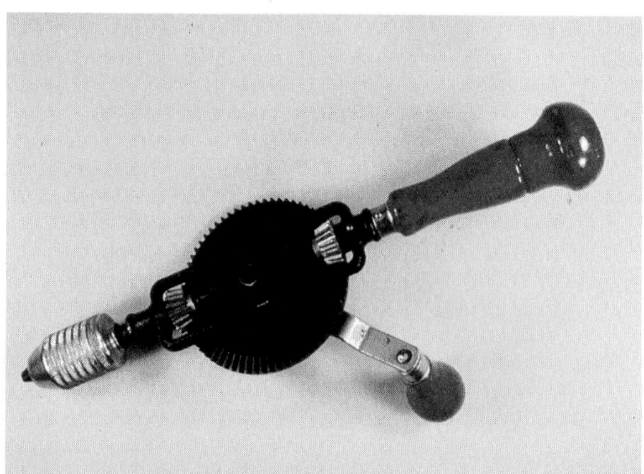

Figure 7.58. Top. Bevel gears are designed to change the direction of the applied force. Bottom. A hand drill uses bevel gears.

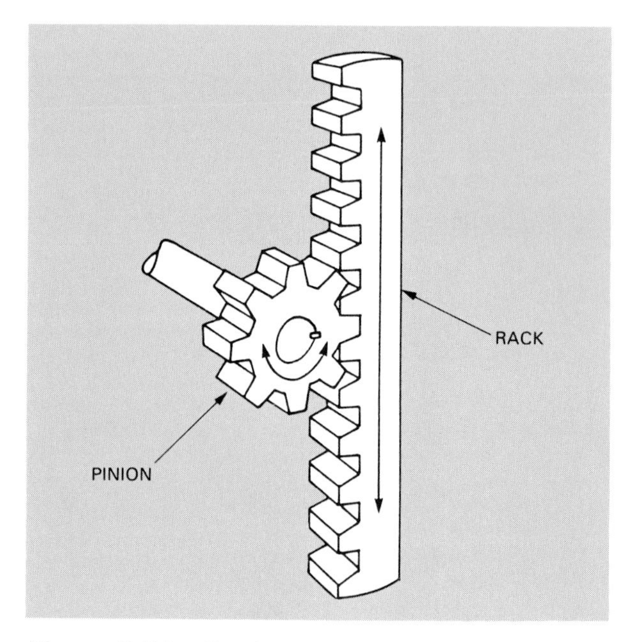

Figure 7.59. Rack-and-pinion gears convert rotary motion to straight-line motion.

Figure 7.61B shows a simple pressure device used to lift an automobile.

To understand how pressure can be increased, think about the area on which the effort presses. Would you rather a woman step on your foot with a small, pointed heel or with a larger, flat heel, figure 7.62?

When a surface area is small, a little effort produces a large pressure. Pressure is the effort applied to a given area. To calculate, divide effort by area. The formula is:

$$\text{Pressure} = \frac{\text{effort}}{\text{area}}$$

For example, if a 120 lb. woman rests her mass on a 4 sq. in. heel, the pressure is 30 psi. On the other hand, if she rests her mass on a 1/4 sq. in. heel, the pressure increases to 480 psi.

In the metric system, effort is measured in newtons. Area is in meters squared (m^2). Pressure is calculated in newtons per meter squared (N/m^2). The metric unit of measure for pressure is the pascal (Pa). Since a pascal is small, kilopascals are generally used (1.0 kPa = 1000 Pa).

Consider another example of how area affects pressure. A knife has a sharp edge. Pressed against a surface, it takes up a very small area. That is why it cuts: the material offers little resistance to such a tiny surface. A dinner fork works in a similar way. The narrow prongs place enough pressure on the food to pierce it easily.

Hydraulics and pneumatics

The study of pressure in liquids is called **hydraulics** and in gases, **pneumatics.** Unlike solids, liquids and gases flow freely in all directions. Pressure, therefore, can be transmitted in all directions. For example, water will flow in a garden hose even when the hose is bent in many directions.

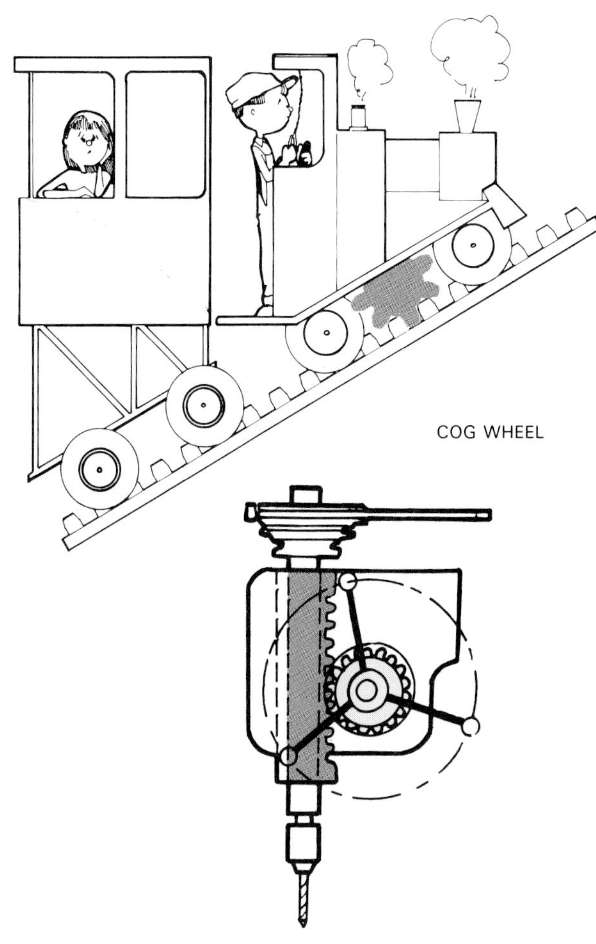

Figure 7.60. An inclined railroad and a feed mechanism for a drill press use the same type of gears.

Figure 7.63 shows a model of a hydraulic lift. The effort is being applied to a piston Pressure produced by the effort is being transmitted by the liquid to a second piston. This piston moves the resistance. The second piston has a larger area and so the pressure presses on a larger area. This produces a larger effort. If the resistance piston has four times the area of the effort piston, the effort on it will be four times greater.

From this example you can see that the effort acting on a piston from a liquid under pressure depends on the area of the piston. The larger the area, the larger the effort. However, the distance moved by the larger

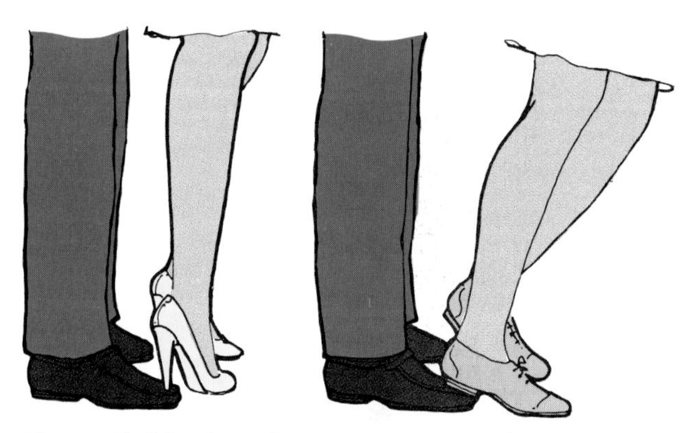

Figure 7.62. Imagine someone stepping on your foot. Which will hurt more, the pointed heel or the flat heel?

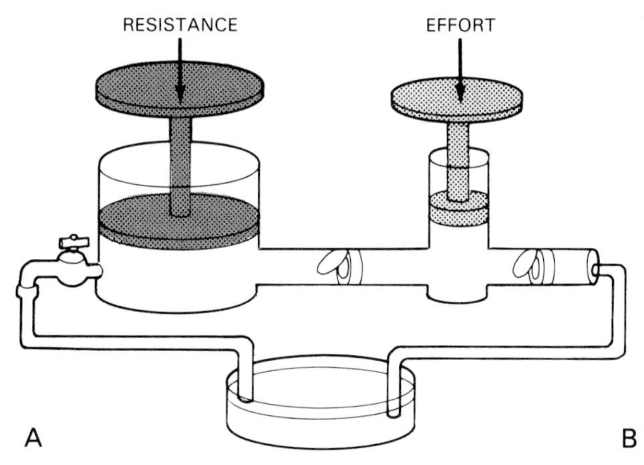

Figure 7.61. Using pressure to increase force. A—A hydraulic system. A fluid transmits effort from where it is applied to where it is used. B—A hydraulic lift is used to support the mass (weight) of an automobile.

piston will be less than the distance moved by the smaller piston. In figure 7.63, the smaller piston moves four times the distance of the larger piston.

The hydraulic brake system on cars operates using the same principles as the hydraulic lift, figure 7.64. Using the brake pedal, a car driver applies a small effort to the piston. Hydraulic fluid is transmitted through the brake lines to a larger piston. The larger piston forces brake pads and shoes against disks and drums. The brake systems of large trucks, buses, and trains are often pneumatically (air) operated.

Among the many common applications of hydraulic power are dentist and barber chairs, door closers, and power steering. Common applications of pneumatic power include a variety of tools such as air drills, screwdrivers, and jackhammers. Sometimes hydraulic and pneumatic systems are combined. For example, air pressure forces hydraulic fluid to raise the lift in a garage, figure 7.61B.

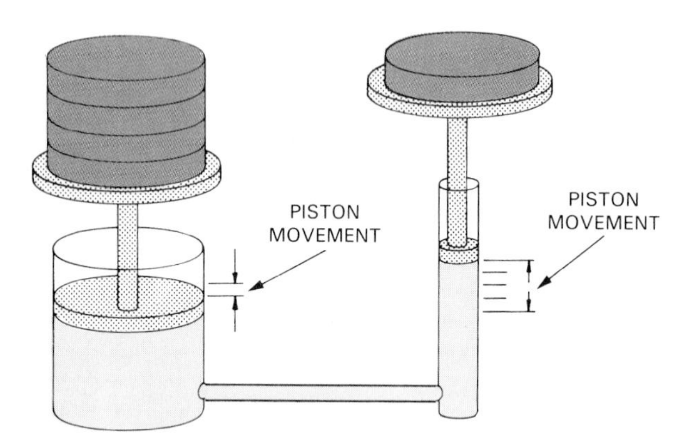

Figure 7.63. Model of a hydraulic lift. Large movement in one piston will create a smaller movement in the other piston. Why?

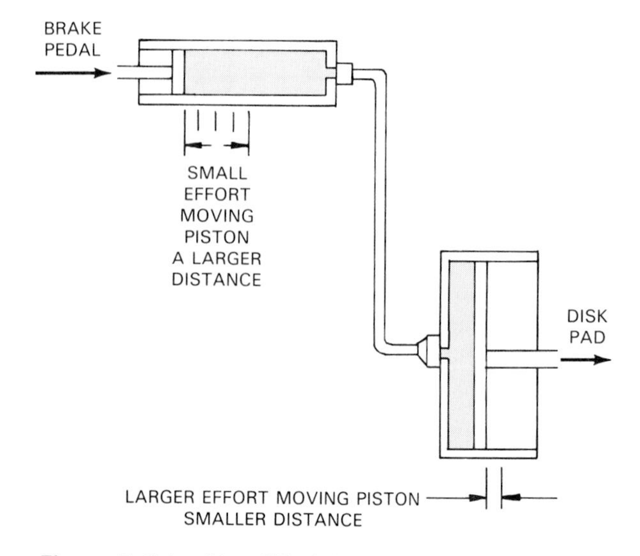

Figure 7.64. Simplified diagram of an automobile brake system.

Because of their many advantages, most industries use hydraulic and pneumatic systems. These advantages include the ability to:

- Multiply a force using minimal space.
- Transmit power to wherever pipe, hose, or tubing can be located.
- Transmit motion rapidly and smoothly.
- Operate with less breakage than occurs with mechanical parts.
- Transmit effort over considerable distance with relatively small loss.

MECHANISM

A mechanism is a way of changing one kind of effort into another kind of effort, figure 7.65. For example, a C-clamp holds two pieces of wood together while glue sets. Rotary motion is changed to linear motion to apply pressure, figure 7.66.

Mechanisms can be combined to form machines. Their advantages include:

- Changing the direction of an effort.
- Increasing the amount of effort applied.
- Decreasing the amount of effort applied.
- Applying an effort to a place otherwise hard to reach.
- Increasing or decreasing the speed of an operation.

Machines change one kind of energy into another and do work. The amount of work done is approximately equal to the amount of energy changed.

$$\text{work} = (\text{energy change}) = \text{effort} \times \text{distance moved in direction of effort}$$

For example, how much work is done to move a 50 lb. resistance a distance of 5 ft.?

$$\text{Work} = 50 \times 5 = 250 \text{ ft./lb.}$$

In metric, how much work would be done to move a 50 N resistance through 4 meters?

$$\text{Work} = 50 \times 4 = 200 \text{ joules}$$

Figure 7.65. How many simple machines make up this mechanism?

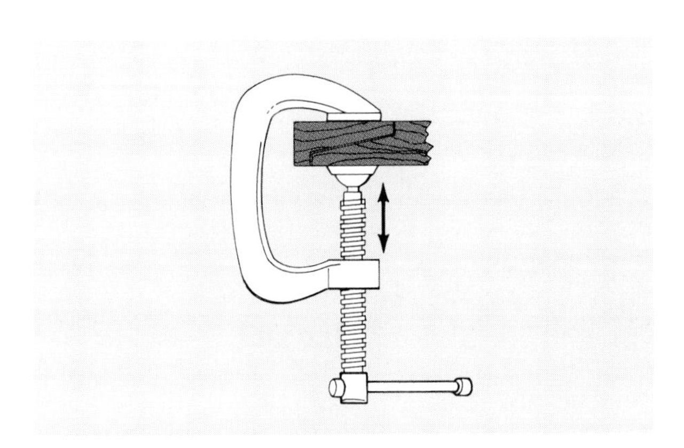

Figure 7.66. A C-clamp is an example of a mechanism.

Machines make it easier to do work. However, no machine does as much work as the energy put into it. If a machine did the same amount of work as the energy supplied, it would be 100 percent efficient. Most machines lose energy as heat or light however. The approximate efficiencies of some common machines are:

Watt's steam engine	3%
A modern steam engine	10%
A gasoline engine	30%
Nuclear power plant	30%
Aircraft gas turbine	36%
Diesel engine	37%
Rocket engine	48%
An electric motor	80%

Mechanisms use or create motion, figure 7.67. The four basic kinds of motion are:

- Linear—straight-line motion.
- Rotary—motion in a circle.
- Reciprocating—backward and forward motion in a straight line.
- Oscillating—backward and forward motion like a pendulum.

Mechanisms are often used to change one kind of motion into another kind. Some examples are shown in figure 7.68.

□ FRICTION

As you pedal your bicycle you are working against friction. Friction is a force that acts like a brake on the movement of moving objects. Your finger will slide without much effort on a pane of glass. But if you do the same thing on sandpaper, you can feel a resistance slowing up your movement.

The moving parts of mechanisms do not have perfectly smooth surfaces. The tiny pro-

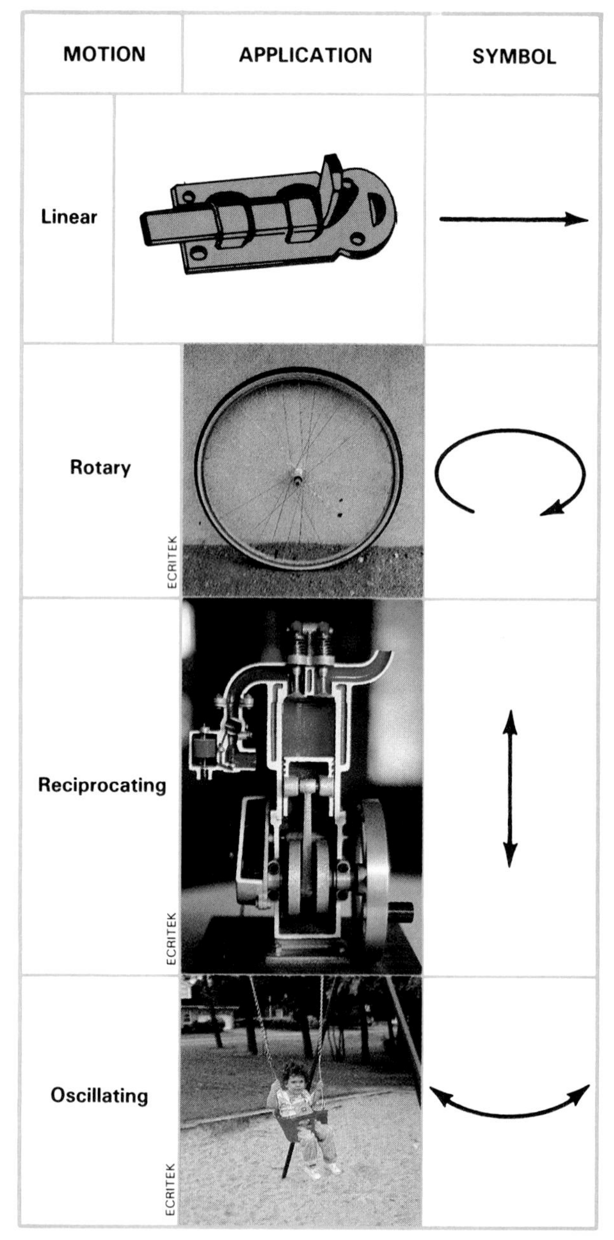

Figure 7.67. Mechanisms are used to create different kinds of motion.

jections on the surfaces rub on one another. This creates friction and results in heat. The friction between the moving parts must be minimized so that:

- Less energy will be needed to work the machine.
- Wear and tear will be reduced.
- Moving parts will stay cooler.

Friction may be reduced in several ways as shown in figures 7.69 through 7.72.

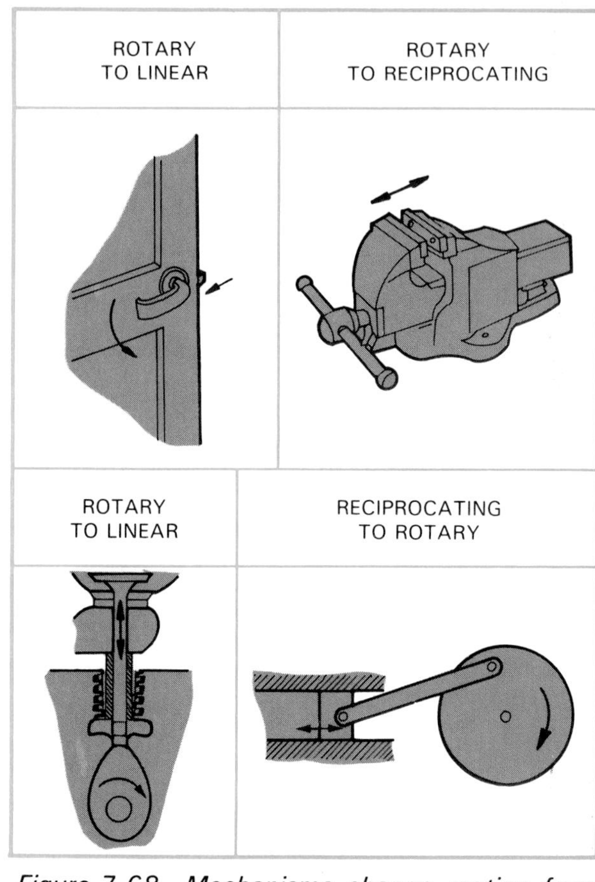

Figure 7.68. Mechanisms change motion from one form to another.

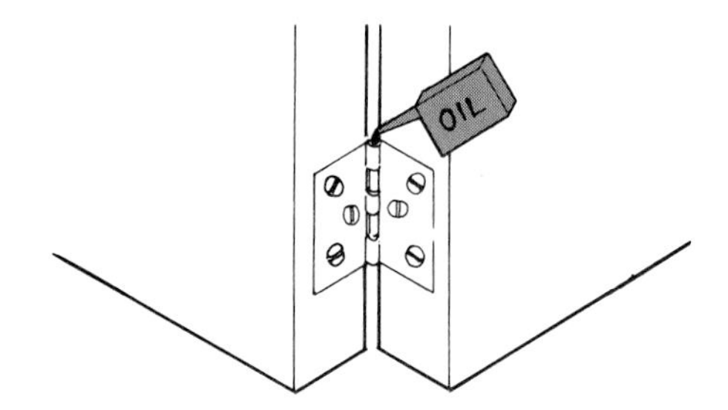

Figure 7.69. Oil lubricates parts to reduce friction.

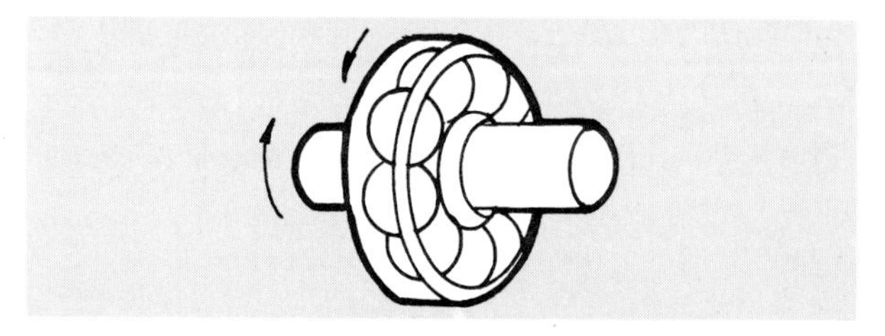

Figure 7.70. Ball bearings allow parts to "roll" over one another, reducing friction.

CANADA COASTGUARD

Figure 7.71. Hovercraft uses a cushion of air to reduce friction.

NFB

Figure 7.72. One of the effects of streamlining is to reduce air drag which is a form of friction.

- Oiling—oil separates two surfaces that would otherwise touch, rub, and wear each other away.
- Ball bearings—steel balls enable surfaces to roll over one another instead of sliding.
- Air or water cushions—compressed air or water separates moving parts.
- Streamlining—the shape of a fast-moving object can be changed to reduce its resistance to air or water.

SUMMARY

Machines can be small like a bottle opener or large and complex like a bulldozer. The six simple machines are the lever, pulley, wheel-and-axle, inclined plane, wedge, and screw.

A lever may be used to increase force, increase the distance a force moves, or the speed at which it moves. They may also reverse the direction of motion.

A pulley is a special kind of lever that acts continuously. It is used for lifting heavy objects vertically. It may also be used to transmit motion, increase or decrease speed, reverse the direction of motion, or change motion through 90 degrees.

The wheel-and-axle is also a special kind of lever. Effort applied to the outer edge of the wheel is transmitted through the axle.

The inclined plane is most commonly seen in the form of a ramp. The more gradual the slope, the less force needed to push an object upward.

The wedge is a special version of an inclined plane. It is two inclined planes back-to-back. A screw is an inclined plane wrapped in the form of a cylinder.

Gears are modified levers used in the transmission of rotary motion. They increase or decrease speed, change the direction of motion, and transmit a force.

Simple machines and gears are able to move a greater resistance with a smaller effort. Pressure can also be used to increase an effort applied. Pressure in liquids is called hydraulics, and in gases, pneumatics.

Mechanisms change one kind of effort into another kind of effort. They can be combined to form a machine. Machines change one kind of energy into another kind and do work.

KEY TERMS

Friction
Gear
Hydraulics
Inclined plane
Lever
Linkage
Pneumatics
Pressure
Pulley
Machines
Mechanical advantage
Mechanism
Moment
Screw
Wedge
Wheel-and-axle
Work

TEST YOUR KNOWLEDGE

Write your answers to these review questions on a separate sheet of paper.

1. List four machines used by early civilizations to make work easier.
2. Identify objects in or around your home that contain one or more simple machines.
3. When you hit a ball with a baseball bat, the bat is an example of a lever. Explain why.
4. Draw three diagrams to illustrate the difference between a Class 1, Class 2, and Class 3 lever.
5. Which of the following objects is a Class 1 lever?
   a. Nutcracker.
   b. Wheelbarrow.
   c. Hockey stick.
   d. Scissors.
6. Which of the following objects is a Class 2 lever?
   a. Nutcracker.
   b. Seesaw.
   c. Fishing rod.
   d. Baseball bat.
7. Which of the following objects is a Class 3 lever?
   a. Scale.
   b. Tweezers.
   c. Nutcracker.
   d. Claw hammer.
8. A laborer using a first-class lever places the load the same distance from the fulcrum as the effort. If the fulcrum is moved closer to the load, the mechanical advantage of the machine ________
   a. Increases.
   b. Decreases.
   c. Remains the same.
   d. Approaches one.
9. A lever is used to move a load of 1500 newtons with a force of 300 newtons. What is the mechanical advantage of the lever?
   a. 30.
   b. 15.
   c. 5.
   d. 3.
10. State the four advantages of levers.
11. The advantage of a single, fixed pulley system is that it:
    a. Increases mechanical advantage.
    b. Decreases the effort required.
    c. Changes the direction of force.
    d. Makes the work harder.
12. The disadvantage of a single, movable pulley system is that it:
    a. Decreases the mechanical advantage.
    b. Makes the operator pull down on the rope.
    c. Increases the effort needed to move a load.
    d. Makes the operator pull up on the rope.
13. What force would be required to raise a load of 15 newtons using a single, fixed pulley?
    a. 30.
    b. 15.
    c. 7.5.
    d. 5.
14. What force would be required to raise a load of 15 newtons using a single, movable pulley?
    a. 30.
    b. 15.
    c. 7.5.
    d. 5.
15. List three examples of a wheel-and-axle.
16. List three practical applications of an inclined plane.
17. A 100 N load has to be moved to the top of an inclined plane 10 m long and 2 m high. What effort is required?
    a. 10 N.
    b. 20 N.
    c. 50 N.
    d. 100 N.
18. A wedge is composed of ________.
19. What is the connection between an inclined plane and a screw thread?
20. State four advantages of simple machines.
21. A gear is a modified form of ________.
22. State three reasons why gears are used.
23. The largest gear in a two-gear train is called the ________.
    a. Driver.
    b. Driven.
    c. Pinion.
    d. Wheel.
24. The wheel in a two-gear train has 60

teeth and the pinion 15 teeth. What is the ratio of the gear train?
a. 4:1.
b. 4:15.
c. 6:15.
d. None of the above.

25. Sketch a simple gear train in which the first and last gears are rotating in the same direction. Use arrows to show the direction of rotation.
26. Pressure is defined as ____________. It is calculated by ________________.
27. The study of pressure in liquids is called _________. The study of pressure in gases is called _________.
28. Give two examples of a hydraulically-operated object and two examples of a pneumatically-operated object.
29. List five advantages of using machines.
30. Name the four basic kinds of motion and give one example of each.

APPLY YOUR KNOWLEDGE

1. Identify three tasks in the home that would be impossible to complete without the assistance of simple machines. Name the simple machines used.
2. Design and build a robotic arm that will move an AA dry cell from one location to another. Use hypodermic syringes and tubing.
3. Design and build a mechanism that will make a loud noise. Your solution must contain at least two simple machines.
4. Design and build a method of weighing (determining the mass of) a series of weights from 1 oz (25 g) to 16 oz (500 g).

BEECH

What differences do you see between this airplane and a commercial airliner?

ALIAS

Design ideas for the car you may drive one day often come from concept cars such as this one.

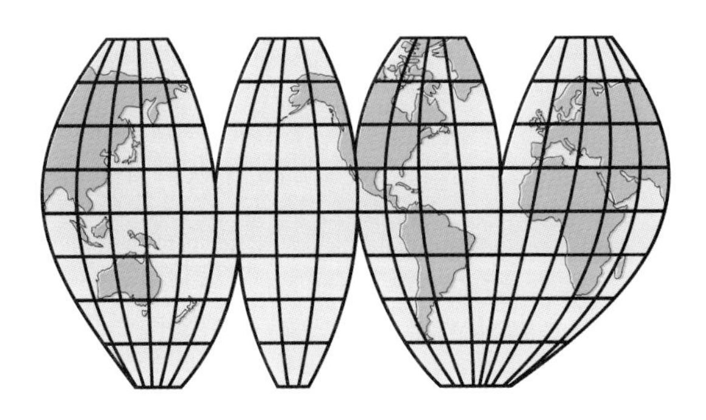

Chapter 8
Transportation

OBJECTIVES

After reading this chapter you will be able to:
- State the advantages and disadvantages of various modes of transportation.
- Explain the principles of various types of engines and motors.
- Explain how transportation technology influences everyday life.
- Discuss the environmental impact of transportation systems.
- Design and build a model of a mode of transportation.

■ MODES OF TRANSPORTATION

When you jump on your bicycle to visit a friend you are using a very simple, but convenient and flexible, form of **transportation.** It is convenient because at any time you can cycle to your friend's house more quickly than you can walk. A bicycle is usable under varying conditions. It can travel over any road, pathway, or reasonably smooth surface.

There are a great many forms of transportation other than the bicycle. Suppose you want to send a parcel to a friend who lives in a distant city. You may transport the parcel to the post office using your bicycle. From the post office the parcel may be taken to the airport by truck. An airplane will deliver it to a city across the continent. Here, another truck takes it to a central depot for sorting. Finally, a mail carrier may use a bus or mail truck to deliver the parcel to your friend's house. The bicycle, truck, airplane, and bus are individual modes of transportation. Together they form a transportation system.

Transportation systems move people or freight from a point of origin to a point of destination. They make it possible to travel from house to house, town to town, country to country and, for a very few people, from earth to the moon. A system may use one or more of the modes of transportation described in figure 8.1.

On-site transportation

On-site transportation transports people and materials from one spot to another. The site may be a mine, gravel, pit, building project, a factory, or a building, to name a few. Elevators, escalators, and conveyors move people or materials in a building. Robots in a factory move materials cheaply and quickly from one spot to another. Some are stationary (sit in one spot) and move parts short distances. Others are called automated guided vehicles or AGVs.They follow a set path and transport parts and material over longer distances. Conveyors, trucks, and pipelines move material at many sites including mines, refineries, wells, and construction sites.

■ ENGINES AND MOTORS

Most forms of transportation need an **engine** to make them move. Engines are machines made up of many mechanisms. They convert a form of energy into useful work.

An engine needs a constant supply of this energy to keep it working. The energy may come from burning fuel such as gasoline, diesel oil, or kerosene. Engines that burn this fuel inside them are called **internal combustion**

MODE OF TRANSPORTATION	ADVANTAGES	DISADVANTAGES
Bicycle	• cheap to operate • flexible • nonpolluting	• transports only one person or small articles • limited to short travel distances
Car	• flexible and convenient • provides privacy	• transports no more than six passengers • causes high levels of pollution • road networks use arable land
Truck	• flexible means of moving freight short and medium distances • goes directly from point of origin to destination	• not suitable for very long distances • causes high levels of pollution
Bus	• relatively flexible and convenient • carries at least 50 people • cheap form of urban mass transit	• causes pollution • not suitable for very long distances
Train	• economical and efficient for large loads of freight over long distances • can carry hundreds of passengers • creates little traffic congestion • very safe • causes very little air pollution	• not as convenient as cars • requires a track system that is expensive to build and maintain
Subway train	• can carry hundreds of passengers • does not cause congestion or pollution	• expensive to build
Ship	• can carry huge loads of freight or large numbers of passengers over long distances	• special facilities needed for loading and unloading • can only be operated on water
Airplane	• can carry passengers and freight over long distances quickly • can pass over all types terrain	• expensive to operate • airport located away from urban centers; needs support of other modes of transportation
Space shuttle	• reusable form of space travel	• extremely expensive • only used for scientific and technological experiments in space

Figure 8.1. Nine different modes of travel and how they compare.

engines. Another type burns its fuel outside the engine. The fuels used as energy sources for such an engine include coal, oil, or nuclear. This kind is called an **external combustion engine.** Internal combustion engines depend on hot, expanding gases for power. External combustion engines may heat water or gases to produce power.

Engines also fall into another kind of grouping: reciprocating piston engines and turbines. Reciprocating engines include **gasoline** and **diesel engines.** These are most commonly used in motorcycles, cars, trucks, and buses. Turbine engines include **turbofan** and **turboshaft** engines. They are used in aircraft, Hovercraft, and helicopters.

A

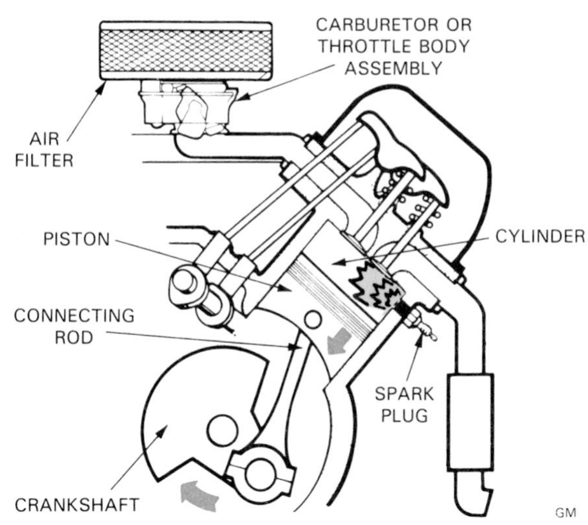

1. Below air filter, a carburetor or throttle body takes in air. It combines with gasoline into a mixture that will burn.
2. Spark plugs make the fuel mixture burn when distributor sends spark to each plug at right time.
3. Fuel mixture burns in cylinder, forcing piston down.
4. Piston slides rapidly up and down, turning crankshaft.
5. Connecting rod joins piston to crankshaft. This shaft turns like the crank on a bicycle.
6. Crankshaft transmits turning power to wheels through drivetrain.

B

Figure 8.2. The four-stroke gasoline engine is a popular power source for transportation. A—Cutaway shows inside parts of a modern engine. B—How the internal combustion engine produces its power.

Four-stroke gasoline engine

Imagine trying to pedal the family car at 55 mph (90 km/h). The driver and one passenger would have to pump their legs up and down about 60 times a second.

Although it is impossible for legs to move this fast, pistons in a gasoline engine slide up and down at this speed. The pistons are short metal drums moving inside cylindrical holes in a metal block. The power to move the pistons comes from a mixture of air and gasoline. This mixture is ignited by an electric spark. Expanding gases push the pistons down to rotate the crankshaft. The crankshaft then transmits turning power to the drive train. The four-stroke gasoline engine is an internal combustion engine, figure 8.2.

The four-stroke cycle is so called because it needs to travel down and up the cylinder four times (down twice and up twice) to produce a complete sequence of events (cycle). This is what happens during one complete cycle of the engine:

First stroke-intake. The intake valve opens, figure 8.3A. The rotating crankshaft pulls the piston downward. The piston sucks fuel and air into the cylinder.

Second stroke-compression. The intake valve closes, figure 8.3B. The crankshaft pushes the piston upward. The mixture is squeezed into a small space to make it burn with an explosion. Just before the piston reaches the top of the stroke, the spark plug ignites the mixture.

Third stroke-power. Both valves are closed, so expanding gases cannot escape, figure 8.3C. The hot, expanding gases force the piston downward with great force. The piston pushes hard on the crankshaft, making it rotate faster.

Fourth stroke-exhaust. At the bottom of the power stroke, the exhaust valve opens, figure 8.3D. The piston comes up again and pushes the gases from the burned fuel out of the cylinder. The cycle starts over again.

The four-stroke gasoline engine has advantages and disadvantages. It adapts easily to changes in speed, provides good acceleration, and has sufficient power for medium size machines, figure 8.4. On the other hand, it is not strong enough for heavy work. It pollutes the air, and uses a relatively expensive fuel.

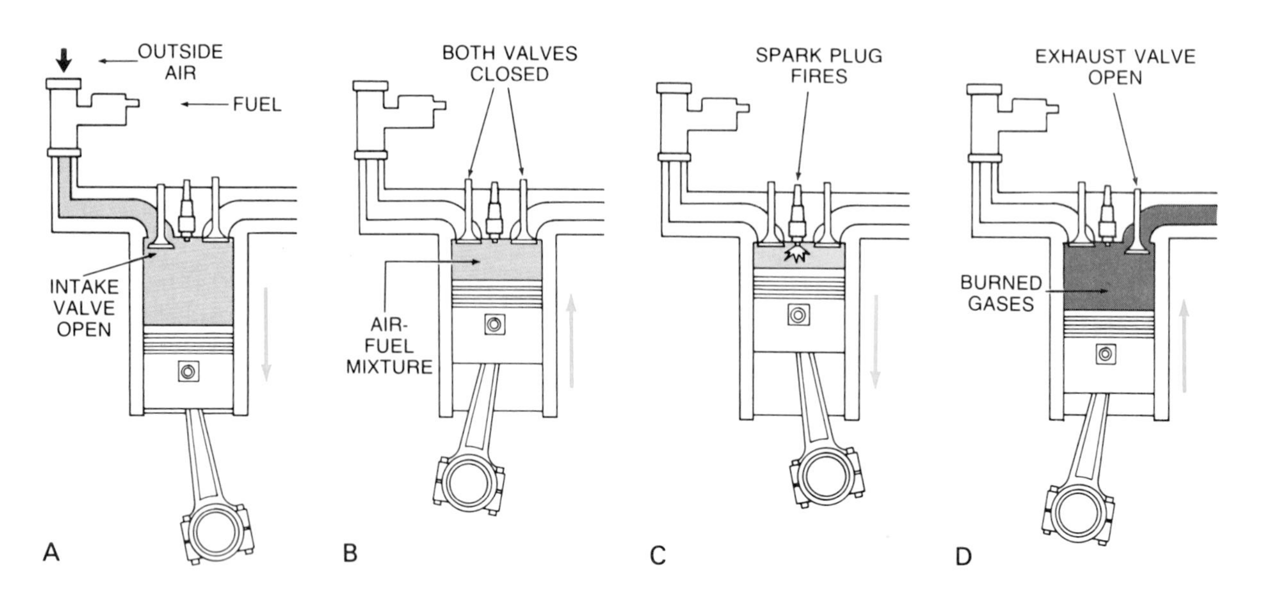

Figure 8.3. How a four-stroke cycle engine works.

Figure 8.4. Gasoline engines power all sorts of vehicles.

The four-stroke diesel engine

Unlike the gasoline engine, the **diesel** engine does not need a spark to ignite its fuel. The fuel catches fire and burns on its own. Pure air is drawn into the cylinders. This air is squeezed to a much higher pressure than the air-fuel mixture in a gasoline engine. At the top of the compression stroke a fine spray of diesel fuel is injected into the cylinder. As the air is compressed, its temperature increases. When the fuel spray meets the hot compressed air, the fuel ignites. The four-stroke diesel engine is an internal combustion engine. Its operation is similar to the four-stroke cycle engine.

- As in a gasoline engine, there is a pair of valves on each cylinder. However, in a diesel, one controls the intake of air only. The other allows exhaust gases to escape from the cylinder.
- The upward stroke of the piston compresses the air.
- An injector squirts fuel into the cylinder.
- The fuel ignites spontaneously and the expansion of gases forces the piston downwards.

The four-stroke diesel engine has several advantages. It stands up well to long hard work, uses less fuel, and lasts longer than a gasoline engine. The disadvantages are that it:

- Is noiser.
- Has slower acceleration.
- Pollutes the air more.
- Must be more heavily built to withstand the high pressures in the cylinders.

Diesel engines are sometimes used in cars. However, they are more commonly used to power large machines and vehicles such as tractors and buses, figure 8.5.

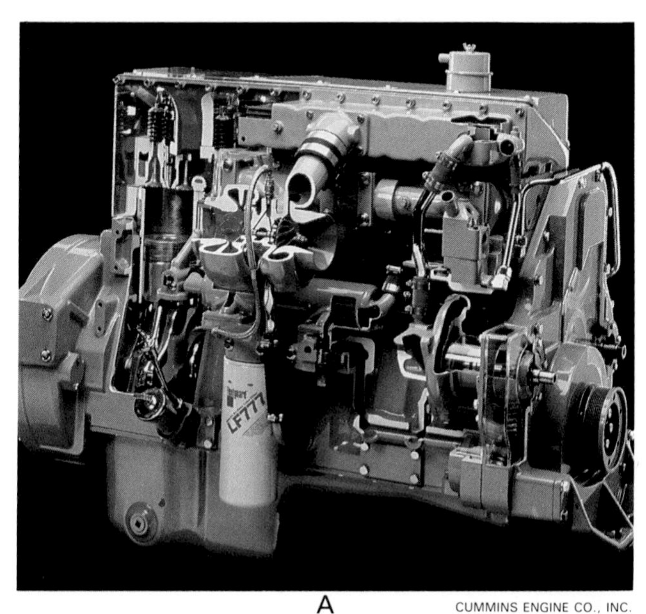

A CUMMINS ENGINE CO., INC.

B TEC

C ECRITEK

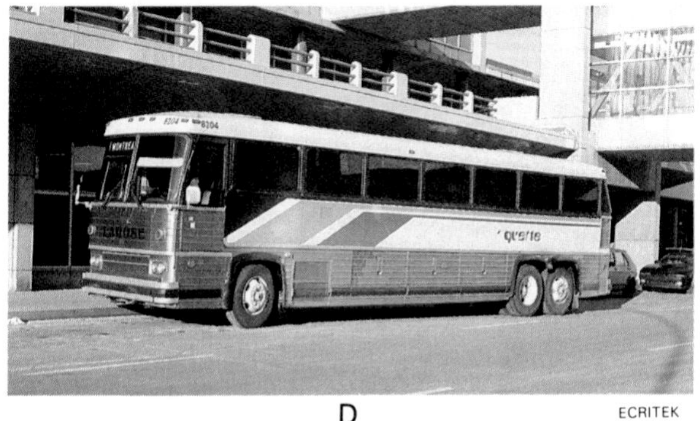

D ECRITEK

Figure 8.5. A—Cutaway of a diesel engine. It is found in many kinds of vehicles and machines. B—Snowcat. C—Ship. D—Bus.

The steam turbine

The term **turbine** originally described machines driven by falling water, figures 8.6 and 8.7. They were also known as water-wheels. There are two types. The overshot wheel is driven with water from above. The wheel rotates clockwise. An undershot wheel is driven with water from below so that the wheel turns counterclockwise. Later, the term turbine was given to heat engines powered by steam. Engineers discovered that a large wheel with blades on it could be turned by a strong jet of steam. A steam turbine's hundreds of blades are set on a long shaft completely enclosed in a strong metal case. The blades on the rotor spin to drive the shaft. Blades on the stator are fixed to the outer casing and cannot spin, figure 8.8. Each stator fan guides the steam flow so that as it moves along the turbine it has plenty of thrust to move the next rotor fan in its path. The spinning fans turn the drive shaft. The shaft, in turn, drives a propeller.

Figure 8.8. Cutaway of a steam turbine. How is it like a water wheel? The stator blades (colored blue) do not move. They guide the steam so it passes over the rotor blades, turning them. Steam turbines are a type of external combustion engine.

After its journey through the turbine, the steam cools and turns into water. This water then flows back to the boiler. Heated to steam again, it returns to work the turbine. The water is heated either by burning fuel (coal or oil) or by a nuclear reactor outside the turbine.

The steam turbine has several advantages. It is smooth running and powerful, has a long life, and is suited to slow, large machines that require an engine bigger than a diesel. This is shown in figure 8.9.

Its disadvantages are that it needs a lot of room since space is also needed for a boiler. It does not adapt as easily as piston engines to changes in speed.

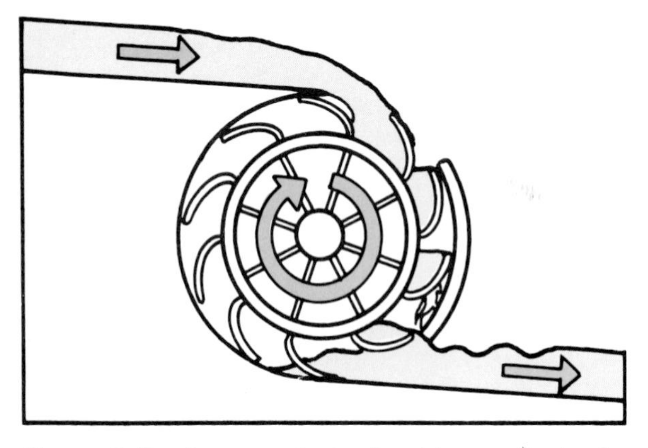

Figure 8.6. An overshot wheel turns clockwise from the mass of the water passing over its blades.

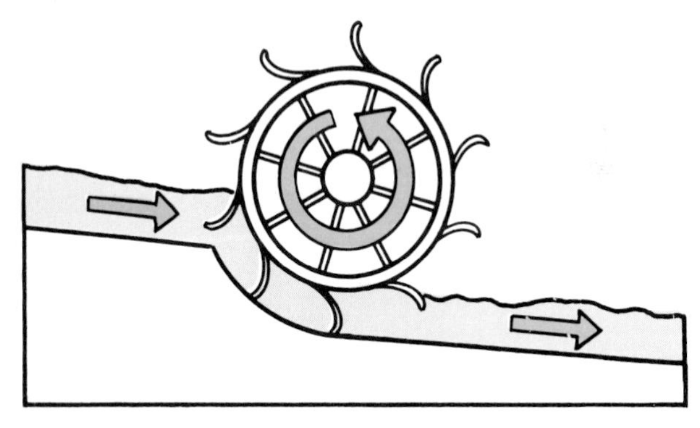

Figure 8.7. An undershot wheel. It is turned counterclockwise from the force or motion of the water striking the blades underneath.

ECRITEK

Figure 8.9. Steam turbines power large sea-going vessels.

The turbofan engine

To get an airplane into the air you need pushing power, or **thrust**. What is thrust? Think of stepping forward off a skateboard, figure 8.10. As you go forward the skateboard moves backward.

Have you ever held a garden hose when someone turned it on full? Suddenly a jet of water bursts out and the hose jumps out of your hand. As the water shoots forward the hose goes backwards. Firemen are sometimes pushed over by this backward force. Blow up a balloon and let it go without tying the neck. The balloon is driven in much the same way as the hose and skateboard.

These are just three examples of one of the basic laws of nature. Isaac Newton discovered it 300 years ago. He said that:

> "For every force in one direction there is always an equal force in the opposite direction."

All jets work on this principle, that for every action there is an equal and opposite reaction. The reaction to the rush of gases out of a jet engine is a thrust that drives the airplane forward. The engine sucks in air at the front, squeezes it, and mixes it with fuel. The mixture ignites and burns quickly. This creates a strong blast of gases. These hot gases expand and rush out of the back of the engine at great speed. Many people believe that the gases "push" against the air to propel the plane forward. This is not true. As the gases shoot out backward, the jet goes forward.

Figure 8.10. An example of thrust. Leaping off the skateboard creates thrust to push the skateboard in the opposite direction.

The **turbofan** is one kind of **jet engine** that powers aircraft, figure 8.11. A jet engine produces a very noisy gas stream. In a turbofan engine, however, the gas stream drives a large fan located at the front of the engine. This creates a slower blast of air. Thrust is as great as a simple jet but the engine is quieter.

Turbofan operation

Air is drawn into the engine by the compressor fans. Its pressure increases. The compressed air mixes with fuel. Ignition takes place. Temperature and pressure increase. The burned mixture leaves the engine through the turbine, which, in turn, drives the compressor and the fan at the front of the engine. Pressure thrusts the engine forward while the exhaust gases rush out of the back in a jet stream. The turbofan is an internal combustion engine.

The turbofan engine has several advantages. It is relatively lightweight, very power-

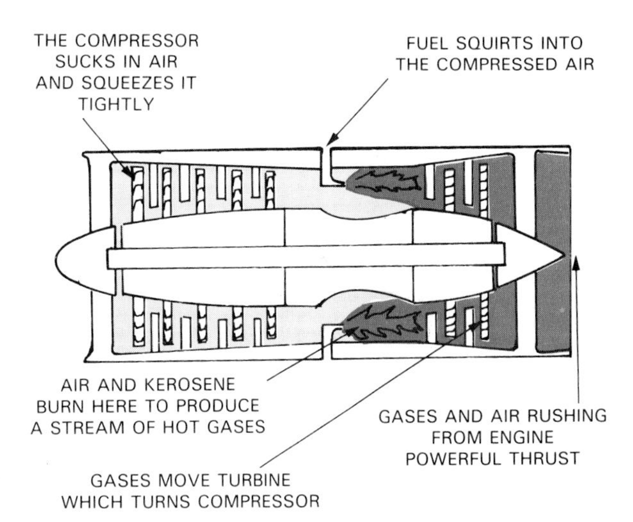

THE COMPRESSOR SUCKS IN AIR AND SQUEEZES IT TIGHTLY
FUEL SQUIRTS INTO THE COMPRESSED AIR
AIR AND KEROSENE BURN HERE TO PRODUCE A STREAM OF HOT GASES
GASES AND AIR RUSHING FROM ENGINE POWERFUL THRUST
GASES MOVE TURBINE WHICH TURNS COMPRESSOR

BRITISH AIRWAYS

Figure 8.11. Top. This is how a turbofan engine operates. Bottom. The Concorde aircraft, using a turbofan engine, is capable of speeds greater than 1300 mph (2100 Km/h).

ful, and uses less fuel than other jet engines. Its disadvantage is that although quieter than other jet engines, all jet engines are noisy.

The turboshaft engine

Like the turbofan engine, the **turboshaft** engine uses a stream of gases to drive turbine blades. The blades turn a shaft, figure 8.12. However, the turboshaft differs from the turbofan in that this shaft is connected to rotors or propellers.

The rocket engine

A firework rocket, an inflated balloon that zooms around the room when it is released, a jet engine, and a rocket engine will work on the same principle, figure 8.13. The compressed air that is forced into a balloon rushes out of the nozzle. This creates thrust. Forward motion results. Inside a rocket, a fuel is burnt to produce hot, high-pressure gas. This escapes from the rocket to provide the thrust.

What pushes rockets forward? Imagine that the fuel is burning and the rocket's exhaust is closed. As the high-pressure gas burns it pushes out in all directions against the inside of the rocket, figure 8.14A. The rocket does not move because the force is equal in all directions. Now imagine that the exhaust is opened. Hot gas will rush through the opening. There is little or no downward force on the bottom of the combustion chamber, but there is upward force on the top. The rocket is pushed up, figure 8.14B.

The difference between a jet engine and a rocket engine is in where they get oxygen to burn their fuel. The jet uses the oxygen in the air. A rocket must carry its own. It operates outside the earth's atmosphere where there is no air.

Rocket engines burn a variety of fuels called propellants. Some propellants are solid, others are liquid. Nearly all space rockets, figure 8.15, use liquid propellants. One of the propellants is the fuel, such as kerosene or liquid hydrogen. The other is liquid oxygen or some substance that can provide the oxygen for combustion.

Figure 8.13. Rocket engines work on the same principle as compressed air rushing out of a balloon. Unequal pressure in the balloon and in the rocket cause them to move.

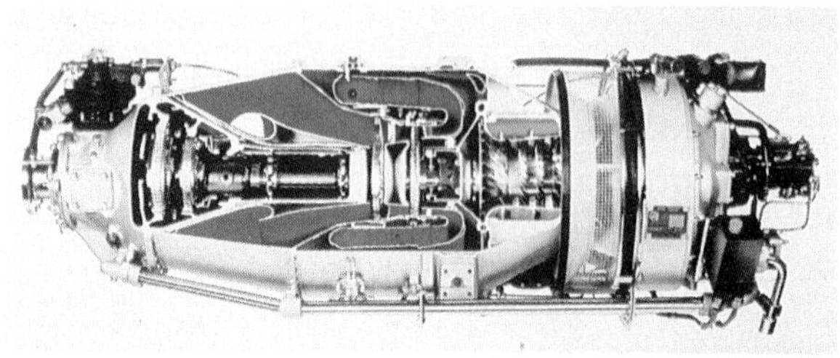

PRATT & WHITNEY, CANADA

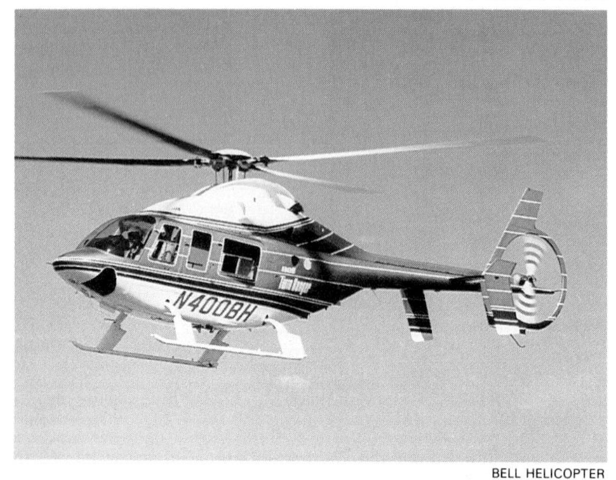

BELL HELICOPTER

Figure 8.12. The turboshaft engine is lightweight and powerful. It is used on all but the smallest helicopters.

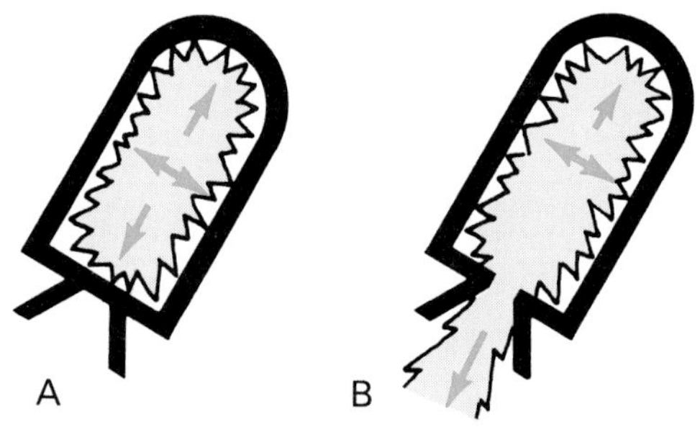

Figure 8.14. Making rockets move. A—This rocket will not move because the pressure inside it is pushing equally in all directions. B—This rocket will move because the pushing force is greatest forward.

NASA

Figure 8.15. Because space has no oxygen, rockets must carry both fuel and oxygen for combustion.

The electric motor

Electric motors change electrical energy into mechanical energy. They provide smooth turning power to drive a shaft. Sometimes the electricity comes from the generators in power stations. This electricity is sent either by overhead cables to trains and trolleys or to the electrified rails used for subway trains.

A limited amount of electricity can be stored in batteries. These provide just enough power to drive short-distance vehicles such as fork lift trucks, golf carts, and experimental cars.

Electrically powered vehicles are reliable and sturdy, figure 8.16. Since there is no combustion process, they do not pollute the environment with exhaust gases. They are, therefore, suitable for transportation vehicles used in closed surroundings, especially subways. They have a very good acceleration at start-up and their speed can be easily regulated. The disadvantage is that electric motors can only be used where there is an electrical power supply. They must be connected to an electric generator or batteries.

The impact of transportation

In general, transportation moves people and materials as rapidly as possible, figure 8.17.

A

B

C

TEC

Figure 8.16. Electric motors power several types of transportation vehicles. A—Subway trains use electric motors because they do not pollute enclosed areas. B—Lift truck motors get their electricity from a battery. C—Many trolley cars run on electricity.

ECRITEK

Figure 8.17. Transportation has both good and bad effects. It is needed to move people and materials. But vehicles cause pollution, noise, and, sometimes, traffic tieups.

ECRITEK

Figure 8.18. One way of avoiding some disadvantages of transportation is to use mass transit such as trains. These commuters park and ride to jobs or shopping.

Even if you only have a week to spend with friends in another country you can get there in less than a day. Products can be delivered from all points of the globe to your local supermarket. Fresh fruit and vegetables are available at any time of the year.

Transportation systems make our lives much more convenient but they do have negative impacts. Several modes of transportation pollute the environment. The pollutants in exhaust emissions from cars, trucks, buses, and airplanes cause many problems. Air pollution causes a bad smell in the air. The pollution level may be so high that people have difficulty breathing. Their eyes water, and they experience a burning sensation in their lungs. Air pollution can also corrode the surface of buildings. The features of historical buildings are eaten away. Noise pollution is another problem. The noise from highways and airports increases the stress level of those living nearby.

Transporting people and products can be dangerous. In general, as the number of vehicles and their speed increases so does the number and severity of accidents. One way to reduce these problems is to make use of mass transit, figure 8.18. Most cars carry a driver and, occasionally, a few passengers. In contrast, one bus will carry more than 50 passengers and a commuter train can seat hundreds.

When no buses or trains are available locally, people should be encouraged to drive to a train station, or to a central car pool, and to take mass transit from that point. In some cities, drastic measures have been taken to reduce traffic problems. Special lanes have been marked on the roads to be used only by full cars or by buses. In some countries, cars are banned from city centers.

Many large communities are examining ways to improve mass transit systems and encourage their use. At the same time, restrictions on the unlimited use of private cars are being considered. Each day tens of thousands of cars stream into and out of many major cities. Each provides convenience and privacy to its occupants but at considerable cost to the environment. Can we continue to pay this cost?

CANADAIR

What are the positive and negative impacts of jet aircraft?

SUMMARY

Transportation systems move people or freight from a point of origin to a point of destination. A system may use one or more modes of transportation: bicycle, car, truck, bus, train, subway, ship, airplane or space shuttle.

Most modes of transportation need an engine to make them move. An engine is a machine composed of many mechanisms that converts a form of energy into useful work. Engines are classified into two groups: internal combustion and external combustion. Internal combustion types include reciprocating piston engines and turbines. Reciprocating piston engines are used in motorcycles, cars, trucks, and buses. Turbine engines are used in aircraft, Hovercraft™, and helicopters.

The most common form of reciprocating piston engine is the gasoline engine. In a gasoline engine a mixture of air and gas is ignited. Pistons are pushed down to drive a crankshaft in a rotary motion. The crankshaft then transmits turning power to the drive train. In a diesel engine air is squeezed to very high pressure. A fine spray of diesel fuel is injected and spontaneous ignition occurs.

In a turbine engine a large wheel with blades is turned by water, a jet of steam, or hot gases. A water wheel uses the force of falling water. A steam turbine uses a jet of steam. A turbofan engine uses a stream of hot gases.

A rocket engine works in a similar way to a turbofan engine. Forward motion is the result of the thrust of high pressure gas escaping from the rocket.

Transportation systems make our lives much more convenient. However, they have negative impacts on the environment due to air and noise pollution. Also, as the number of vehicles and their speed increase, so do the number of accidents. One way to reduce these problems is to increase the use of public transportation.

KEY TERMS

Electric motor
Engine
External combustion engine
Diesel engine
Gasoline engine
Internal combustion engine
Jet engine
Thrust
Transportation
Turbine
Turbofan
Turboshaft

TEST YOUR KNOWLEDGE

Write your answers to these review questions on a separate sheet of paper.

1. Describe the ways in which the modes of transportation affect the way your family lives.
2. State the best mode of transportation for each of the following situations:
   a. A student traveling five blocks to the sports field.
   b. Carrying a huge load of grain from one continent to another.
   c. Transporting passengers across town without congestion or pollution.
   d. Providing a cheap and flexible form of urban mass transmit.
3. List the modes of transportation that you have used in the past. For each also list its disadvantages.
4. Name, in the correct sequence, the four strokes of an internal combustion engine.
5. What is the function of an intake valve?
6. What is the function of a spark plug?
7. What is the function of an exhaust valve?
8. What is the major difference between a gasoline engine and a diesel engine?
9. Make a sketch to show how a jet of steam can be used to turn a shaft.
10. What is the basic principle of a jet engine?
11. Turbofan engines are used to power _________ and turboshaft engines are used to power _________.
12. Use sketches to show how rockets are pushed forward.
13. Explain why a rocket engine must carry its own oxygen.
14. Electric motors change _________ energy into _________ energy. They do not pollute the environment because _________________________.
15. Several modes of transportation pollute the environment. State one way in which this pollution can be minimized.

APPLY YOUR KNOWLEDGE

1. List the modes of transportation that you use in a period of one week. State the advantages and disadvantages of each.
2. Name the most appropriate mode of transportation to move:
   a. A prize bull from a farm in Texas to a farm in Alberta.
   b. 10,000 commuters from a suburb to a city center.
   c. A transplant organ from one city to another, 500 miles (800 km) away.
   d. A large supply of fresh fruit from South America to North America.
3. Identify one problem in your town caused by a mode of transportation. Describe the problem and suggest ways to resolve it.
4. Design and build a vehicle that will travel 15 ft. (5 m) across a flat surface. The only power source permitted is a large elastic band and a propeller.
5. Design and build a balloon rocket using a balloon, drinking straw, clear tape, and string. The rocket must travel in a straight line for a distance of at least 15 ft. (5 m).

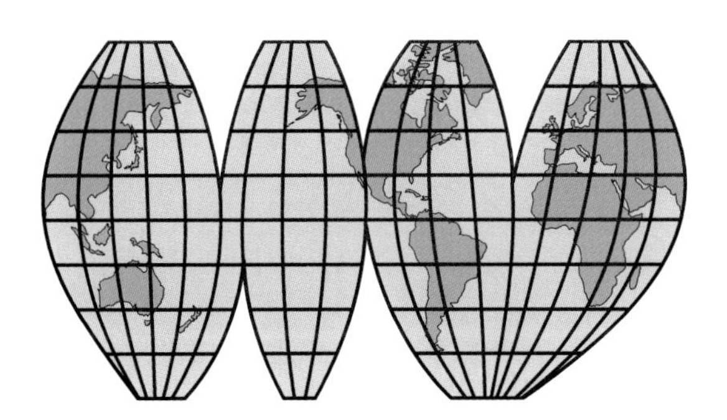

Chapter 9

Energy

OBJECTIVES

After reading this chapter you will be able to:

- Explain society's dependence on energy.
- Describe the difference between potential and kinetic energy.
- Identify the various forms of energy and their applications.
- Describe how energy can be changed from one form to another.
- Distinguish between nonrenewable and renewable sources of energy.
- List the advantages and disadvantages of each source of energy.

Every day we use energy in one form or another. When you ride your bicycle to school you are using your own physical energy to turn the pedals. When you fly a kite, wind gives energy to keep the kite in the air. The family car uses the energy in gasoline.

At home the sun shines in your windows and warms up your house. The sun provides the energy. In winter, the energy stored in wood, coal, or other fuels can heat your home. When you turn on a light you are using electrical energy.

What is energy? In a few words, it is the ability to do work. To understand it more fully, we need to look at two categories: **potential energy** and **kinetic energy**.

Think about lifting a sledge hammer to drive a post into the ground. As you hold the hammer in the air it has potential energy. Energy is stored in the hammer until it is dropped and the energy is released to hit the post. When you wind up a spring in a toy you are giving the spring potential energy. Once again, energy is being stored. A stretched elastic band also has potential energy. Potential energy is often called "stored energy."

When the head of the hammer is dropped to hit the post, or the spring is released to drive the toy, both the hammer and the toy gain kinetic energy. Kinetic energy is the energy an object has because it is moving, figure 9.1.

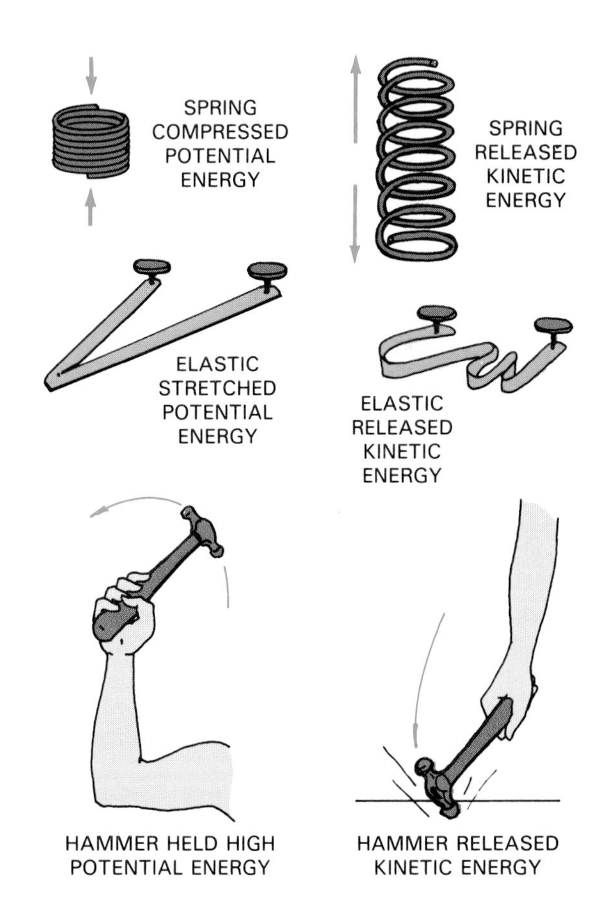

Figure 9.1. These are examples of potential and kinetic energy.

■ FORMS OF ENERGY

Look at figures 9.2—9.9. They show that energy is available in many different forms. Each picture describes a different form of energy and the way energy gets things moving so that work is done.

A great deal of **chemical energy,** figure 9.2, is locked away in different kinds of substances. Such energy is found in the molecules making up food, wood, gasoline, and oil. The energy often is released by burning the chemical. Burning rearranges the substance's molecules and releases heat.

Objects always tend to move toward the lowest possible level. This is due to **gravitational** attraction or pull of the earth. It causes objects to fall. It is why water runs or objects roll downhill, figure 9.3.

Certain materials that can be stretched or compressed have a tendency to return to their original shape. This is known as **strain energy** or the energy of deformation. It is the kind of energy found in a rubber band, a fishing pole, or bow and arrow, figure 9.4.

Electrical energy is the movement of particles from one atom to another. Atoms are made up of several particles: neutrons, protons, and electrons. Protons are positively charged, while electrons are negatively charged. Electrons will jump from one atom to another if attracted by protons. This energy is enough to provide light and operate electrical devices such as motors and heaters, figure 9.5.

Heat energy occurs as the atoms of a material become more active. If you could look at them under a strong enough microscope,

- Chemicals in the body provide the muscles with energy to do work.
- Gasoline provides the chemical energy to keep the motorcycle moving.
- Chemical energy can be stored in batteries.

Figure 9.2. Chemical energy is often called stored energy.

- A skateboard at the top of a hill has gravitational energy. This is energy available because of the pull of gravity.
- Gravitational energy is potential energy before the object moves and kinetic energy when the object is moving.

Figure 9.3. Gravitational energy has energy because of its position.

- Strain energy can be stored in a material that stretches, such as elastic, nylon, or a spring.
- When an arrow is fired from a bow, the potential (strain) energy of the bow and string changes into the kinetic energy of the arrow.

Figure 9.4. Strain energy or energy of deformation. Some materials are elastic and try to return to their original shape or size.

- Electrical energy is a flow of electrons from one atom to another within a conductor.
- Electrical energy can move from place to place and readily changes into other forms such as heat, light, and sound.
- Electrical energy can be stored in batteries or produced in a generating station.

Figure 9.5. Electrical energy. How many of the devices you use everyday need electrical energy?

you would see that the atoms move about. The faster the movement, the warmer the material, figure 9.6.

Light energy is related to heat energy. Another name for it is radiant energy. It travels as a wave motion. This is a type of energy that is coming to us from the sun. Light is part of the electromagnetic spectrum. It travels at 186,000 miles (300,00 km) per second, figure 9.7.

Sound energy is a form of kinetic energy, figure 9.8. It moves at about 1100 ft. (331 m) per second. This is much slower than light energy.

Nuclear energy occurs as atoms of certain material are split. This action, called nuclear fission, creates huge amounts of energy, figure 9.9. Most of it is in the form of heat.

Energy conversion

One of the most important laws of science states that energy can neither be created nor destroyed. It can only be changed from one form to another. The energy you use to pedal your bicycle comes from the food you eat. Your body has made a change in the form the energy takes.

There are many examples of energy changing its form. In the example of a cyclist the chemical (potential) energy in the muscles is changed to kinetic energy of the bicycle. When the brakes are applied, this kinetic energy is

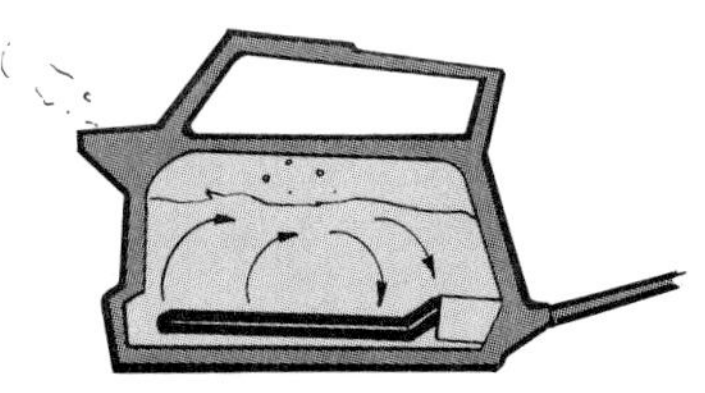

Heat energy travels through matter in three ways:

- Convection occurs when expanded warm liquid or gas rises above a cooler liquid or gas. When liquid is heated, it expands and its volume increases. The amount of material (its mass) does not change. Since its mass is more spread out, hot liquid is less dense than cold liquid around it. In a mixture of hot and cold liquid, the cold liquid will sink to the bottom and the hot liquid will rise to the top.

- Conduction occurs when heat energy passes from molecule to molecule in a solid. There is movement of heat energy without any obvious movement of the material.

- Radiation occurs when heat energy is moving in the form of electromagnetic waves. For example, when you stand in the sunshine or in front of an electric heater, heat is transmitted without involving a material between you and the source of the heat.

Figure 9.6. Heat energy can move through matter by convection, conduction, and radiation.

changed into heat energy as a result of friction between the brake shoes and the wheel. When a flashlight is switched on, the chemical energy in the dry cell (battery) is changed to electrical energy. Electrical energy is changed to light energy when the bulb is lit. A diver has gravitational energy because of the height above water. This energy is changed to kinetic

- Light travels in straight lines.
- TV pictures, lamps, and the sun are seen because of the light they send out.
- Most other objects are seen because they reflect light.

Figure 9.7. Light energy. Most objects have no light of their own. We see them by the light they reflect.

- Sound is produced when matter, such as a tuning fork or human vocal cords, vibrates.
- The vibrating object has kinetic energy due to movement.
- The string on a guitar has potential energy when pulled back. When relesed, it has kinetic energy.

Figure 9.8. Sound energy. Sound waves carry vibrations from a source to our ears and make our eardrums vibrate.

energy during the dive. If the horn of a car blows, electrical energy becomes sound energy.

Losses during conversion

When switching on a light bulb, you may expect to change all of the electrical energy to light energy. Not so! Only a portion of the electrical energy is converted into light energy. The rest is converted into heat energy. You can feel the heat produced by holding your hand near to the light bulb, figure 9.10. In all energy changes, some of the energy is used as intended, but some is wasted. However, remember that although there has been a change in the form of energy the total amount of energy remains the same: energy is neither created nor lost.

- Nuclear energy comes from the **fission** (splitting) of radioactive atoms or the **fusion** (joining together) of atoms.
- Energy is released mainly in the form of heat.

Figure 9.9. Nuclear energy. Uranium is used in nuclear power stations to produce heat and turn water into steam.

Figure 9.10. A light bulb is one example of energy being converted. Electrical energy converts to light energy.

■ WHERE DOES ENERGY COME FROM?

Almost all of the energy we use comes from the sun. The sun's heat keeps us warm. Heat from the sun causes wind and rain. Most plants need light energy from the sun for growth. These plants provide humans and animals with the energy they need to do work. Over millions of years some of these plants have been changed into petroleum (oil and gas) and coal. These fossil fuels may be used to provide energy for machines.

All sources of energy make up two groups: **nonrenewable** and **renewable.** Nonrenewable energy sources will eventually be used up and cannot be replaced. They include coal, oil, and natural gas. Renewable energy sources will always be available. They include the sun, wind, and water. These two major groups are summarized in figure 9.11.

RENEWABLE

NONRENEWABLE

OIL

Figure 9.11. Energy comes from a source which is either renewable or nonrenewable.

■ NONRENEWABLE SOURCES OF ENERGY

The most important nonrenewable sources of energy are coal and petroleum (oil and gas). All were most probably formed from the remains of living matter.

Coal developed from the remains of plants that died millions of years ago. For this reason it is often referred to as a fossil fuel. The coal-forming plants probably grew in swamps. As the plants died, they gradually formed a thick layer of vegetable material. Sometimes ancient seas covered this layer. Sediments (small pieces of soil, clay, sand, and stone) buried the plant layers. As this process was repeated, the layers of vegetable material became squeezed under great pressure and heat for a long time. The result was coal, figure 9.12.

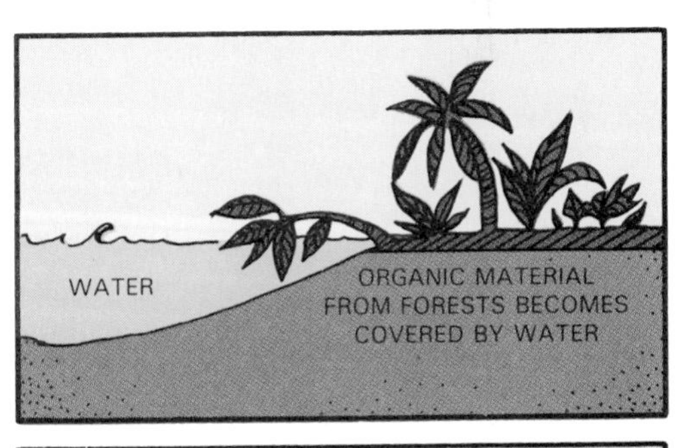

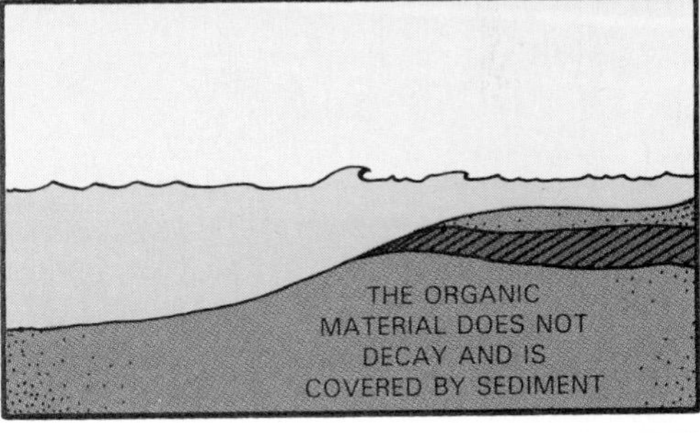

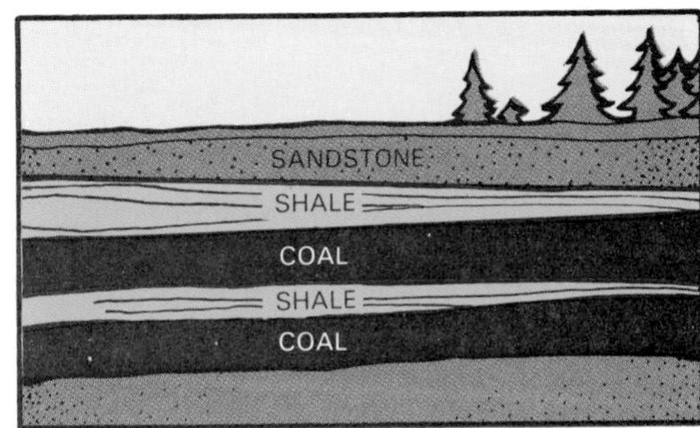

Figure 9.12. After millions of years of heat and pressure, organic material becomes coal.

Removing coal from the ground is called mining. Coal mines are of two types: surface mines and underground mines. Surface mining involves stripping away the soil and rock that lie over a coal deposit. The coal can then easily be dug up and hauled away, figure 9.13. Underground mining involves digging tunnels into the coal deposit. Miners go down a shaft in a large elevator, then ride through the tunnels in cars. The cars take them to the coal face where large machines rip coal from its million-year-old home, figure 9.14. Coal is primarily used as a fuel for electrical power generating stations. It is also used to power industrial processes, particularly those manufacturing steel.

SEEDS

Figure 9.13. Open pit or surface mining. Dragline removes overburden (soil and rock) to expose a coal deposit.

SEEDS

Figure 9.14. An underground coal mine. The tracks are for coal cars that carry coal and miners to the coal face.

Petroleum (oil and gas) was formed from the bodies of countless billions of microscopic plants and animals that lived in the seas millions of years ago. As these animals and plants died, they sank to the bottom to mix with mud and sand on the sea floor. As more and more layers were added, pressure increased and the remains of the sea life in the deeply buried layers slowly transformed into oil and gas, figure 9.15.

Oil and gas are removed from the ground by drilling deep holes. Holes are made either by drilling rigs located on land or by drilling platforms situated on the ocean, figure 9.16.

In its natural or crude form, oil removed from the ground is useless. The crude oil is processed in an oil refinery. The process consists basically of heating the crude oil. This separates the components. The lighter components rise to the top and the heavier components remain at the bottom, figure 9.17.

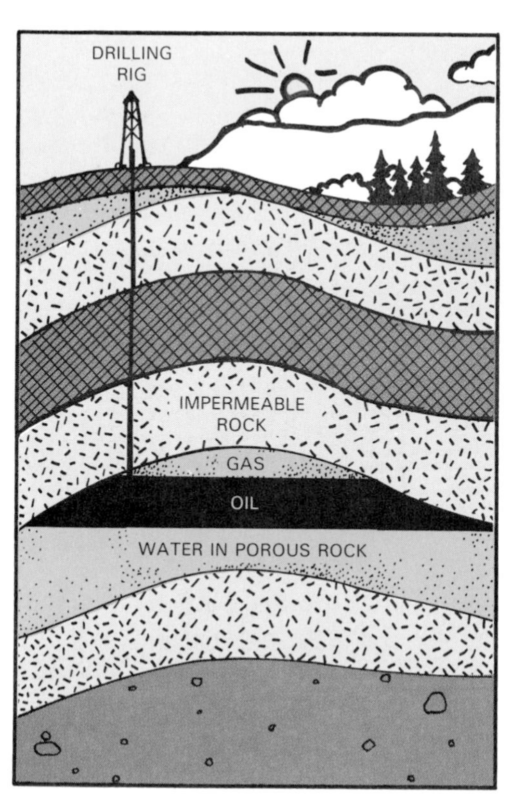

Figure 9.15. Typical rock formation containing oil and natural gas deposits. When a well is sunk, the oil may come out in a gusher. This is due to underground pressure. Sometimes, however, it must be pumped to the surface.

Figure 9.16. Off-shore drilling platform is needed to drill oil wells under the oceans.

■ RENEWABLE SOURCES OF ENERGY

Renewable energy is energy that does not get used up. The energy provided by the source can be renewed as it is used. In the past, most energy has been obtained from burning nonrenewable fossil fuels. Currently, a great deal of electrical energy is produced by hydroelectric generating stations. But as the nonrenewable sources of energy become scarce, alternative sources (**solar, wind, tidal, geothermal,** and **biomass**), are being developed.

Solar energy. Solar energy, that is, using the energy from the sun, is the most important of the alternative sources of energy. The idea of collecting energy from the sun may seem to be a very good one. The main drawback is that this type of energy is not always available. In the winter and on cloudy days there may be too little. At night, there is none. Yet, these are the times when it is most needed. Energy from the sun can, however, be collected and stored for later use.

A solar panel collects heat from the sun's rays. The heat is carried away to provide hot water or to heat buildings. In one kind of solar panel, water flows through pipes or channels under a plate of glass. These pipes or channels are painted black to absorb heat better. This heat transfers to the water. Pipes carry it to the hot water system where the heat is released. Solar panels are usually placed on the roof of a building, figure 9.18.

Using solar cells, we can convert the sun's energy directly into electrical energy. These cells are silicon wafers. They generate electricity by the photovoltaic method. In this method, solar energy dislodges electrons from the wafers. The loose electrons move through conductors. This creates an electric current.

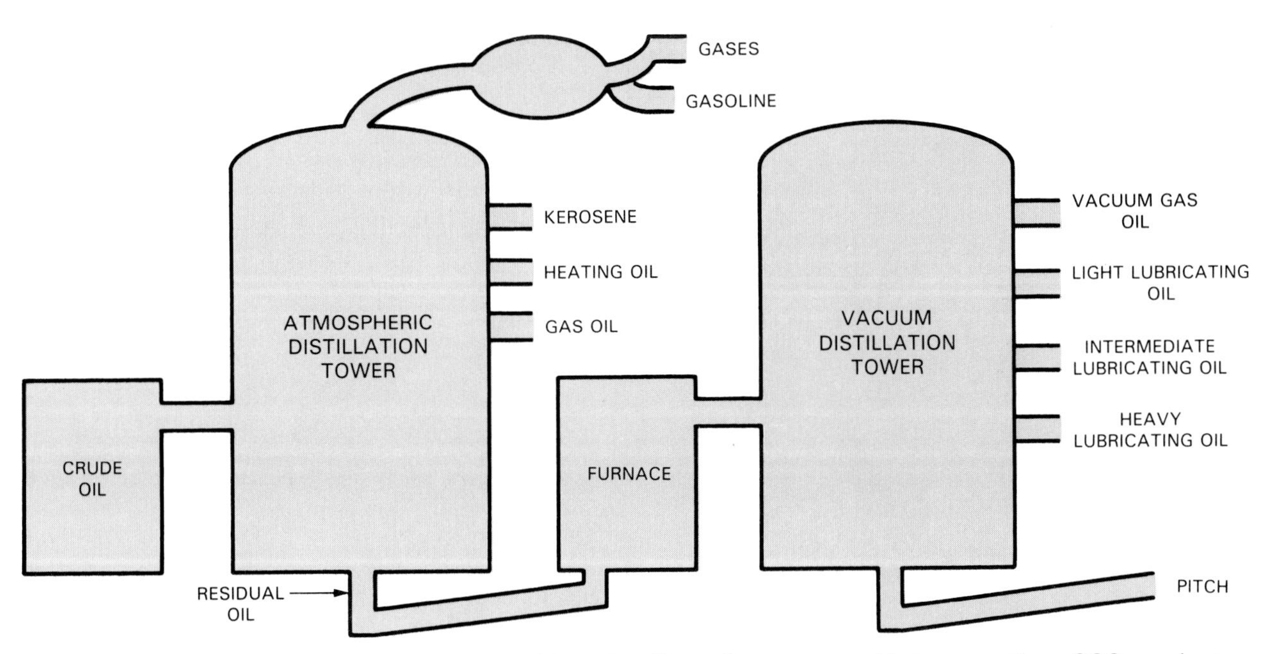

Figure 9.17. By distillation or "cracking," crude oil can be converted into more than 800 products.

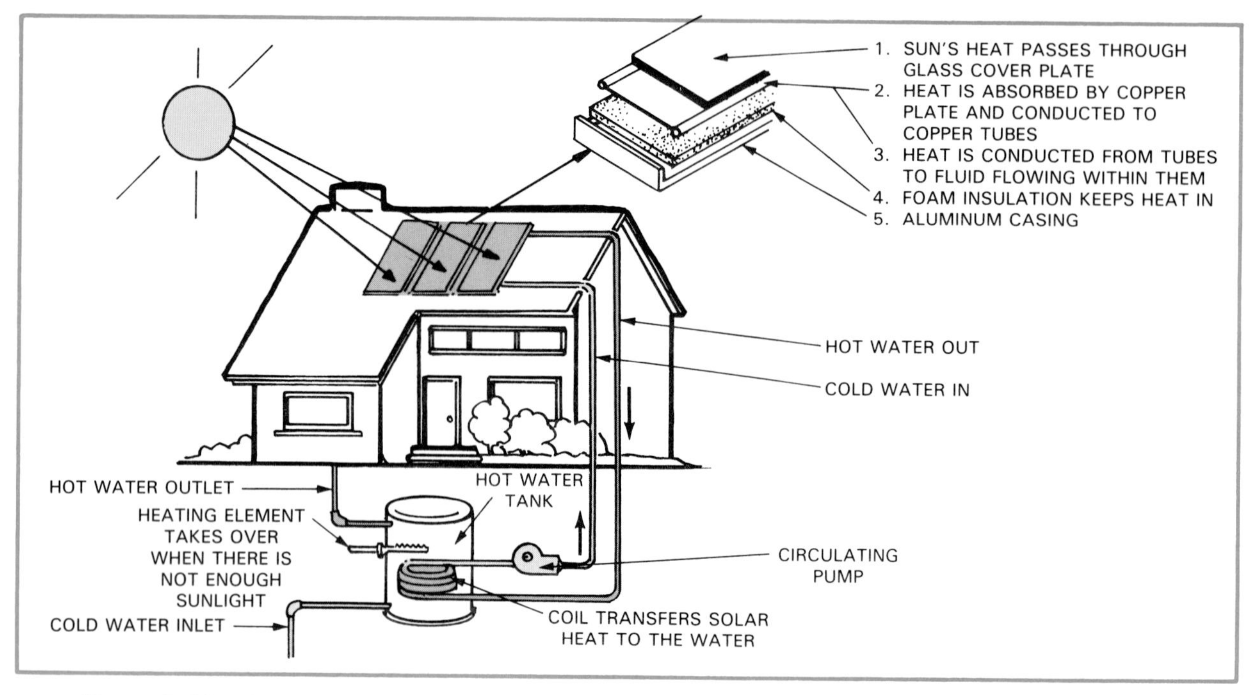
1. SUN'S HEAT PASSES THROUGH GLASS COVER PLATE
2. HEAT IS ABSORBED BY COPPER PLATE AND CONDUCTED TO COPPER TUBES
3. HEAT IS CONDUCTED FROM TUBES TO FLUID FLOWING WITHIN THEM
4. FOAM INSULATION KEEPS HEAT IN
5. ALUMINUM CASING
HOT WATER OUT
COLD WATER IN
HOT WATER OUTLET
HEATING ELEMENT TAKES OVER WHEN THERE IS NOT ENOUGH SUNLIGHT
COLD WATER INLET
HOT WATER TANK
CIRCULATING PUMP
COIL TRANSFERS SOLAR HEAT TO THE WATER

Figure 9.18. Solar panels collect heat from the sun. This heat is used to provide hot water.

The photovoltaic method has recently received great publicity because of its use in space. For example, the Skylab space station used more than 500,000 silicon cells covering an area of 2422 sq. ft. (225 m^2). At present, only small amounts of electricity are generated in this way. Portable devices, such as some pocket calculators, make use of this energy source.

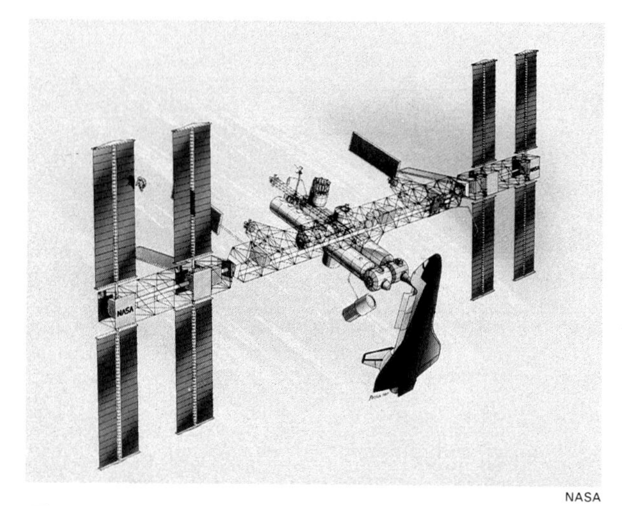
NASA

Figure 9.19. This is a concept for a space station. Its solar panels will produce 75 kW of power.

Wind energy. As long as the sun continues to shine, the wind will continue to blow. Wind currents occur because of differences in temperature between different parts of the earth. More heat from the sun reaches the equator than the poles, figure 9.20. This is because the sun's rays hit the earth directly at the equator (A). Near the poles (B) they hit at an angle. The heat is spread over a wider area.

The air above the equator expands most and rises by convection. When it reaches about 30° latitude, the warm air cools and falls. At about 60° latitude, it meet cold air from the poles and has to rise over it. This air move-

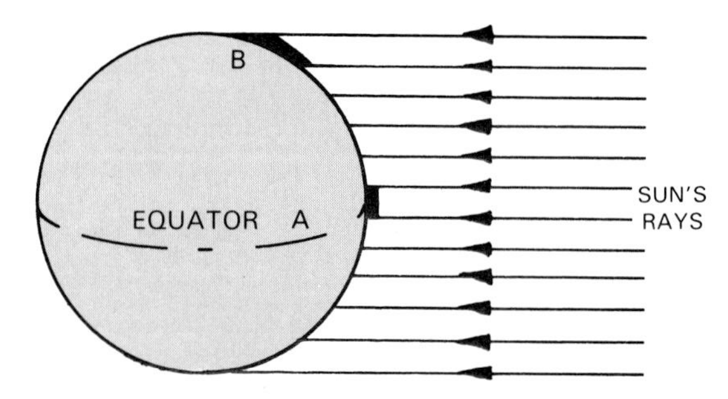
B
EQUATOR
A
SUN'S RAYS

Figure 9.20. More solar heat is delivered at the equator than at the poles. Can you explain why?

ment creates still more convection currents, figure 9.21.

Convection currents are also created as a result of the difference in temperature between the land and the sea. During the day, the land heats up more quickly than the sea. Warm air rises from the land. Cooler air from the sea moves in to take its place, figure 9.22. At night the land cools more quickly than the sea. The process reverses. Warm air rises from the sea. Cooler air from the land moves in to replace it, figure 9.23.

Wind is one of the oldest sources of energy. For many centuries, wind has turned wheels to grind grain and pump water. Today, wind is being looked at to do other work. It can spin wind turbines that will generate electricity, figure 9.24.

The wind is free. Unfortunately, it is also unreliable. The wind does not blow constantly. Also, no one can predict when it will blow. Therefore, some method is needed to store some of the electricity generated during times

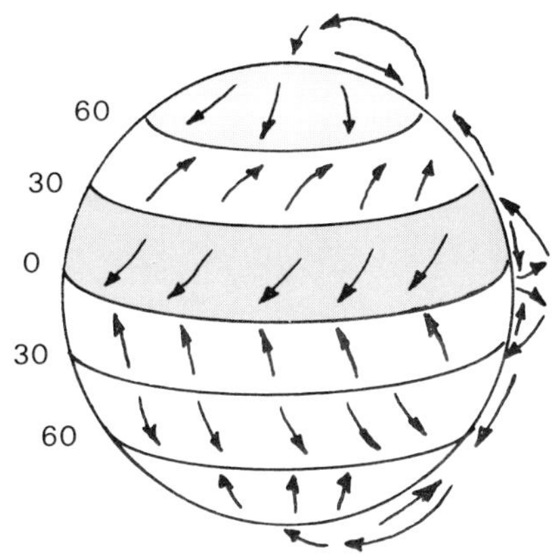

Figure 9.21. Air currents follow a certain pattern over the entire earth.

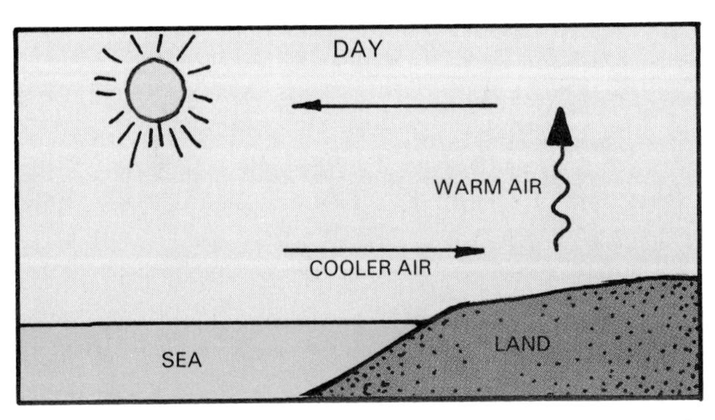

Figure 9.22. Sea breezes occur as warmer air over land rises and cooler sea air moves in to replace it.

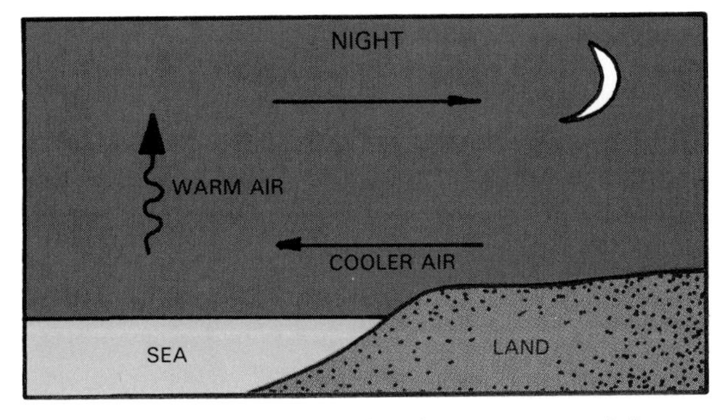

Figure 9.23. At night, land masses cool faster than oceans. The breezes blow from land to sea.

Figure 9.24. Eole wind turbine, on Isles-de-la-Madeleine in the Gulf of St. Lawrence, is 354 ft. (108 m) high. This turbine was named after the Greek god of the winds.

of high winds. Usually a number of batteries serve this purpose. On windless days the batteries provide electrical energy.

Energy from moving water. The energy of moving water is found in rivers **(hydroelectricity)**, estuaries **(tidal energy)**, and oceans **(wave energy)**. Why do rivers flow? Look at figure 9.25 for the answer.

The sun evaporates water (turns to vapor) mainly from the sea and also from rivers, lakes, and plants. This water vapor rises to form clouds. The clouds move with the land breezes. When they reach high ground they are forced to rise. This causes them to cool, and they cannot hold as much water. Water falls as rain on high ground and forms rivers.

Hydroelectricity. Water from rivers can be stored behind a dam, figure 9.26. To generate electricity, water flows through very large pipes called penstocks. The penstocks direct water onto turbine blades, spinning them. The turbines are connected to generators as figure 9.27 shows.

Hydroelectricity is cheap compared to other electric power systems. After the initial expense of building the dam and generating sta-

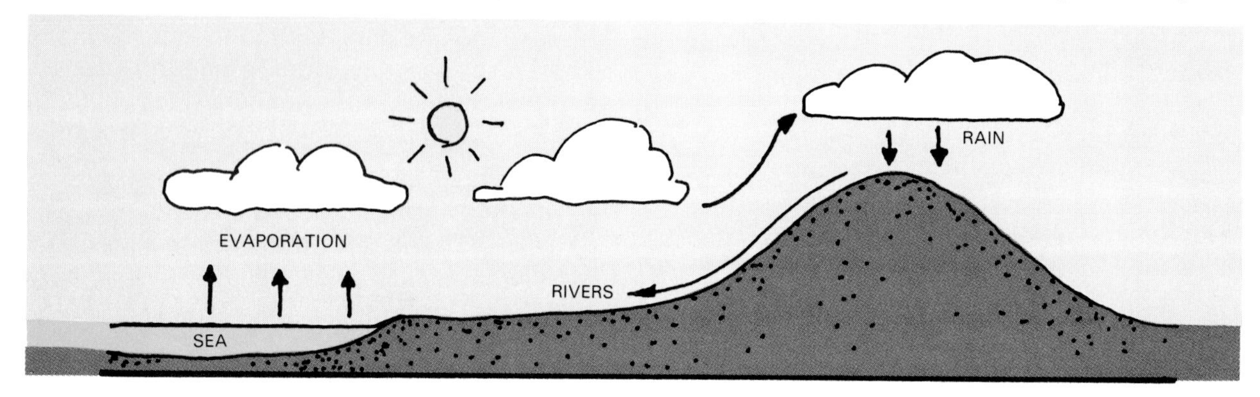

Figure 9.25. Rivers flow because of gravitational energy. Rains deliver the water to high ground. Rivers flow to the lower ground.

Figure 9.26. Dams store water which is used to produce electricity.

tion, the cost of producing electricity is small. No fuel is needed apart from the energy provided by the sun.

Tidal energy. Tides from the sea are yet another alternate source of energy. The tides are regular and inexhaustible. The force of tidal currents can be used to produce electricity. The method is much the same as the way waterfalls or streams and rivers are used. The force of tidal water, however, can be captured when it is rising as well as when it is falling.

To understand this method, think of a damlike structure being placed across the mouth of a bay. As the tide rises, the water flow through a tunnel in the dam. It turns a turbine inside the tunnel. As the tide falls, the water flows back towards the ocean. Once, again, it turns the turbine, figure 9.28.

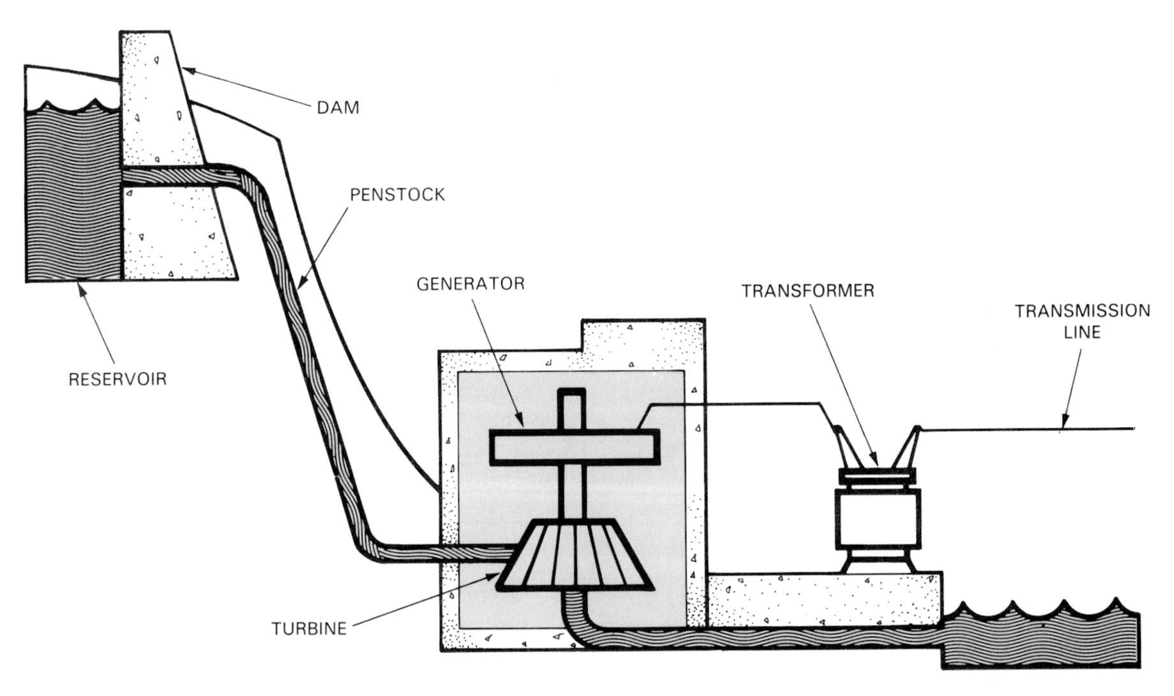

Figure 9.27. Stored water runs through the turbine with great force causing it to spin rapidly. The turbine drives an electric generator, producing electricity.

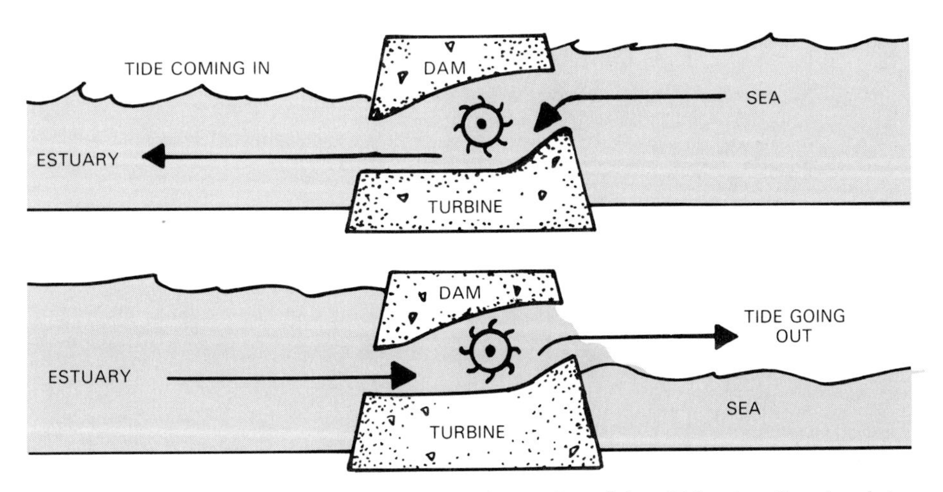

Figure 9.28. Tides can be used to produce electricity. Whether flowing into or out of the estuary, the water spins a turbine.

Only a few places on earth have tides great enough to make tidal energy practical. Tidal electricity generating plants are working in France. One is working experimentally in the Bay of Fundy, Nova Scotia.

Wave energy. Another experimental method of generating electricity is to harness the motion of waves. Waves moving up and down can be used to drive turbines.

Nuclear energy. There are two types of nuclear energy: **nuclear fission** and **nuclear fusion.** Nuclear fusion is the same energy source that powers our sun and the stars. It requires high temperatures. So far, technologists have not been able to produce it on earth. At present, nuclear power stations use only the fission process.

What is fission? Remember, all solids, liquids, and gases are composed of chemical elements. The smallest unit of each element that still retains the properties of that element is an atom. Although atoms are very small, they are made of even smaller subatomic particles called protons, neutrons, and electrons. At the center of each atom is a tiny nucleus, figure 9.29.

Most atoms have a stable nucleus. This means that they do not change. In a few atoms the nucleus is unstable. These unstable nuclei try to become stable. They throw off particles or rays. These rays are called **radiation.** The atoms are radioactive.

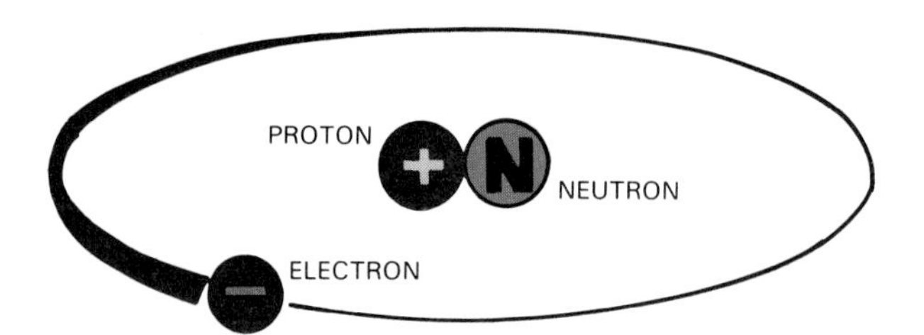

Figure 9.29. Nucleus of an atom contains protons and neutrons. Electrons move around outside the nucleus.

Uranium is a metal. Its atoms have very large nuclei. Very large nuclei are often particularly unstable. When an atom of uranium is hit by a neutron, the nucleus of the atom splits, figure 9.30.

The atom splits into two parts, called fission fragments. Together, the fragments weigh slightly less than the original atom. The loss in mass turns into energy. The fissioning atom produces, on average, one, two, or three neutrons. These neutrons may hit other uranium nuclei. When they do, these nuclei may also split. This, in turn, gives out more neutrons. A chain reaction occurs.

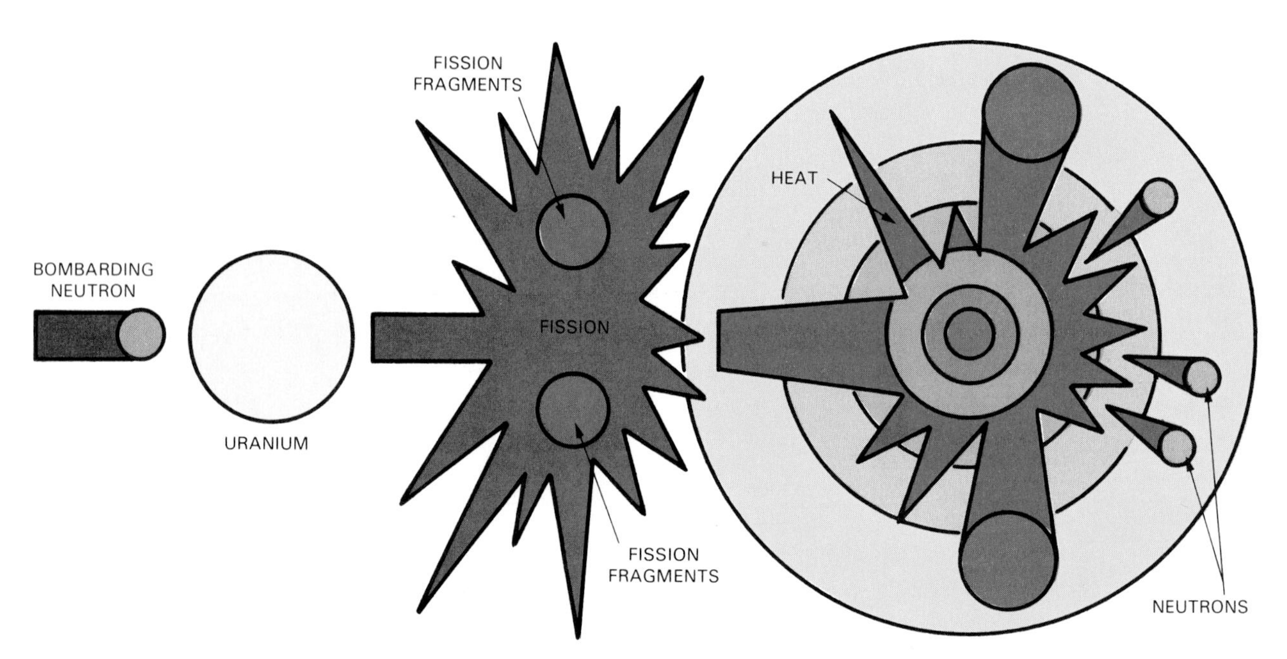

Figure 9.30. Nuclear fission occurs when an atom of uranium splits.

The heat produced by nuclear fission is used to heat water. This produces steam. This steam is, in turn, used to drive turbines. Generators attached to the turbines produce electricity.

The fission process is noted for the large amount of heat energy released. The fissioning of 2.2 lb. (1 kg) of uranium produces about the same amount of heat as burning 2.9 million lb. (1.3 million kg) of coal!

It was once thought that nuclear energy would become the major energy source. Before this can happen, however, major problems remain to be solved. These include:

- The difficulty of finding appropriate power station sites with large amounts of cooling water.
- The need to site power stations away from cities and towns due to the danger of radioactive leaks.
- The need to be sited in places that are not subjected to earthquakes.
- The safe disposal of highly radioactive waste. These will remain lethal for thousands of years.

Geothermal energy. The rocks deep within the earth are hot. Underground water changes to steam when it comes near these rocks. This is called **geothermal energy.** Technologists are not absolutely sure how geothermal energy originates. It is generally assumed that about 80 percent of the heat is due to the decay of naturally occurring radioactive materials such as uranium. The remainder is heat left over from the original formation of the earth.

Hot springs and geysers are evidence of the tremendous heat still in the earth's core. Most of the houses in Reykjavik, Iceland, are heated by water that has been warmed by the natural heat of the earth.

If wells are drilled in an active area, steam can rush to the surface with enough force to drive turbines and generate electricity. Geothermal generating stations are already working in Iceland, Japan, Mexico, New Zealand, and the United States.

Another way of using the heat in the earth is to drill two parallel holes into the hot rock. An explosive charge is placed at the bottom of the holes. The resulting fracture joins the holes. Cold water is then pumped down one hole. It heats up and returns through the other hole as steam. The steam generates electricity, figures 9.31 and 9.32.

One problem with geothermal power concerns pollution. Geothermal steam brings strong chemicals with it. If this condensed

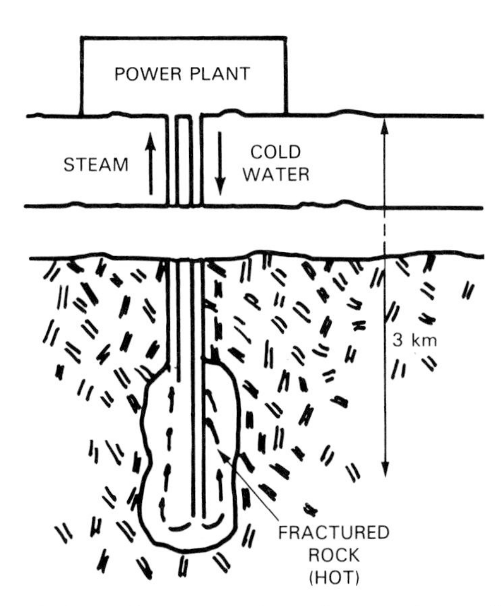

Figure 9.31. Cross section of a geothermal well with a power station. Cold water is sent down through one well to be heated by the super hot rocks. Steam rises up the other well.

SEEDS

Figure 9.32. These pipes carry geothermal heat to a power station to make electricity.

steam is dumped into streams and rivers, their temperature can rise to dangerous levels. The chemicals will also pollute the water.

Other renewable sources of energy. Other renewable sources of energy are being developed in different parts of the world. The most important are:

- Energy from garbage.
- Energy from plants.
- Energy from rotting matter.
- Hydrogen.

Together, these four sources produce only a small part of our energy needs.

Energy from garbage. People produce a lot of garbage. The average North American throws away about 4 lb. (2 kg) of garbage a day. Think of the population of your town or city. How much garbage do you suppose is produced in one week?

About 50 percent of garbage is combustible (will burn). These combustible materials may be used as fuel. They will produce steam to turn the turbines in an electric generating plant. This is a partial solution to the disposal of large amounts of urban garbage.

Energy from plants. Energy from plants is called **biomass energy.** Some fast-growing plants can be burned as fuels. Wood has been used as a fuel for thousands of years. It is still the most commonly used fuel in the developing world. There, four out or five families depend on it as their main energy source.

When trees are harvested in North America, about 50 percent of the tree is converted into lumber or pulpwood. The remaining 50 percent, mainly branches, twigs, and bark, is discarded. This is wasteful. All parts of the tree contain stored solar energy.

Recently technologists have tried to use this previously wasted energy. Branches, twigs, and bark are ground up into wood chips and wood pellets. This waste wood is then used as fuel for a steam boiler. Many pulp and paper mills use wood-fired generators to make their own electricity, figure 9.33.

Biomass fuels may also be in liquid form. For example, sugar cane grows quickly in many regions. Sugar produced from the cane can be fermented to make alcohol. This alcohol may be mixed with gasoline to produce gasohol, figure 9.34.

Energy from rotting matter. On farms, manure can be collected. Farms also have plant wastes. There are pasture plants that have not been eaten, leftover feedstock and fruits, vegetables, and grains that are damaged or unsold.

When manure and organic wastes are put into closed tanks, bacteria will digest them. This produces methane gas. Called biogas, it can be used for cooking, lighting, and running engines. This is a common method of producing energy in many parts of the world. In China there are over 7 million biogas digesters supplying energy for 35 million people.

Digesters are designed to be batch load or

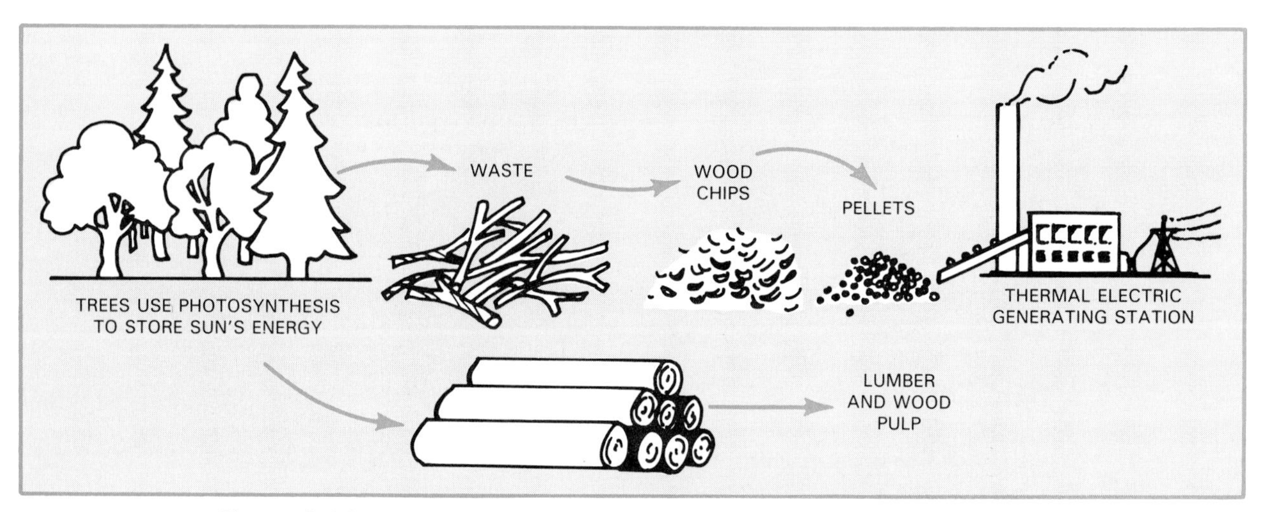

Figure 9.33. Waste wood products provide energy for producing electricity. Wood is a biomass source of energy.

continuous load. In a batch load type, the digester is loaded with a soupy mix of wastes. The mix is called a slurry. The digester is then sealed. It will not be emptied until the materials stop producing gas.

A continuous load digester will accept a small amount of fresh slurry continuously. This type will produce gas as long as slurry is being fed into it. Figure 9.35 diagrams two different kinds of continuous load digesters.

Hydrogen. Hydrogen is one of the two elements in ordinary water. Since 70 percent of the earth's surface is covered by water, the supply of hydrogen is practically inexhaustible. Hydrogen is produced by the electrolysis of water. A direct electrical current passes through a solution of water and a catalyst causing the water to separate into hydrogen and oxygen. (A catalyst is material that speeds up the process.) When hydrogen is separated from water it is a very combustible gas. It can be used as a fuel in engines. In its gaseous form, hydrogen takes up a large amount of space. Storage tanks would be too large to be

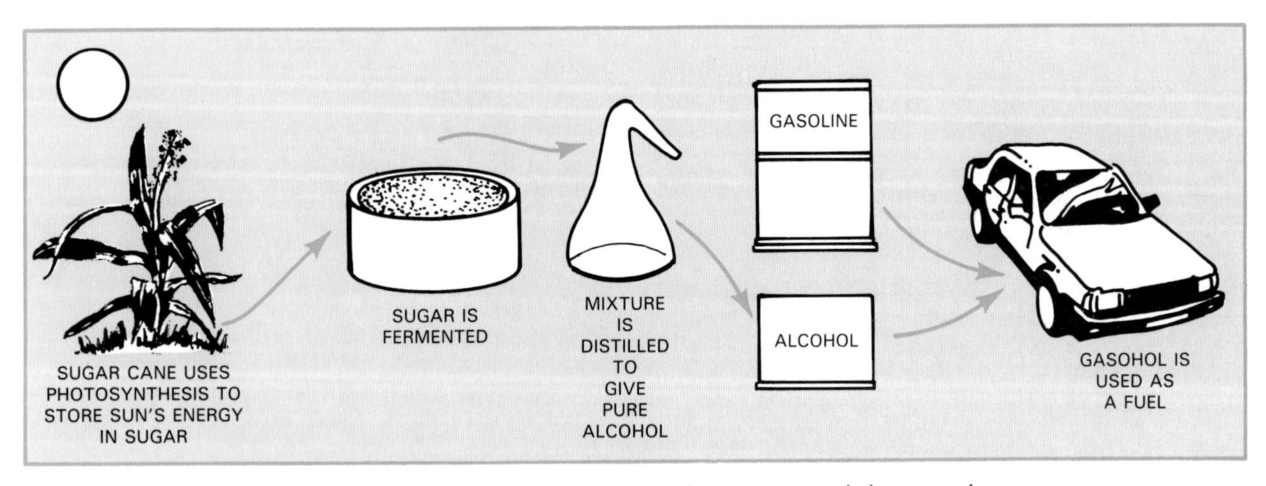

Figure 9.34. Sugar can be processed from cane and then used to produce alcohol. Blends of alcohol and gasoline are used to fuel automobiles.

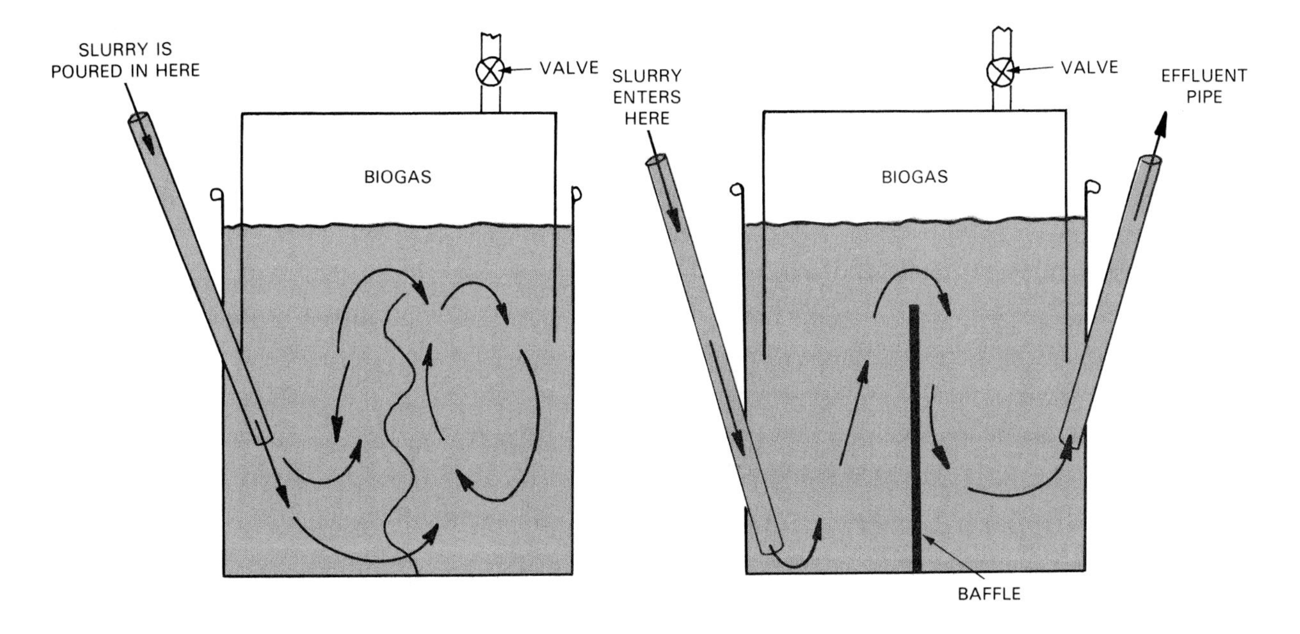

Figure 9.35. Two types of biogas digester. Waste is made into a soupy mixture called a "slurry."

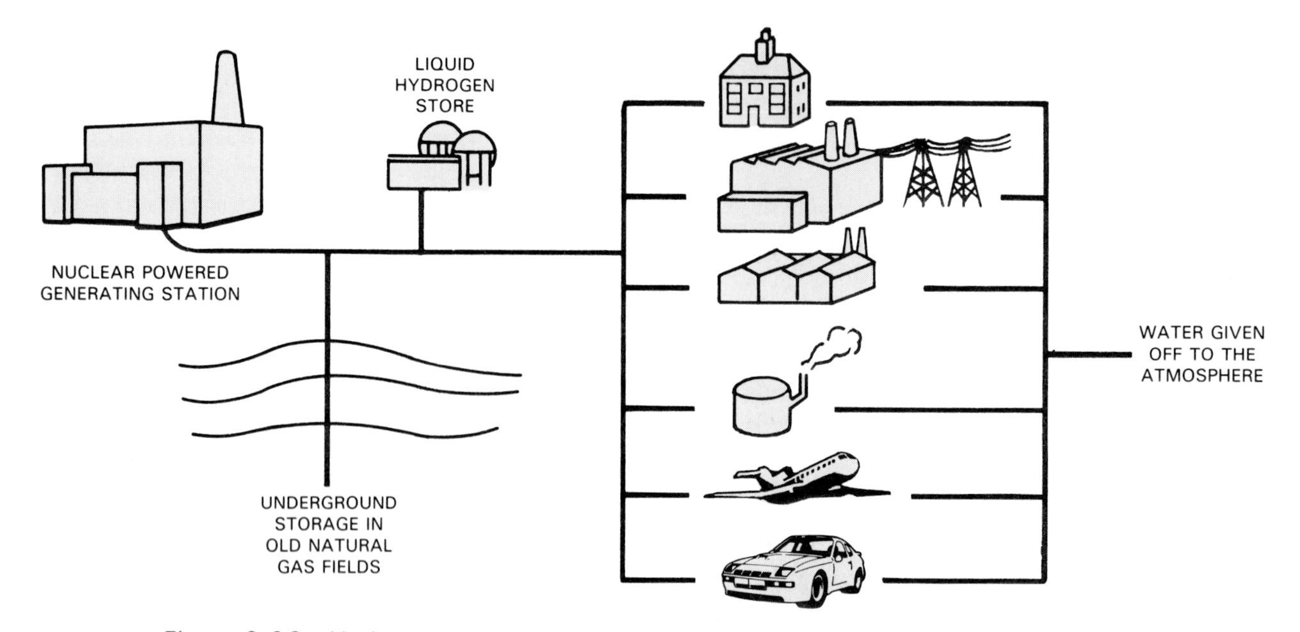

Figure 9.36. Hydrogen can be used as a fuel for homes, factories, aircraft, and automobiles.

practical. Therefore, hydrogen gas is converted to liquid hydrogen. This is done by cooling it to $-425\,^{\circ}F$ ($-254\,^{\circ}C$). Liquid hydrogen is stored in special insulated tanks.

Large amounts of electrical energy are needed for electrolysis. Thus, hydrogen may have to be produced using electricity generated by nuclear power stations. Hydrogen could be stored in gaseous form in underground spaces that once held natural gas. Also, it may be stored as liquid hydrogen in above-ground containers, figure 9.36. One day hydrogen may replace natural gas to heat homes and factories and power vehicles.

SUMMARY

Energy is available in many different forms: chemical, gravitational, strain, electrical, heat, light, sound, and nuclear. Energy can neither be created nor destroyed. It can only be changed from one form to another.

Almost all energy originates from the sun. Sources of energy can be divided into two groups: renewable and nonrenewable. Nonrenewable energy, coal, oil, and natural gas, cannot be replaced. Renewable energy, sun, wind and water, will always be available.

Coal developed from the compressed remains of plants that died millions of years ago. It is removed from the ground by surface-mining or underground mining. Petroleum (oil and gas) is formed from the compressed layers of plants and animals that once lived in the seas. Oil and gas are removed from the ground by drilling wells.

Solar energy uses energy from the sun. Solar panels are usually placed on the roof of a building. They can extract heat from the sun's rays and use it to provide hot water or to heat buildings. The sun's energy can also be converted directly to electrical energy using solar cells.

Wind is one of the oldest forms of energy used to do work. In the past, wind was used to turn windmills. Today it powers wind turbines.

The water from a river can be stored behind a dam and used to drive turbines that generate electricity. The force of tidal currents can be used to generate electricity in much the same way. Experiments are also being conducted to harness the energy of waves.

Nuclear power stations use nuclear fission. In this process the nucleus of an atom of uranium is split into two parts. Together these parts weigh slightly less than the original. The loss in mass turns into energy.

Geothermal energy is produced when underground water comes into contact with hot rocks deep within the earth and is changed into steam. Wells are drilled and the steam rushes to the surface with enough force to drive turbines.

KEY TERMS

Biomass energy
Chemical energy
Conduction
Convection
Electrical energy
Geothermal energy
Gravitational energy
Heat energy
Hydroelectricity
Hydrogen
Kinetic energy
Light energy
Nonrenewable energy
Nuclear energy
Nuclear fission
Nuclear fusion
Potential energy
Radiation
Renewable energy
Solar energy
Sound energy
Strain energy
Tidal energy
Wave energy
Wind energy

TEST YOUR KNOWLEDGE

Write your answers to these review questions on a separate sheet of paper.

1. A wound spring has ________ energy, and a falling boulder has ________ energy.
2. List the eight forms of energy and give one example of each.
3. Give examples from your daily life where you have experienced the three types of heat transmission.

Type of Heat TRANSMISSION	SITUATION
Conduction	
Convection	
Radiation	

4. Give three examples to show that energy can neither be created nor destroyed. It is only changed from one form to another.
5. Three examples of nonrenewable energy sources are ________, ________, and ________ ________. Three examples of renewable energy sources are ________, ________, and ________.
6. Which of the following is a 'renewable' source of energy?
   a. Coal.
   b. Gasoline.
   c. Natural gas.
   d. Solar energy.
7. Describe two methods of collecting energy from the sun.
8. What is the major disadvantage of wind turbines as a source of electrical energy?
9. Make a sketch to show how hydroelectricity is produced. Indicate, using arrows, the direction of flow of water.
10. Which alternative energy source is similar to hydropower, and why?
11. What are the major problems in using nuclear energy to produce electricity?
12. Describe how geothermal energy is used to produce electricity.
13. How can energy be produced from garbage?
14. Energy from plants is called ________ energy.
15. Why is biogas a useful form of energy in a rural area?

APPLY YOUR KNOWLEDGE

1. Think back to a time when a power failure occurred in your area. List the devices you were unable to use. Imagine that the power stayed off for 48 hours. What alternative sources of energy could you use?
2. Describe three situations that involve potential energy and three that involve kinetic energy.
3. Make a list of eight forms of energy shown in figures 9.2 to 9.9. For each, describe a further example of how the energy form is used to do work.
4. List five devices in your home that use energy. Describe the energy change(s) that take place when each is used.
5. List the ways in which you use energy each day (a) around the house, (b) in traveling, (c) at school, and (d) for leisure. State whether the energy comes from a renewable or nonrenewable source.
6. State the advantages and disadvantages of three sources of energy.

Huge hydroelectric projects flood large areas of land. What effect does this have on the area's people or on wildlife?

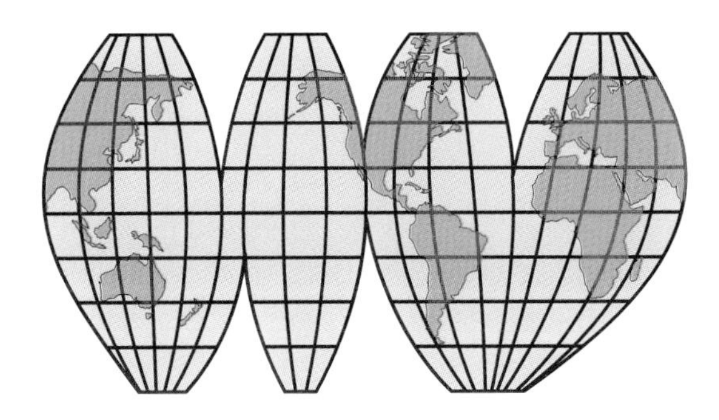

Chapter 10

Electricity and Magnetism

OBJECTIVES

After reading this chapter you will be able to:

- O Describe the different ways electricity can be produced.
- O List the advantages and disadvantages of each method of generating electricity.
- O Describe the principal components of a network for the transmission and distribution of electricity.
- O Explain the nature of electricity by referring to the movement of electrons.
- O State the laws of magnetism.
- O Explain how the laws of magnetism are used in the generation of electricity.
- O Describe how an electric motor operates.

Imagine your town or city without electricity. It can happen. At seventeen minutes after five o'clock in the afternoon of November 10, 1965, the lives of 30 million people were suddenly interrupted.

- 800,000 riders were trapped in the New York subway.
- All nine television channels in the metropolitan area of New York were forced to go off the air.
- Kennedy International and LaGuardia airports were shut down and airplanes found themselves circling, unable to land.
- 5,000 off-duty police officers were summoned to duty.
- 10,000 national guardsmen were called up to help protect the city.
- Militiamen were alerted in Rhode Island and Massachusetts.
- Broadway theaters and movie houses were closed.
- Thousands of peoples hiked across the Brooklyn and Queensboro bridges.
- Highways were jammed with traffic for more than five hours.
- Thousands of New Yorkers were trapped in elevators in the city's skyscrapers.
- In less than 15 minutes the power failure spread across more than 49,000 sq. miles (about 128,000 km^2). New York state, New England, and parts of New Jersey, Pennsylvania, Ontario, and Quebec had no electricity.

It was the largest power failure in history. The first signs of trouble appeared at the power company's control center. An operator noticed problems on the center's interconnecting system with upstate power companies. By then the blackout was only seconds away. It was too late for action. The demand for reserve power went so high that automatic switches shut the system down to protect it.

In our daily lives we take electricity for granted. To most people, it is merely something that always arrives at the home. Only when it is gone do we realize how we depend on it, figure 10.1.

■ GENERATING ELECTRICITY

The electrical energy supplied to your home comes from a **generating station.** Making electricity is called generating.

There are several types of generating stations. They are named after the power source used. Remember the names, hydroelectric and thermal-electric.

Figure 10.1. How would a blackout this evening affect you?

TEC

Figure 10.2. Dams are built to store water.

Hydroelectric

"Hydro" is another word for water. Hydroelectric generating stations use the energy of flowing or falling water. The station is located at a waterfall or at a dam, figure 10.2. As the water drops to a lower level its mass spins a turbine, figure 10.3. A turbine is a finned wheel. When the falling water strikes the fins the turbine turns rapidly. The turbines are connected to **generators.** A generator is a device that produces an electric current as it turns.

Thermal-electric generating stations

Thermal-electric generating stations use steam to drive turbines. A heat source produces the steam. The steam is directed onto the blades of a turbine. The turbine spins rapidly. As in hydroelectric systems, the turbines drive generators. The spinning generators produce electricity.

Heat for powering thermal-electric turbines comes from one of two sources. The first is by burning fossil fuels, figure 10.4. Fossil fuels come from once-living animals and plants. They include coal, oil, and natural gas. The second source is nuclear fission. Fission is the splitting of uranium atoms. The process releases enormous amounts of heat. In a nuclear station, figure 10.5, the nuclear reactor does the same job as the furnace in fossil-fuel stations.

TEC

Figure 10.3. This is the turbine room of a hydroelectric generating station.

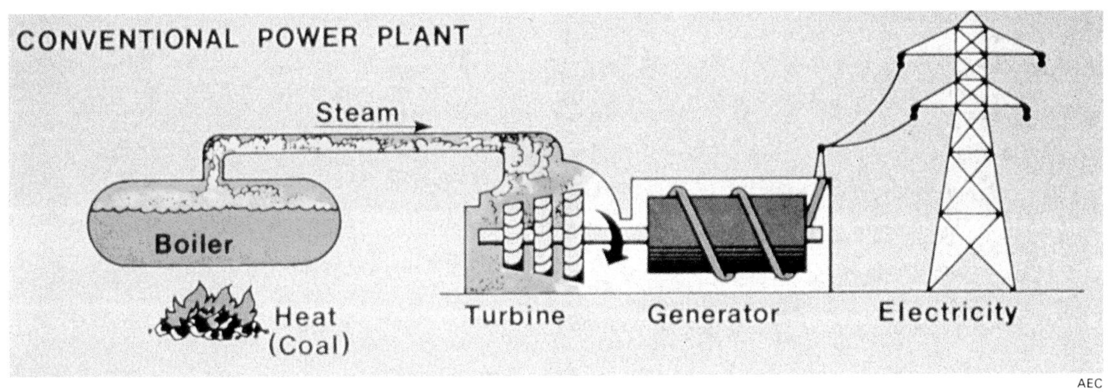

Figure 10.4. This is a simple diagram of a system for generating electricity with fossil fuels.

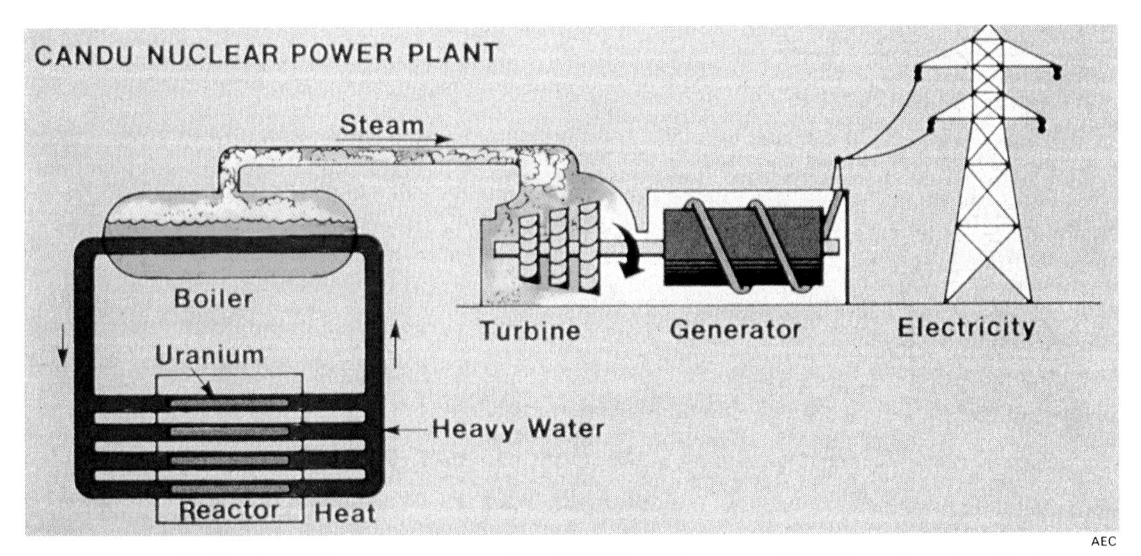

Figure 10.5. A nuclear power station. Splitting atoms, rather than burning fuels, creates the heat.

Any device that changes one form of energy to another is called a converter, figure 10.6. Hydroelectric generating stations change the potential (stored) energy of water behind a dam. As it falls into the turbine it becomes kinetic energy. Thermal electric generating stations convert the heat energy stored in fossil fuels and uranium into kinetic energy. In both cases, the kinetic energy is converted to electrical energy by generators.

Each method of generating electricity has advantages and disadvantages. Look at figure 10.7. If a generating station had to be built near your home, which would you choose? Why?

Most of the electricity used in homes and factories is produced either in hydroelectric or thermal-electric generating stations. There are, however, other methods. Friction, chemical action, light, heat, and pressure also generate electricity, figure 10.8.

■ TRANSMISSION AND DISTRIBUTION OF ELECTRICITY

Generating stations are rarely found close to where the electrical energy is used. The electricity that comes to your home may have traveled a great distance.

After leaving the generating station, the electricity is fed into a network of **transmission lines** and distribution lines. These lines transport the electricity to wherever it is needed, figure 10.9.

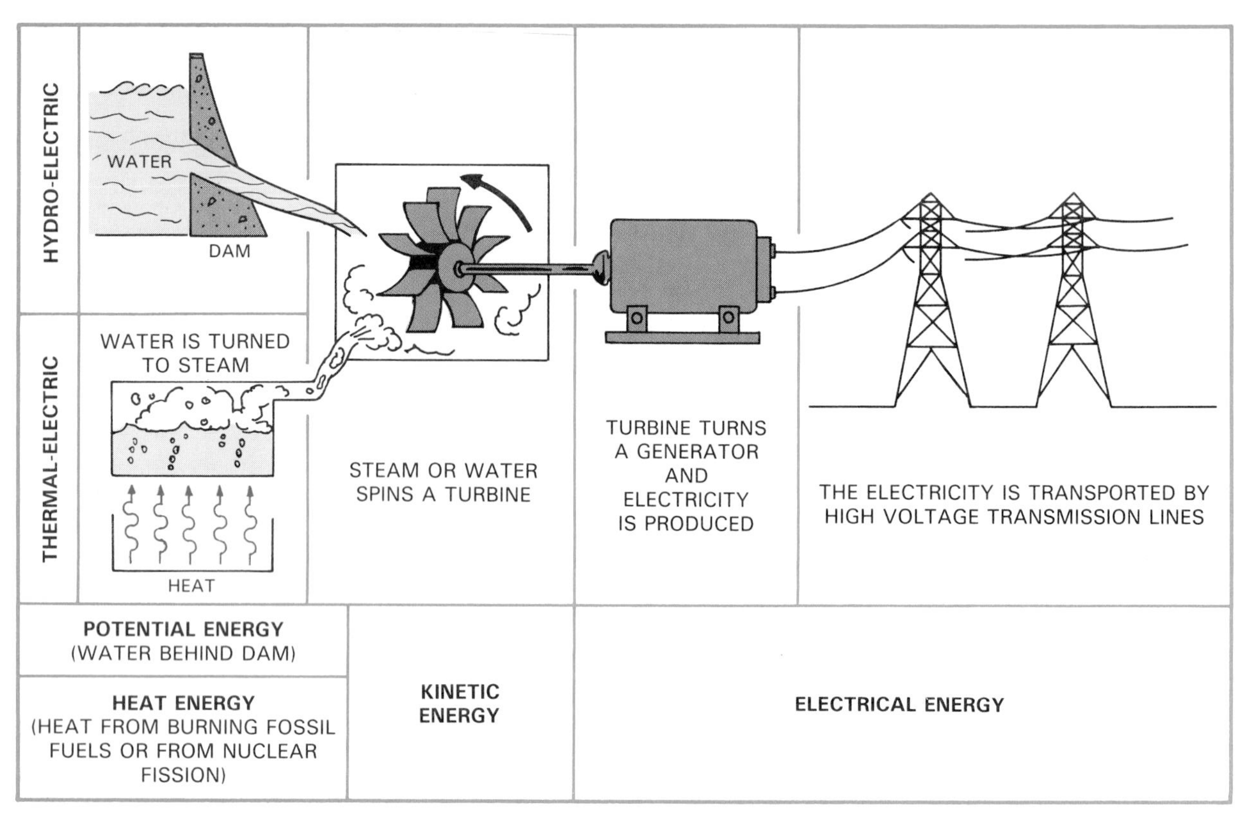

Figure 10.6. All generating stations are energy converters.

	ADVANTAGES	DISADVANTAGES
HYDRO	Cheapest method overall. Most environmentally safe method. Cheap to operate — raw material is free. Low maintenance and operational costs. No harmful combustion products. No harmful wastes of any kind. Water used is not polluted. Flow of water, and, therefore, the amount of electricity generated is adjustable.	Sites are normally a long way from cities so long wires are needed to bring the electricity to the consumers. Transmission towers are unsightly. Large amounts of land are taken up by transmission corridors. Transmission lines emit electromagnetic radiation that may be a health hazard. Dams disrupt rivers and, therefore, the marine life. The reservoir behind the dam covers a large expanse of land, thereby displacing people and animals.
THERMAL-FOSSIL	Uses fuel that is often available locally or can be easily transported. Small generators can be built to supply local needs.	Uses a nonrenewable resource. Oil-fired thermal plants are becoming too costly and must be converted to natural gas or coal. Coal is bulky, heavy, costly to move, and dirty. Burning coal produces a large amount of ash. Rain that penetrates ash heaps or buried ashes will pollute streams or ground water. Harmful particles and gases are released into the air and combine with water vapor in the air to form acid rain that damages trees, lakes, and buildings. Mining coal is a dangerous occupation. Oil extracted from under the sea may sometimes leak into the sea water.
THERMAL-NUCLEAR	Large amounts of electricity can be generated using a small amount of material. Can be built wherever there is a supply of water for cooling. No acid rain is created.	Uses a nonrenewable resource. Radioactive nuclear wastes must be disposed of. Deep burial sites must be found for waste that will remain radioactive for thousands of years. Mining the uranium fuel is expensive; it is also hazardous to miners who are exposed to cancer-causing gas. Reactors are expensive to build and maintain. A reactor could overheat and release radioactive substances into the environment.

Figure 10.7. There are three types of electricity generating stations. Which type would you select?

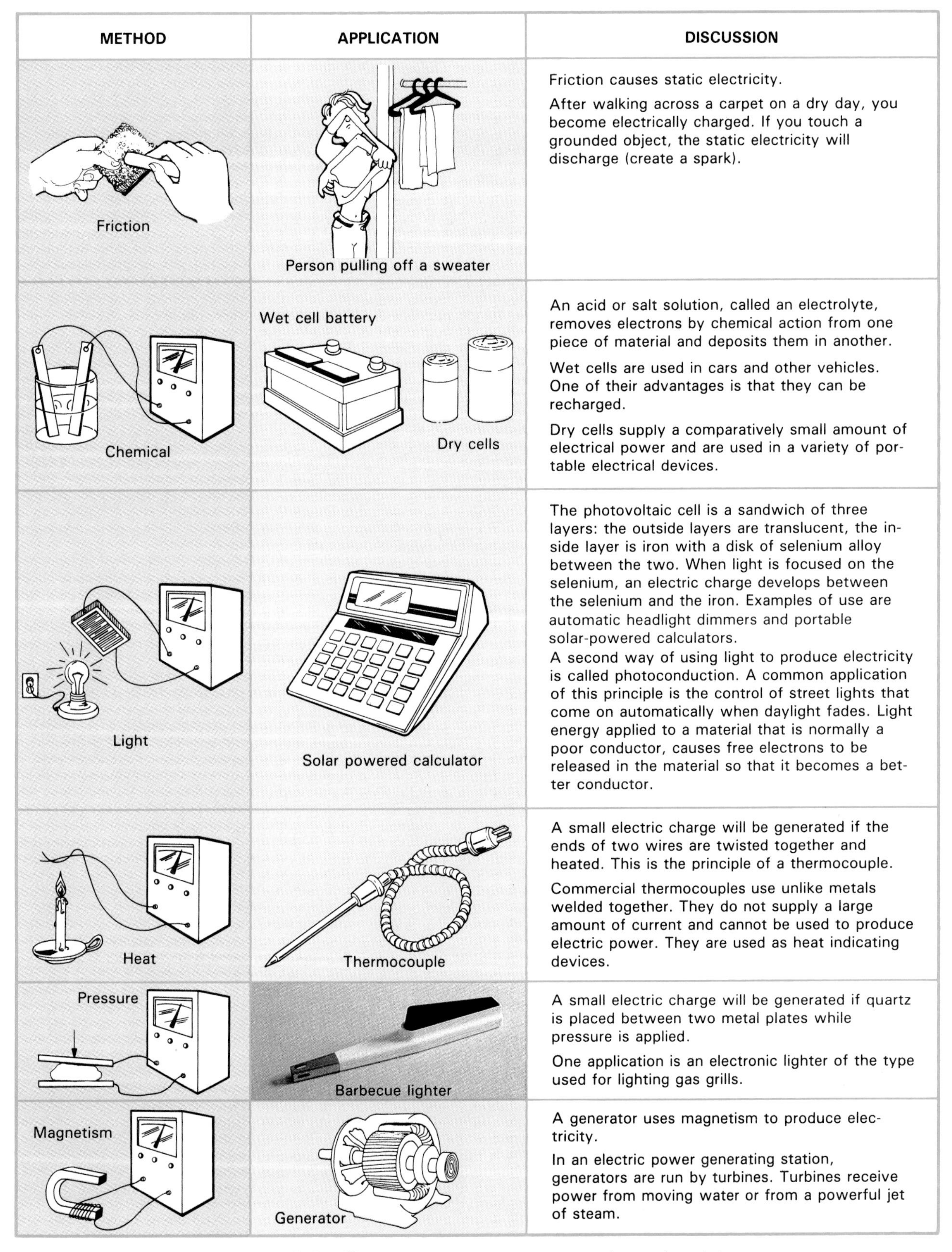

METHOD	APPLICATION	DISCUSSION
Friction	Person pulling off a sweater	Friction causes static electricity. After walking across a carpet on a dry day, you become electrically charged. If you touch a grounded object, the static electricity will discharge (create a spark).
Chemical	Wet cell battery Dry cells	An acid or salt solution, called an electrolyte, removes electrons by chemical action from one piece of material and deposits them in another. Wet cells are used in cars and other vehicles. One of their advantages is that they can be recharged. Dry cells supply a comparatively small amount of electrical power and are used in a variety of portable electrical devices.
Light	Solar powered calculator	The photovoltaic cell is a sandwich of three layers: the outside layers are translucent, the inside layer is iron with a disk of selenium alloy between the two. When light is focused on the selenium, an electric charge develops between the selenium and the iron. Examples of use are automatic headlight dimmers and portable solar-powered calculators. A second way of using light to produce electricity is called photoconduction. A common application of this principle is the control of street lights that come on automatically when daylight fades. Light energy applied to a material that is normally a poor conductor, causes free electrons to be released in the material so that it becomes a better conductor.
Heat	Thermocouple	A small electric charge will be generated if the ends of two wires are twisted together and heated. This is the principle of a thermocouple. Commercial thermocouples use unlike metals welded together. They do not supply a large amount of current and cannot be used to produce electric power. They are used as heat indicating devices.
Pressure	Barbecue lighter	A small electric charge will be generated if quartz is placed between two metal plates while pressure is applied. One application is an electronic lighter of the type used for lighting gas grills.
Magnetism	Generator	A generator uses magnetism to produce electricity. In an electric power generating station, generators are run by turbines. Turbines receive power from moving water or from a powerful jet of steam.

Figure 10.8. There are many ways to produce electricity.

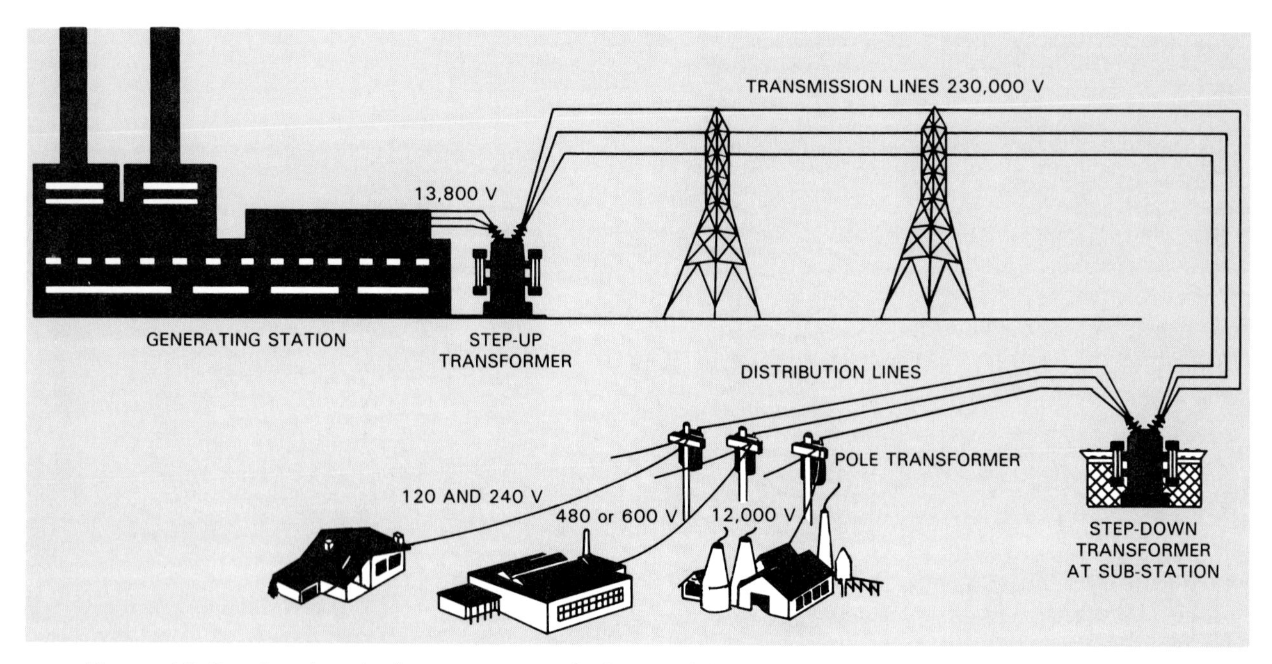

Figure 10.9. An electrical power transmission and distribution system. Voltage is greatly increased before the electricity is transmitted over long distances.

The transmission lines, figure 10.10, resist the flow of electrical energy. Thus, some of the energy is lost along the way. Increasing **voltage** and reducing current (amperage) will greatly reduce this loss. Voltage is a measure of electrical pressure. Amperage measures the amount of current.

TEC

Figure 10.10. Transmission lines are supported by large metal towers. Notice the large insulators that support the transmission lines where they attach to the tower.

TEC

Figure 10.11. Distribution substations step down the voltage.

A **transformer** changes voltage and amperage. It consists of a pair of coils.

Neither your home nor a factory can use electricity at high voltage. It would destroy the wiring, appliances, and machines. The voltage must be reduced before current enters distribution lines. Once more, a transformer is used. The first drop in voltage occurs when transferring electrical energy to distribution lines, figure 10.11. Another reduction occurs when transferring electrical energy from distribution lines to service lines. This transformer may be located on a pole or on the ground, figure 10.12. Electricity enters your home through a customer service line. It also passes through a meter, and a main switch, figure 10.13.

How transformers work

Transformers, therefore, are of two types. Step-up transformers increase voltage. Step-down transformers reduce it.

Basically, a transformer consists of two coils (windings) of wire around an iron core. One coil has more turns than the other. Figure 10.14 shows the construction of step-up and step-down transformers.

TEC

Figure 10.12. A pole transformer further reduces the voltage in the distribution lines.

TEC

Figure 10.13. Electricity enters your home through a customer service line, a meter, and a main switch.

■ WHAT IS ELECTRICITY?

When you flip a light switch, electricity flows through wires and lights a bulb. What exactly is it that flows through the wire when the switch is turned on? There is no perfect answer. A scientist would say that electric current consists of a flow of **electrons.** What does this mean?

One explanation is called the electron theory. Everything around us is made from very small particles called atoms. Atoms are made of even smaller particles called protons and electrons. The protons have a positive charge. The electrons have a negative charge.

An atom normally has the same number of electrons as protons. The negative and positive charges cancel each other. Such atoms are electrically neutral, figure 10.15.

In many materials, some electrons are only loosely held to their nuclei. These electrons

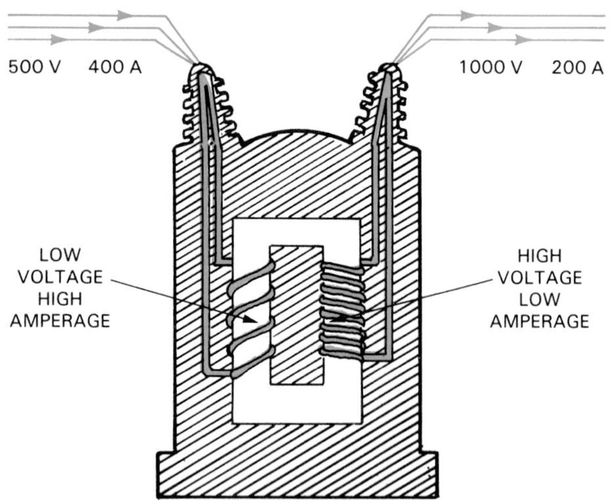
500 V 400 A
1000 V 200 A
LOW VOLTAGE HIGH AMPERAGE
HIGH VOLTAGE LOW AMPERAGE
A STEP-UP TRANSFORMER. WHEN THE INPUT VOLTAGE IS CONNECTED TO THE COIL WITH THE LEAST NUMBER OF TURNS THE OUTPUT VOLTAGE IS INCREASED.

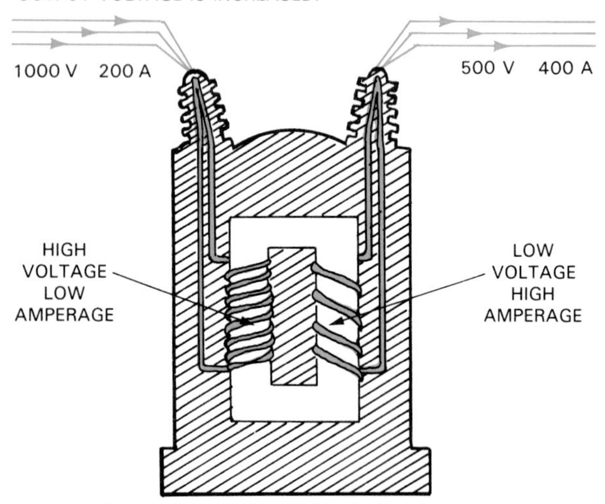
1000 V 200 A
500 V 400 A
HIGH VOLTAGE LOW AMPERAGE
LOW VOLTAGE HIGH AMPERAGE
A STEP-DOWN TRANSFORMER. WHEN THE INPUT VOLTAGE IS CONNECTED TO THE COIL WITH THE GREATEST NUMBER OF TURNS THE OUTUPT VOLTAGE IS DECREASED.

Figure 10.14. The "inside" of step-up and step-down transformers will help explain how they work.

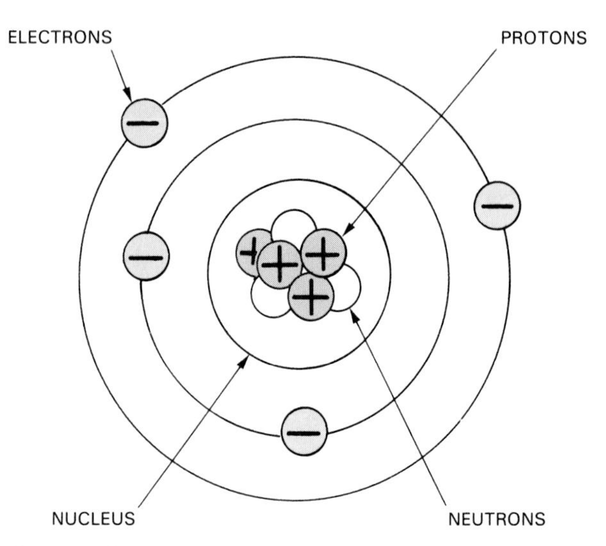
ELECTRONS
PROTONS
NUCLEUS
NEUTRONS

Figure 10.15. Protons and neutrons make up the nucleus of the atom. Electrons orbit around the nucleus.

break away. They can be made to flow through the material. This flow of electrons is what is called an **electric current,** figure 10.16.

Suppose that electrons are made to flow from one end of a wire to the other. The end that loses electrons becomes positively charged. It is called the positive terminal. The end that gains electrons becomes negatively charged. It is called the negative terminal. Electrons can move through the wire that is called a **conductor.**

Why are electrons able to move through a wire? There are two reasons.

- A force pushes them along a path.
- There is a closed path, called a **circuit,** around which they can move.

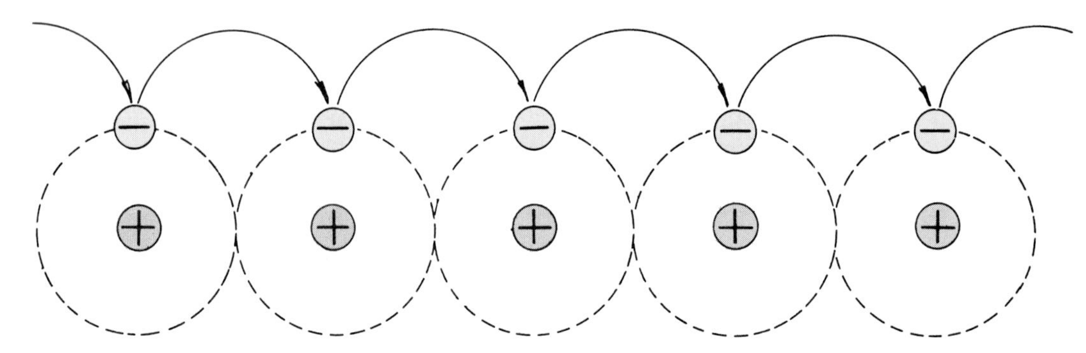

Figure 10.16. The movement of electrons. When a force is applied to one end of a wire, the free electrons move from one atom to the next. This process is repeated along the whole length of the wire. This causes what we call an electric current.

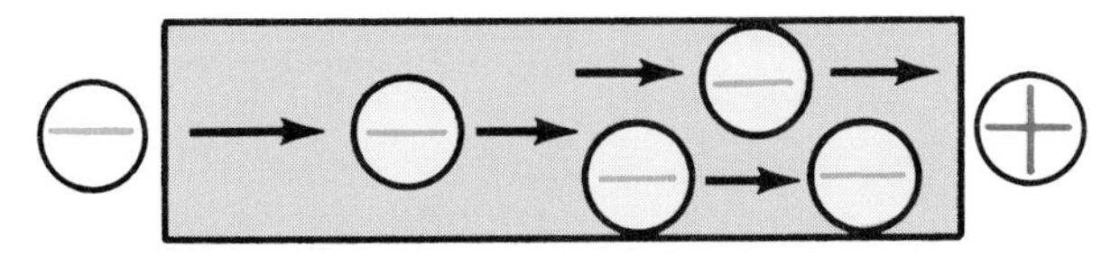

Figure 10.17. The terminal with a surplus of electrons is called "negative." The terminal with a scarcity of electrons is called "positive."

Think of the force as like that created by a pump pushing water through pipes. It is thought that electrons move because an electromotive force (EMF) pushes them through a conductor (wire). We also call the force voltage.

Just as we cannot say what an electron really is, we cannot describe an electromotive force. However, it is possible to describe some of the devices that are capable of producing electromotive force.

The two most common sources of EMF are:

- The mechanical movement (turning) of generators.
- Chemical reaction (changes) in dry cells and batteries.

These two sources provide most of the electricity that we use. Other sources of EMF are friction, light, heat, and pressure. (Look back at figure 10.8.)

□ MAGNETS AND MAGNETISM

The production of electricity depends upon magnets and magnetism. **Magnetism** is the ability of a material to attract pieces of iron or steel.

Materials that are attracted by magnets are called magnetic materials. Among them are iron, steel, and nickel. Nonmagnetic materials are not attracted by magnets. Examples are aluminum, copper, glass, paper, and wood, figure 10.18. Magnets fall into three different groups: natural magnets, artificial or permanent magnets and electromagnets.

Natural and artificial magnets

Natural magnets, such as lodestone, occur in nature. Lodestone is a blackish iron ore (magnetite). Its weak magnetic force varies greatly from stone to stone.

Artificial or permanent magnets are made of hard and brittle alloys. Iron, nickel, cobalt,

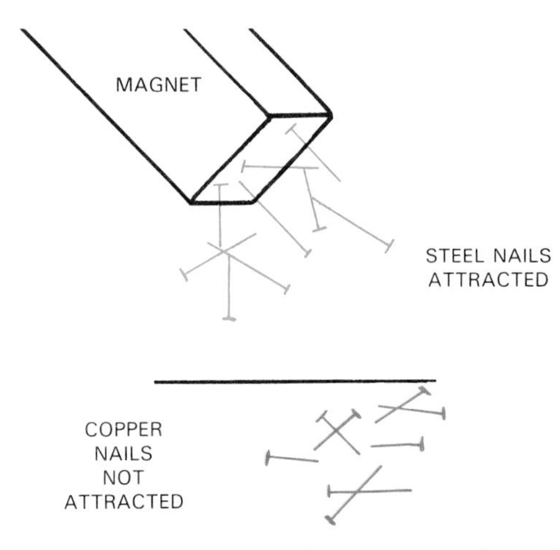

Figure 10.18. Which materials are magnetic and which are nonmagnetic?

and other metals make up the alloys. The alloys are strongly magnetized during the manufacturing process.

Permanent magnets come in many shapes and sizes. The most common are horseshoe magnets, bar magnets, and those used in compasses, figure 10.19. Natural and artificial magnets can retain their magnetism indefinitely.

Electromagnets

Electromagnets are so named because they are magnetized by an electric current. They consist of two main parts. One is a core of special steel. The other is a copper wire coil wound on this core, figure 10.20. Electromagnets, unlike permanent magnets, can be turned on or off. Their magnetic force can be completely controlled.

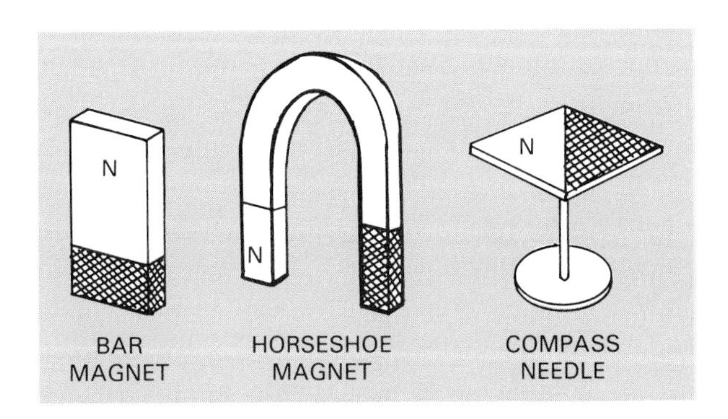

Figure 10.19. There are three common artificial magnets.

Figure 10.20. An electromagnet can be turned on and off.

How magnets act

Figure 10.21 shows a bar magnet suspended from a loop of thread. Held this way, the magnet will twist until it is lined up in a north-south direction. The end that points toward the north is called the north-seeking pole. The end that points toward the south is called the south-seeking pole.

Suppose that two magnets are suspended so that the north pole of one is brought close to the south pole of the other. The magnets will attract (move towards) one another, figure 10.22. This is the first law of magnetism. It states that unlike magnetic poles attract each other.

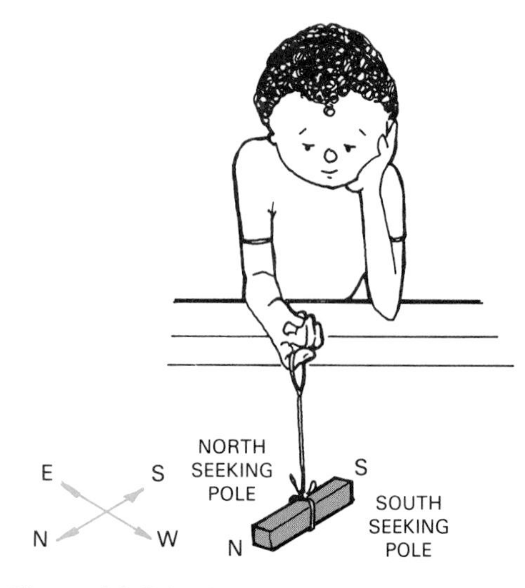

Figure 10.21. The north pole of a bar magnet that is free to swing will always point north.

Now, suppose that the north poles (or south poles) come close to one another. The magnets will push or move away from each other. This is the second law of magnetism. It states that like magnetic poles repel one another, figure 10.23.

Lines of force

Like poles repel and unlike poles attract. This suggests that around a magnet there are invisible lines of force. Although you cannot see them they can be shown to exist. Place a sheet of paper over a magnet and sprinkle iron filings on the paper. When the paper is tapped gently, the small iron particles form a distinct pattern, figure 10.24. The lines of force shown by the iron filings take the shape shown in figure 10.25.

Figure 10.22. Unlike magnetic poles attract.

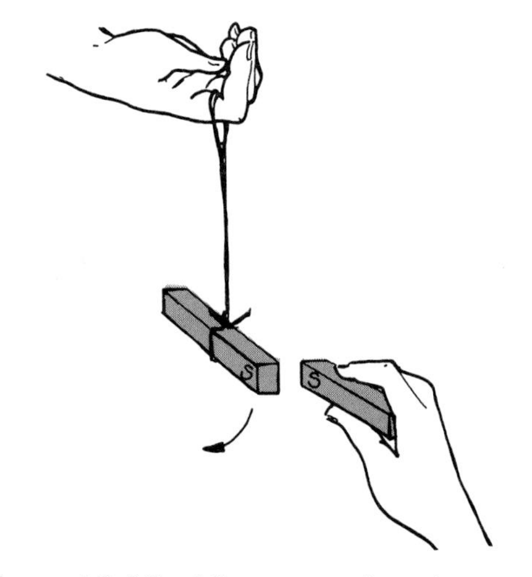

Figure 10.23. Like magnetic poles repel.

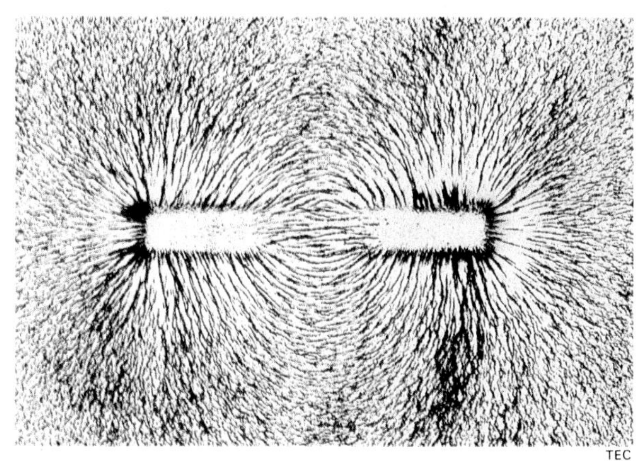

Figure 10.24. Iron filings will show the lines of magnetic force around a magnet.

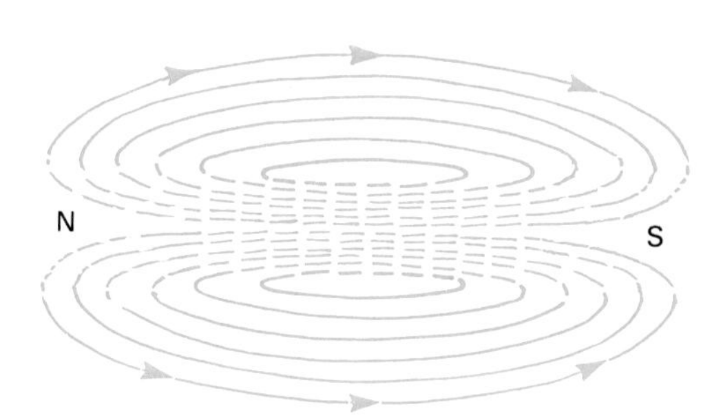

Figure 10.25. Note the pattern and direction of magnetic lines of force.

Now, suppose that two magnets are laid end to end, and the experiment with iron filings is repeated. The lines of force will demonstrate the two laws of magnetism, figure 10.26.

Magnetism and electric current

An electric current passed through a wire will also create a magnetic field around the wire. Magnetism produced by this means is called electromagnetism. This principle is used to make the electromagnet in figure 10.20.

SAFETY WARNING: The demonstration shown in figure 10.27 should only be done by your teacher. A carbon-zinc cell should be used. NEVER use an alkaline cell. It may explode.

If the wire in figure 10.27 is wound to form a coil, it becomes a magnet with poles, figure 10.28. The magnetic strength of this coil can be controlled. It depends upon the strength of the current and the number of loops in the coil.

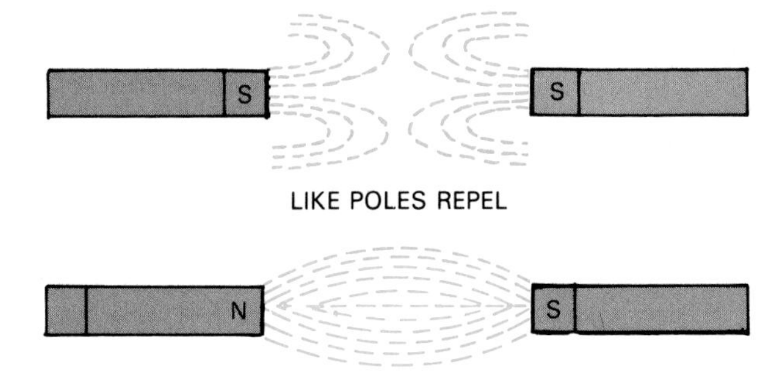

Figure 10.26. Try this experiment with iron filings and two magnets.

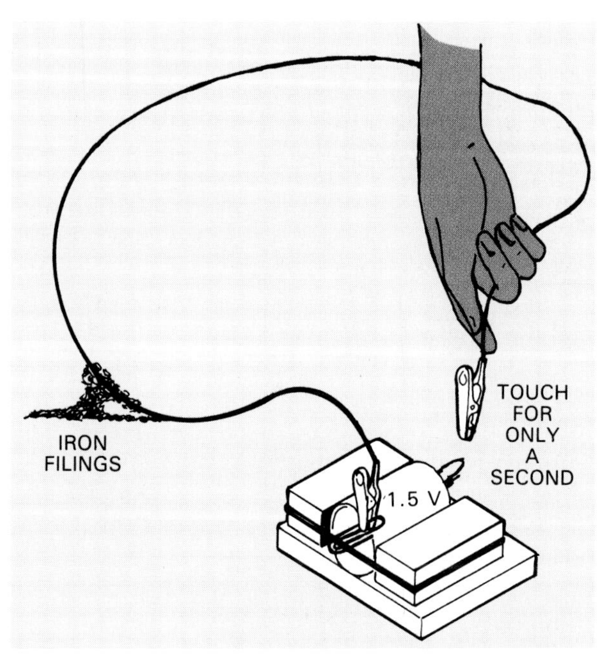

Figure 10.27. Producing a magnetic field. Connecting a wire to a dry cell will make the wire a magnet without poles.

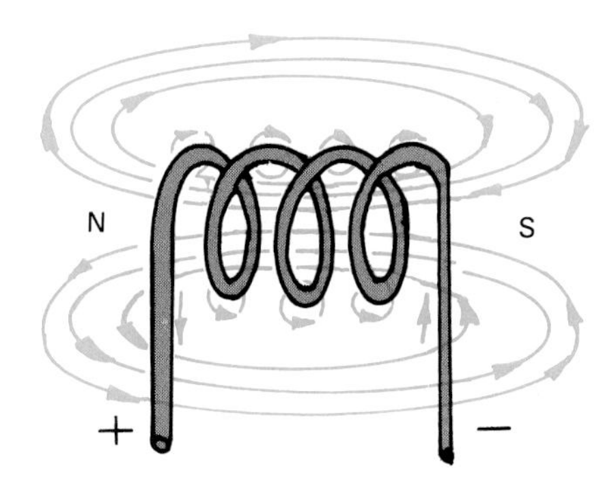

Figure 10.28. Coiling the wire of figure 10.27 will create a magnet with poles.

If a wire is coiled around a core of magnetic material, such as a soft iron nail, the nail becomes an electomagnet. It remains strongly magnetic only as long as there is current in the wire, figure 10.29.

□ THE GENERATION OF ELECTRICITY USING MAGNETISM

So far, you have learned that electric current is the flow of electrons in a circuit. Causing electrons to flow is called ''generating electricity.'' Electrical energy is not created. It is converted from other forms of energy.

A generator is the most practical and economical method today of producing electricity on a large scale. It uses magnetism to cause electrons to flow.

To see this in action, connect a length of copper wire to a milliammeter (sensitive current meter). As shown in figure 10.30, move part of the wire loop through a magnetic field. A small current will flow while the wire is cutting across the magnetic field.

The strength of the current (the number of electrons that flow) depends on two things. One is the strength of the magnetic field. The other is the rate at which the lines of force are cut. The stronger the magnetic field, or the faster the rate at which the lines of force are cut, the greater the current.

The direction of electron flow depends on the direction in which the lines of force are cut.

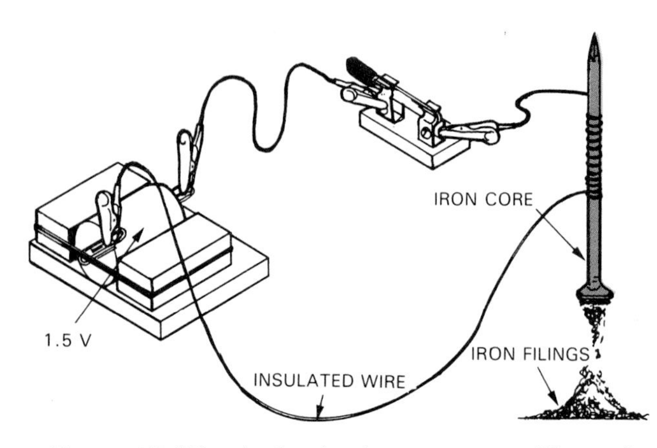

Figure 10.29. A simple electromagnet. The nail remains magnetized as long as there is current in the coil.

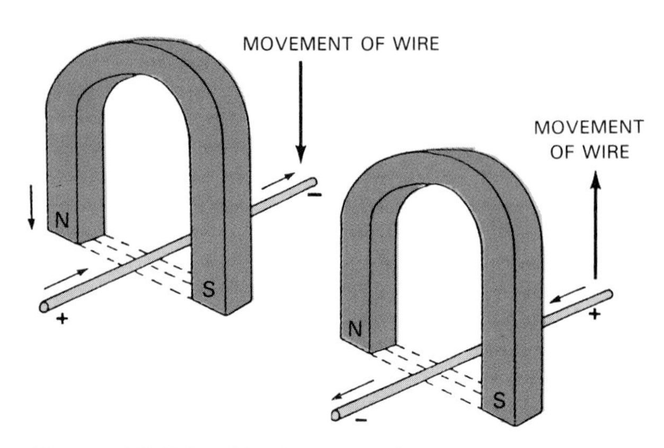

Figure 10.30. Moving a wire loop through a magnet will create a small current in the wire.

Look at figure 10.30 again. When the wire moves down through the lines of force of the magnet, electrons flow in one direction. When the wire moves up, electrons flow in the opposite direction.

The end that loses electrons becomes positively charged. The end that gains electrons becomes negatively charged.

Alternating current

Alternating current (AC) is electron flow that reverses direction on a regular basis. It is the type of current you use in your home. It is the type of current produced by power stations.

How is it produced? This will become clear as the basic operation of a generator is explained.

Figure 10.31 shows a simple generator. It is no more than a loop of wire turning clockwise between the poles of a magnet. Remember what was explained earlier. Current is produced only when a wire cuts through lines of magnetic force.

Now refer once more to figure 10.31. With the loop (wire) in position A, no lines of force are cut. The generator produces no current. As the loop continues turning, it reaches position B. At this point, one side of the loop moves downward through the lines of force. At the same time, the other side of the loop is moving up through the lines of force. Because the wire is a closed loop, current travels through it in one direction.

As the loop reaches position C, half a revolution is completed. As in A, there is no current. Why? No lines of force are being cut.

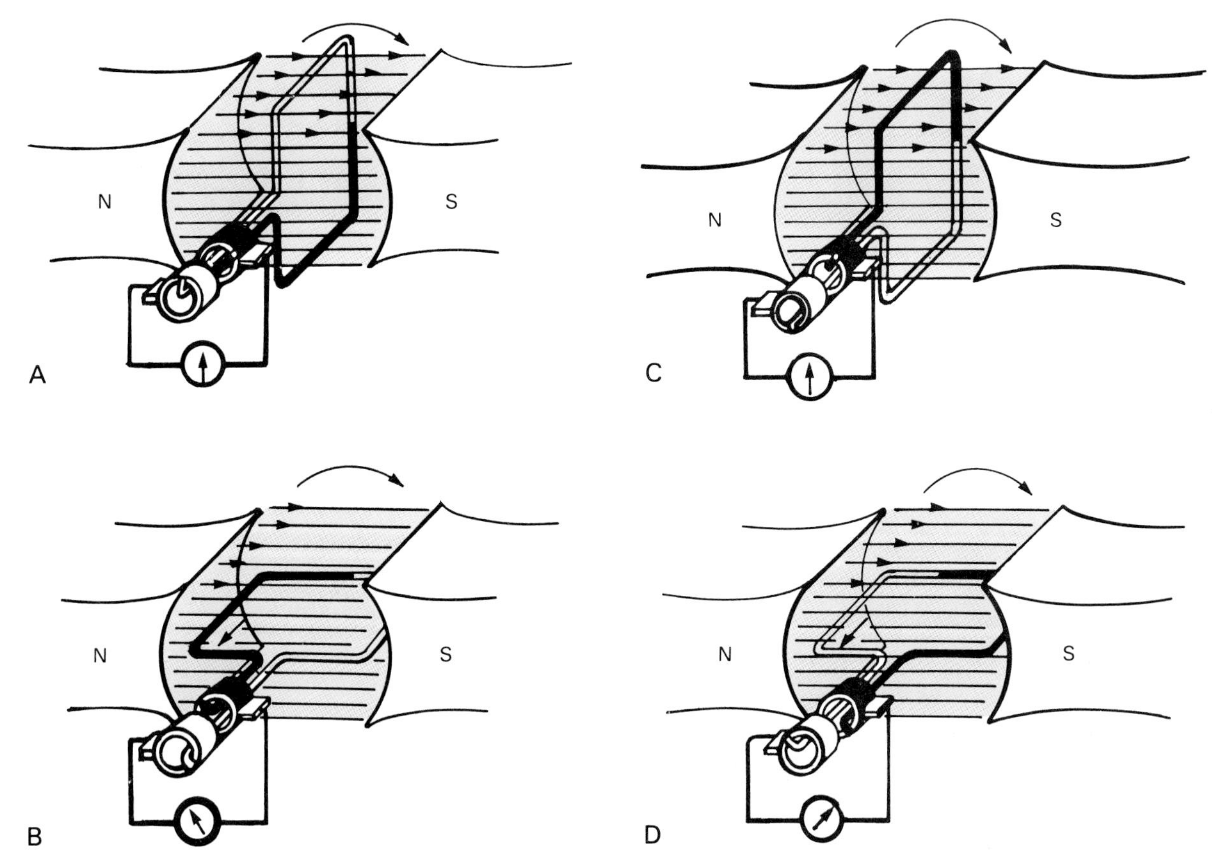

Figure 10.31 A basic generator. Note that an ammeter is connected across the terminals of the wire loop. At positions B and C the loop or wire is cutting across lines of force to create current in the wire. The current moves through the wire into the ammeter. At B it is going one direction. At D, current reverses.

The loop continues to turn. It reaches position D. The two sides once more cut lines of force. There is a difference, however. The side that moved downward before is now moving upward. Likewise, the side that moved upward before is now moving downward. What happens? The electron flow reverses. Because the direction of flow alternates as the loop turns, the current produced is called alternating current.

Electricity produced by the generator must have a path along which it can flow. You know the path as a **circuit.** Therefore, the terminals (ends) of the loop must always be in contact (in touch) with an outside wire. This outside wire is stationary (does not move). The contact is made with slip rings and brushes.

A separate slip ring is permanently fastened to each terminal of the wire loop. Each slip ring turns with the loop. A brush is placed against each slip ring. As the slip rings turn, the brushes maintain rubbing contact with them. The wire forming the stationary part of the circuit is attached to the brushes. Electrical devices, such a light bulb, are connected to the external part of the circuit, figure 10.32.

Current produced in the loop of the generator flows from the generator through a slip ring and brush into the external circuit. It travels in the external circuit through the electrical device. Then it returns to the generator through the other brush and slip ring.

As already noted, the direction of electron flow keeps changing or alternating. About 90 percent of the electricity produced in the world today is alternating current. It is easier to generate in large quantities than direct current. Even more important, it is easier to transmit from one place to another.

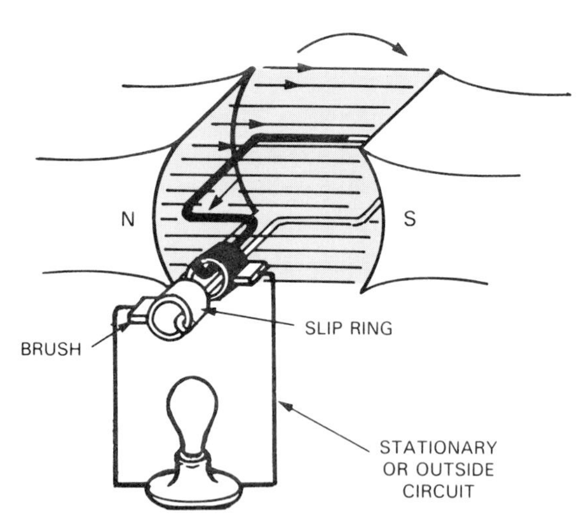

Figure 10.32. A simple generator and its external circuit. What would happen if one end of the wire becomes detached from its brush?

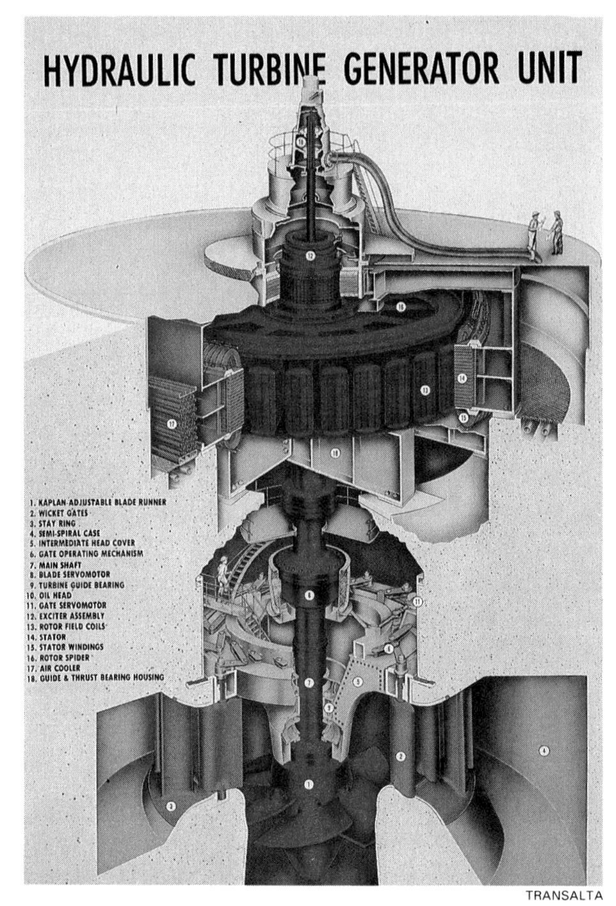

TRANSALTA

Figure 10.33. A large, commercial generator. Which part is the rotor and which the stator?

In North America, with few exceptions, alternating current makes 60 complete cycles each second. A cycle is a flow or pulse in one direction and a pulse in the opposite direction. In many European countries the alternating frequency is 50 cycles per second. Cycles are given in Hertz rather than cycles per second. A Hertz is equal to one cycle per second.

Generators at any of the large generating stations are more complex than the simple loop generator shown in this chapter. Their basic principle is the same, however. It is known that generated current can be increased in two ways:

- By increasing the rate at which the lines of force are cut.
- By strengthening the magnetic field.

Therefore, many loops of wire are used instead of one. Powerful electromagnets supply the magnetic field.

For practical reasons, the loops are mounted around the inner surface of the generator housing. They remain stationary and are called the stator. The electromagnets are mounted around a rotating shaft. This assembly is called a rotor. It is placed inside the stator. In this way, current is created by having lines of force cutting across conductors instead of by having a conductor cutting across lines of force, figure 10.33.

Direct current

Direct current (DC) is current that does not change direction in an external circuit. The direct current generator uses a single, split ring. It replaces the two split rings of an alternating current generator. Current in the loop still alternates. However, the split ring, called a commutator, sends current only one way through the circuit. The brushes and commutator of a DC generator are shown in figure 10.34.

Each half of the commutator is attached to one of the wire loop's terminals. As the current changes direction, the rotating commutator switches the terminals from one brush to the other every half revolution. Figure 10.35 shows a small DC generator.

Direct current is used in portable and mobile equipment. The list includes flashlights and car accessories. It is also used in electronic and

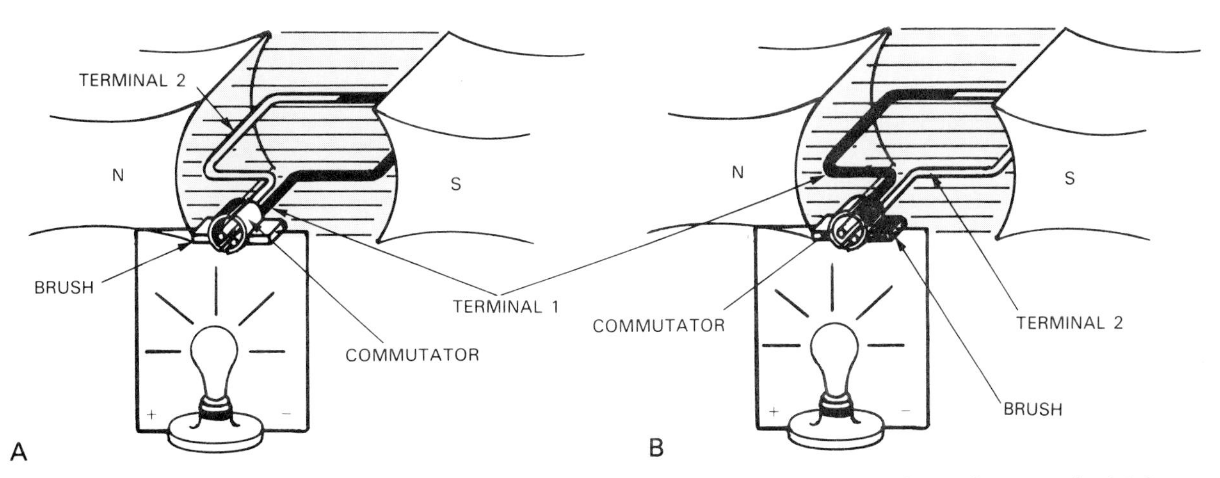

Fig. 10.34. How a commutator produces direct current. A—At this point of rotation, terminal 1 is contacting the brush connected to the negative side. B—Half a turn later, as current changes direction, terminal 1 is in contact with the brush connected to the positive side. Current through the external circuit continues in the same direction.

sound reproduction equipment. A disadvantage of DC current is that it is difficult to move over long distances.

□ ELECTRIC MOTORS

In many ways, an electric motor is like a generator. However, while the two have similar parts, their purposes are different. A generator converts kinetic energy, or movement, to electrical energy. An electric motor changes electrical energy to kinetic energy.

Both a generator and an electric motor apply the laws of magnetism. Both contain magnets and a rotating coil of wire. The coil of wire of an electric motor is placed in a magnetic field, figure 10.36. The motor will spin when a current is applied to the coil of wire.

What makes an electric motor run? Electrons flow through the coil of wire of an electric motor. They cause a magnetic field around the coil. Remember the first law of magnetism? Unlike poles attract. Like poles repel. When current is introduced in the coil the coil's magnetic field reacts with the magnets in the motor. The coil spins as it is either attracted or repelled by the motor's permanent magnets.

Unfortunately, the coil will only spin for part of a turn from the effects of magnetism. The rotation would stop except for the effects of a split copper ring that rotates with the coil.

ECRITEK

Figure 10.35. A bicycle's light uses direct current supplied by the small generator. What advantage does this generator have over a light operated by a battery?

For an instant, the current stops and the coil coasts. When current starts up again, magnetic force keeps the coil turning.

The split copper ring is called a commutator, figure 10.37. Current passes into and out of the coil through brushes that press against the

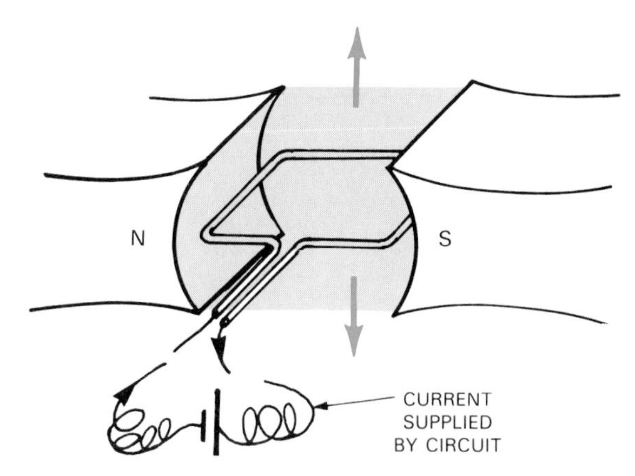

Figure 10.36. A simple electric motor is much like a generator.

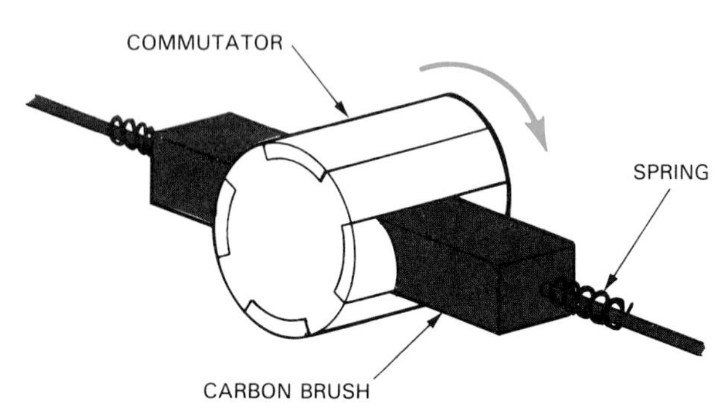

Figure 10-37. Brushes of an electric motor are always rubbing against the spinning commutator. This allows the electricity to flow into the commutator and into the coil.

commutator. In this way, current always passes down on the right side and back on the left side of the coil. The effect of this is to switch the poles in the coil's magnetic field. The rotation then continues in one direction.

The brushes also serve a second purpose. Since they do not rotate, they prevent the wires from twisting.

Brushes are usually made from carbon. It is a good conductor and produces less friction than metal. The brushes are spring-loaded. Pressure from the spring ensures continuous contact with the commutator.

□ CELLS AND BATTERIES

What most of us call a battery is not a battery at all. It is really a cell. The energy source for a car, however, is rightly called a battery. It is made up of several cells. How they differ and how they work will be explained.

Voltaic cell

The simplest of cells is the **voltaic cell.** Two rods, one copper and one zinc, are immersed in a container filled with a solution of water and sulphuric acid. (The mixture is known as an electrolyte.) The acid attacks and corrodes both of the metals, figure 10.38.

Some of the atoms from the metals pass into the solution. Each atom leaves behind a pair of electrons. However, the zinc rod tends to lose atoms to the solution faster than the copper rod. Since the zinc rod builds up more electrons than the copper rod, it becomes negative. If the two electrodes are connected by a conductor, excess electrons tend to flow along the conductor from the zinc to the copper. This flow of electrons produces an electric current. The current will illuminate (light up) a light bulb. **Cells** are classified as either **primary** or **secondary.**

Primary cell

A primary cell is one whose electrode is gradually consumed (used up) during normal use. It cannot be recharged. Primary cells are used in flashlights, digital watches, and cameras.

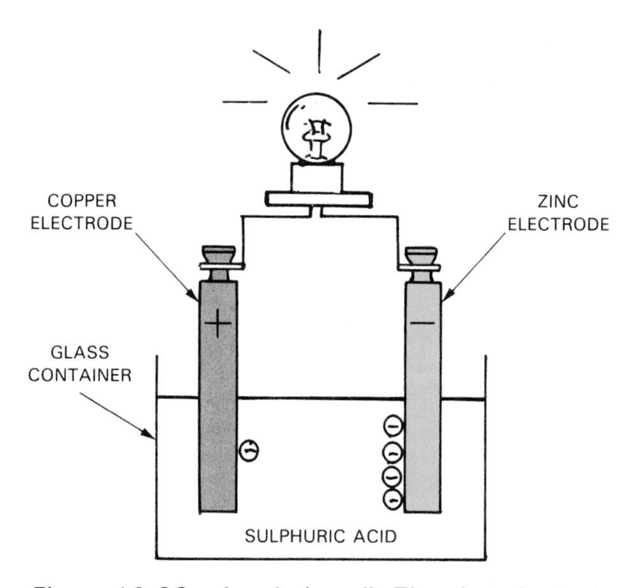

Figure 10.38. A voltaic cell. The zinc electrode is negative. The copper electrode is positive.

The primary cells used today employ the same principles as the voltaic cell in figure 10.38. There are many different types of primary cell. They all have three main parts: the electrolyte and two electrodes. The electrolyte is usually a very active chemical such as an acid or an alkali. (Acids are compounds that react with a base such as metal. An alkali is a substance capable of neutralizing acids.) Inside the battery, figure 10.39, two chemical reactions take place. One is between the electrolyte and the negative electrode (cathode). The other is between the electrolyte and the positive electrode (anode). These reactions change the chemical energy stored in the cell into electrical energy. When the chemicals have all reacted, the cell has no chemical energy left. It can give no more electricity.

Carbon-zinc. The carbon-zinc is the most common primary cell. Carbon-zinc cells are produced in a range of standard sizes, figure 10.40. These include 1.5 V AA, C, and D cells, as well as 9 V rectangular batteries. Carbon-zinc cells are the least expensive but are short-lived.

Alkaline. Alkaline cells are produced in the same sizes. However, they can supply current longer.

Mercury. Both the carbon-zinc and alkaline cells are much too large for some uses, such as digital watches, hearing aids, calculators, and miniature electronic equipment. For these applications, mercury cells are used, figure 10.41. Mercury cells develop 1.34 V.

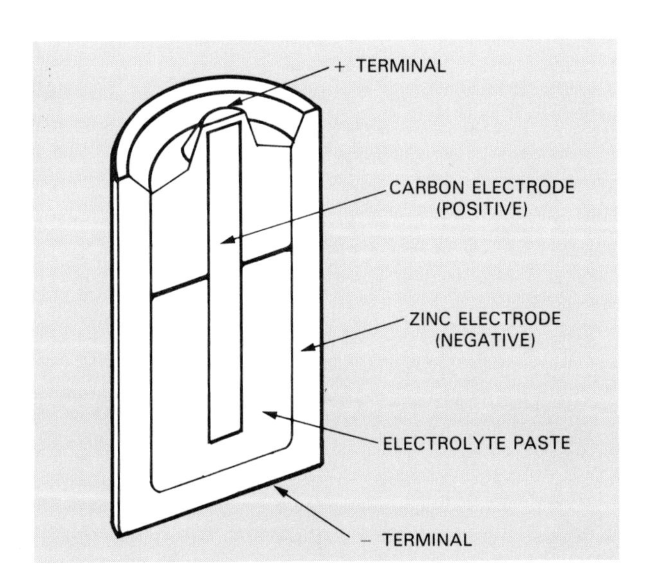

Figure 10.39. Cross section of a dry cell. It provides chemical storage of electricity.

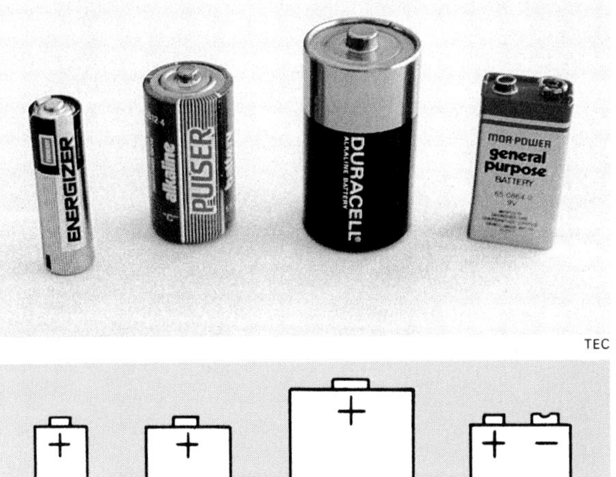

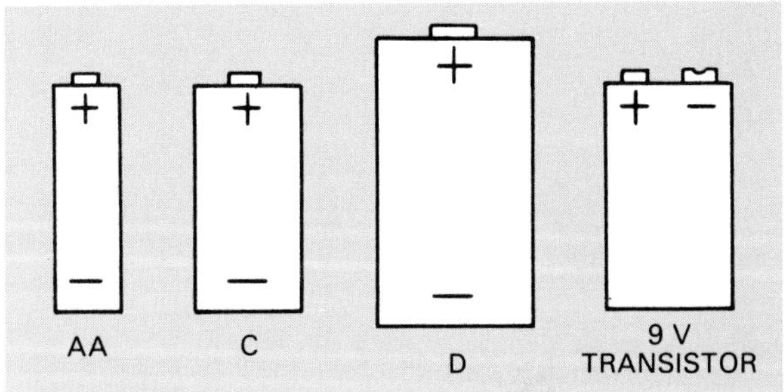

Figure 10.40. Top. Typical primary cells. Bottom. Names of each size. Which have you used?

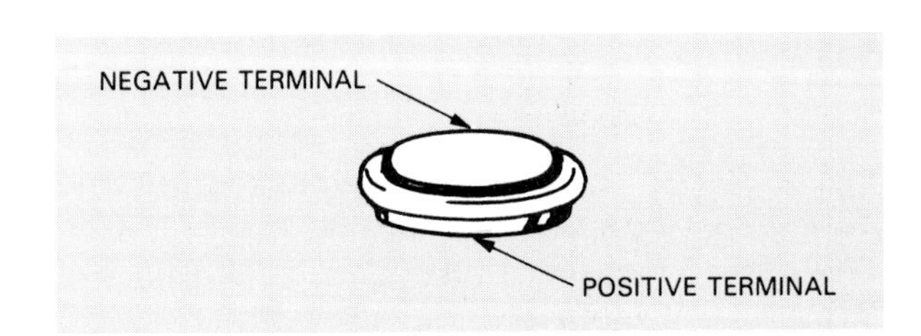

Figure 10.41. Mercury cells are smaller and shaped differently from carbon-zinc and alkaline cells.

Nickel-cadmium. The three types of cell described so far must be thrown away when they run down. Nickel-cadmium primary cells can be recharged. These cells are very expensive and require a special transformer for recharging.

Secondary cells

A secondary cell, figure 10.42, is one that can store electrical energy fed into it. Then, as needed, the electricity can be drawn from the cell in the form of an electric current. Lead plate electrodes are placed in a solution of

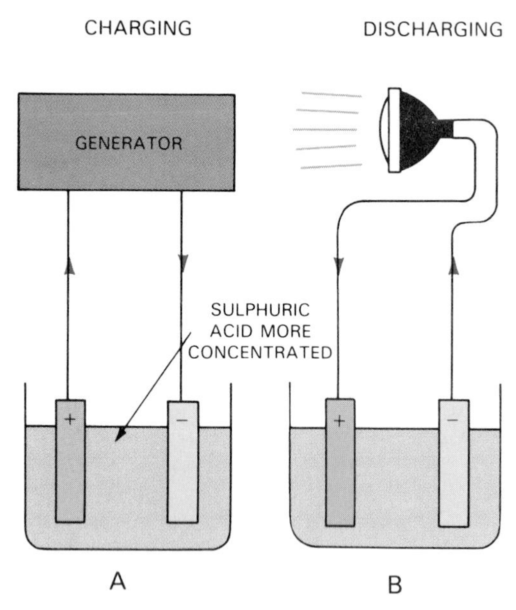

Figure 10.42. A typical secondary cell. A—It can store energy fed into it during charging. B—As needed, the stored energy produces an electric current.

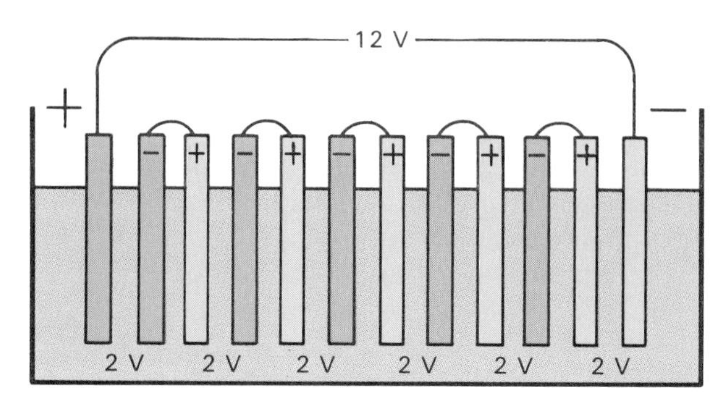

Figure 10.43. A modern automobile battery is made by connecting six cells to produce 12 V.

TEC

Figure 10.44. A 9 V battery combines several 1.5 V cells. What other sizes (voltages) of batteries have you used?

sulphuric acid. A current passing through the lead plates produces chemical changes. The sulphuric acid solution gets stronger and the cell becomes capable of producing an electric current. This is called charging a cell. When charged, the cell can produce a current in a circuit.

As electricity is drawn from the cell, the chemical change that took place during charging reverses. However, the materials in the cell are not used up, only changed. Therefore, the entire process can be repeated.

Each pair of electrodes in a secondary cell can produce about 2 volts. Most motor vehicles require 12 volts to operate the starter motor. Therefore, six pairs of electrodes, or cells, must be connected together. Figure 10.43 shows a number of cells connected together to form a battery.

Primary cells can also be connected to form a battery. For example, six 1.5 V cells can be put together to form a 9 V battery, figure 10.44.

SUMMARY

Most of the electricity we use is produced at hydroelectric or thermal-electric generating stations. Other ways to produce electricity include friction, chemical action, light, heat, and pressure.

Electric current consists of a flow of electrons. Electrons move because an electromotive force pushes them. This force is provided by a generator or by chemical change in a dry cell or battery.

Most electricity is produced by generators using the principles of magnetism. An electric current flows in a wire if it is moved through a magnetic field. The current generated may be AC or DC.

An electric motor is similar to a generator. However, a generator converts kinetic energy to electrical energy; an electric motor changes electrical energy to kinetic energy.

Small amounts of electricity may be stored in cells or batteries. Primary cells are consumed gradually and cannot be recharged. Secondary cells can be recharged.

KEY TERMS

Alternating current	Generator
Circuit	Magnetism
Conductor	Primary cell
Direct current	Secondary cell
Electric current	Transformer
Electromagnet	Transmission lines
Electron	Voltage
Generating station	Voltaic cell

TEST YOUR KNOWLEDGE

Write your answers to these review questions on a separate sheet of paper.

1. If the electricity supply were cut off at 6:00 p.m. tonight, how would your community be affected?
2. What is the meaning of the word ''hydro'' in the phrase ''hydroelectric generating station?''
3. Compare hydroelectric generating stations and thermal-electric generating stations by stating:
   a. In what way they are similar.
   b. In what way they are different.
4. After each entry below, describe where, in a transmission and distribution system, you would find the voltages listed.
   a. 315,000 V.
   b. 13,800 V.
   c. 240 V.
   d. 120 V.
5. A transformer can increase or decrease __________ and __________.
6. Electric current may be described as ______________________________.
7. Name the three different groups of magnets.
8. What is the main different between a permanent magnet and an electromagnet?
9. State the two laws of magnetism.
10. Make a sketch to show how you would make an electromagnet.
11. The most practical and economical method of producing electricity is ______________.
12. If a wire is moved through a magnetic field created by a horseshoe magnet, ________ are caused to flow.
13. The alternating frequency in North America is ________ cycles (Hertz) per second.
14. Describe the difference between alternating current and direct current.
15. How is the purpose of an electric motor different from that of a generator?
16. What is the disadvantages of most dry cells?
17. How does a dry cell produce electrical energy?
18. Over a period of a year a portable stereo system uses a large number of dry cells (often referred to as batteries). In order to reduce the amount you spend, what kind of cells could be used and why?
19. Describe the difference between a dry cell and a battery.
20. To produce 24 V a lead acid battery needs ________ cells.

APPLY YOUR KNOWLEDGE

1. Where and how is the electricity used in your home produced?
2. Make a model to illustrate one method of generating electricity.
3. Describe the components of a network for the transmission and distribution of electricity. How many of these components can you see in your neighborhood?
4. Make sketches with notes to illustrate (a) an atom, and (b) electron flow.
5. Describe how the laws of magnetism are used to generate electricity.
6. Repeat the experiment illustrated in figure 10.24. Iron filings can be made by cutting steel wool into tiny pieces using an old pair of scissors. You can fix the pattern of iron filings in place using hair spray.
7. List five objects in your home that use an electric motor.

SOFSOURCE, INC.

How have computers affected your life?

Chapter 11

Using Electricity and Electronics

OBJECTIVES

After reading this chapter you will be able to:
- Design, draw, and build different types of circuits.
- List examples of insulators and conductors.
- Use Ohm's Law to calculate current, voltage, or resistance.
- Name and state the function of common electronic components.
- Design and construct a power-driven prototype.

Think about the things that electricity does. It operates motors found in many large and small appliances. The motors run electric mixers, blowers, pumps, dishwashers, washing machines and many other appliances. They power street cars, buses, and golf carts. Perhaps you have had an electric toy with a small electric motor.

Electricity is used to help us communicate. Think of the ways we keep in touch with one another. We use telephones, radios, tape recorders, television, and computers.

Factories use electricity and electrical circuits for starting and stopping machines automatically. Electricity controls assembly lines and robots. Whole factories can be run electrically. A few people working at computers can control machines, lights, assembly lines, packaging and loading of products. Indeed, it is hard to imagine what we would do without electricity.

■ ELECTRIC CIRCUITS

An **electric circuit** is a continuous path for electric current. The path starts from a source—a cell, for example. It continues through a resistance such as a lamp before it returns to its source. To make a simple circuit, connect a lamp, two pieces of wire, and a 1.5 volt cell. In the simple circuit shown in figure 11.1, a conductor (the wire) is connected to the negative terminal of the cell. Electrons flow from the negative terminal. They continue through the wire and the lamp to the positive terminal.

Figure 11.1 is a pictorial of this circuit. However, it takes too long to make a picture of each component. This is especially true when the circuit is complicated. Therefore, it is easier to use something simpler. The circuit diagram in figure 11.2 uses symbols rather than pictures. Symbols may be drawn quickly and are understood everywhere.

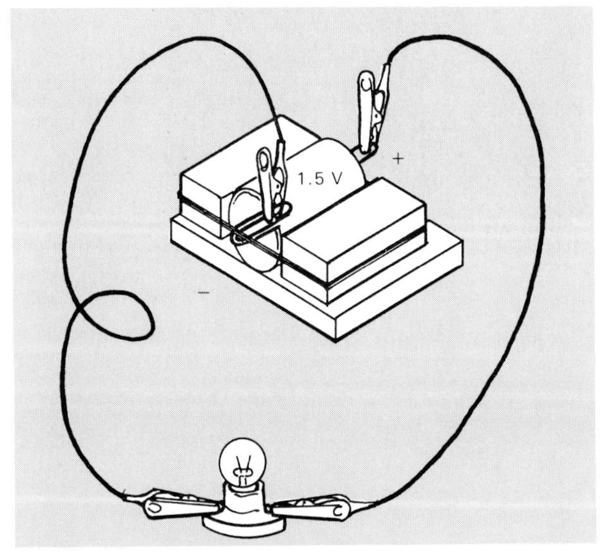

Figure 11.1. Pictorial drawing of a simple circuit. The circuit is used to bring current to the light bulb.

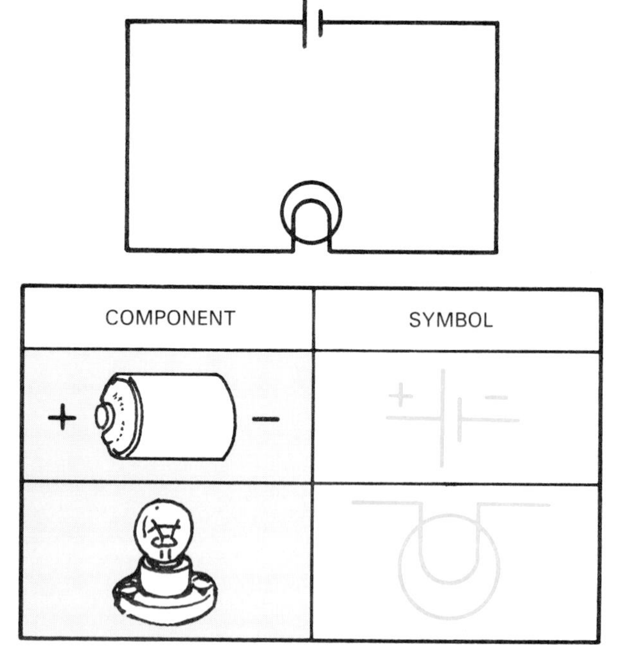

Figure 11.2. Top. A circuit diagram using symbols. It is the same circuit pictured in figure 11.1. Bottom. The components and their symbols.

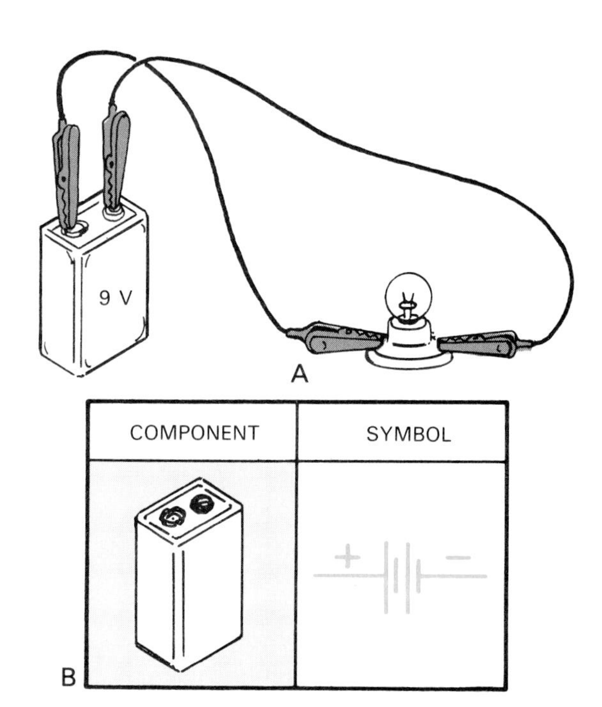

Figure 11.3. A—You could turn off the light by releasing any of the four clips. B—A battery and its symbol.

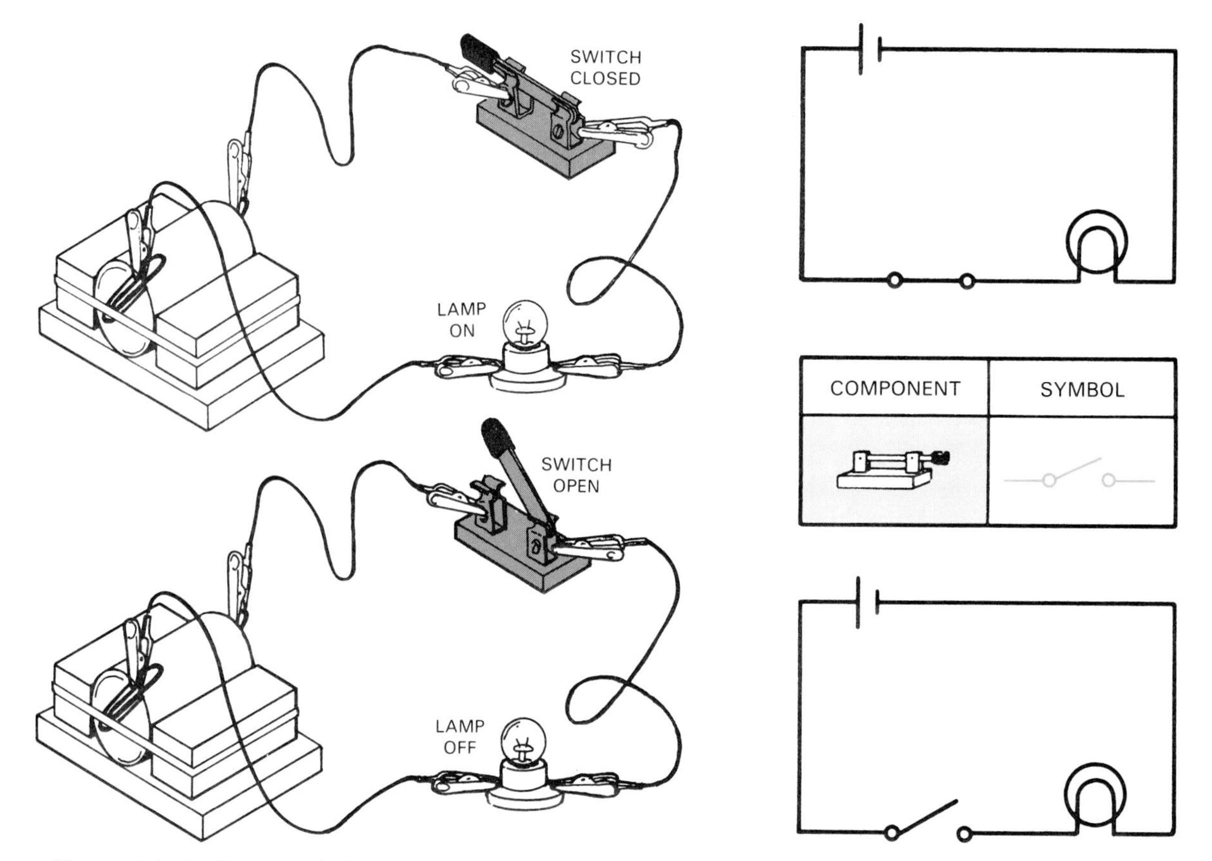

Figure 11.4. Open and closed circuits. What is the difference between the two circuit diagrams?

The circuit in figure 11.2 has voltage, resistance and current. Voltage is supplied by the cell. The resistance is the glowing element in the light bulb. The current is the flow of electrons in the wires. If there is only one resistance, such as a light bulb in the circuit, it is a simple circuit.

Suppose you wished to turn the bulb on and off. You could do it by connecting and disconnecting a wire at any one of the four places shown in figure 11.3.

Most electric circuits have a switch. This is a device that enables the circuit to be turned on and off. With the switch closed, electrons flow. The light bulb lights. When the switch is open, the flow stops. The bulb turns off. Mechanical switches are also used to direct current to various points. The simplest is a single-pole, single-throw (SPST) knife switch, figure 11.4. There are many types of switches. Figure 11.5 shows five.

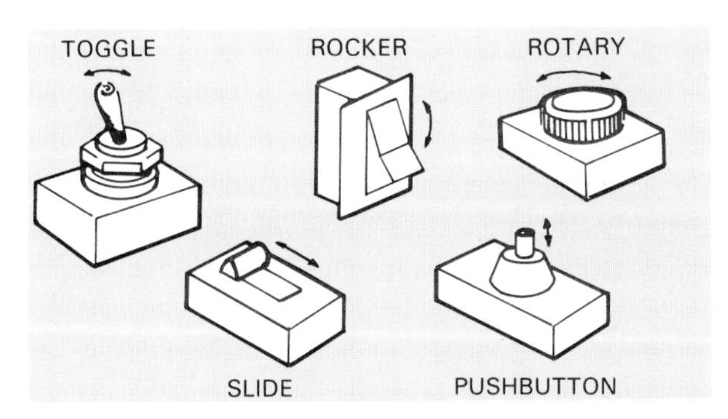

Figure 11.5. Where have you seen these different switches used?

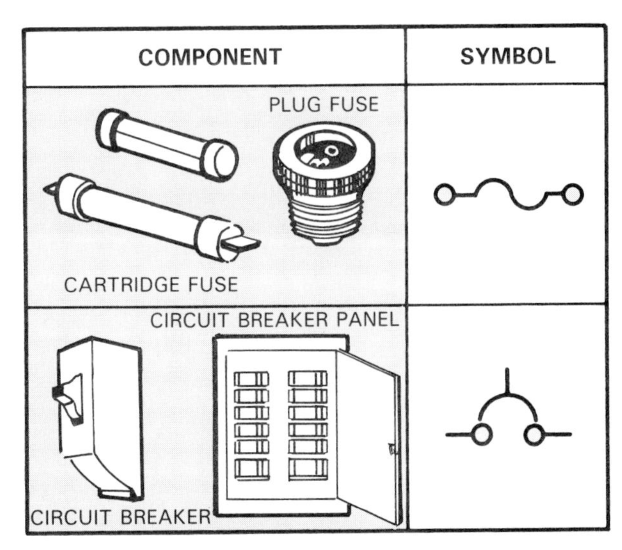

Figure 11.6. Overload devices protect the wiring in a circuit.

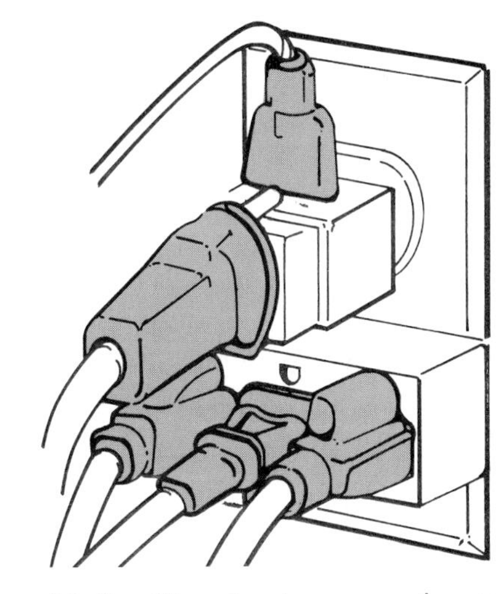

Figure 11.7. Plugging too many lamps and appliances into one outlet will cause an overloaded circuit.

Protecting circuits

Too much current can overheat and damage circuits. To prevent this, a fuse or a circuit breaker is added, figure 11.6. These are devices that open the circuit when current is too high. The fuse "blows" (burns out). The circuit breaker trips a contact. In either case, when the circuit is open, it is no longer complete. Current stops.

When applied to using electricity, the term **load** means something in the circuit that uses up the electric current. The bigger this load, the more current it needs.

High current can be caused by an **overload** or a **short circuit**. An overload occurs when lights or appliances in the circuit demand more current than the circuit can safely carry, figure 11.7. A short circuit occurs when bare conducting wires accidentally touch, figure 11.8.

How fuses and circuit breakers work

When too much current attempts to pass through a fuse, figure 11.9, a thin wire, called a filament, melts. Current stops before the wires can be damaged. When high current enters a circuit breaker, it heats a bimetal (two metals) strip. The strip bends as one metal expands more than the other. This opens con-

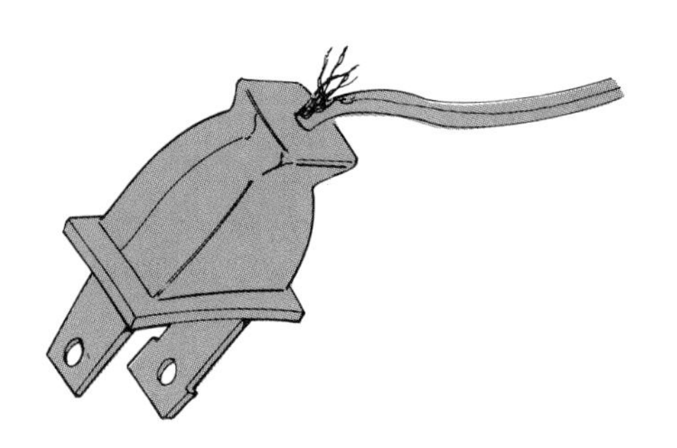

Figure 11.8. A short circuit is a fault in a circuit. It allows current to return to its source without traveling through the entire circuit.

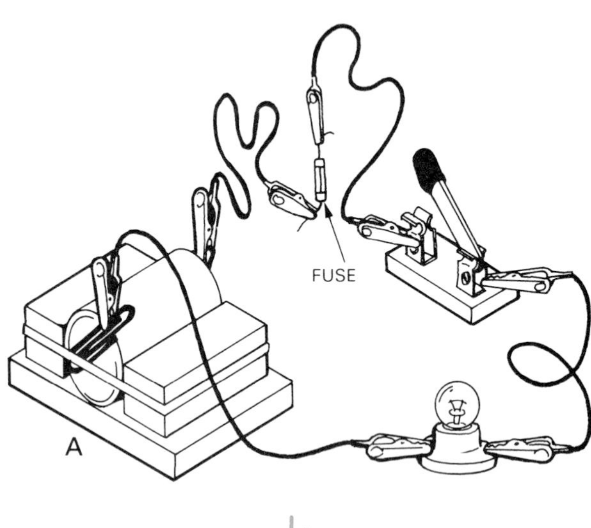

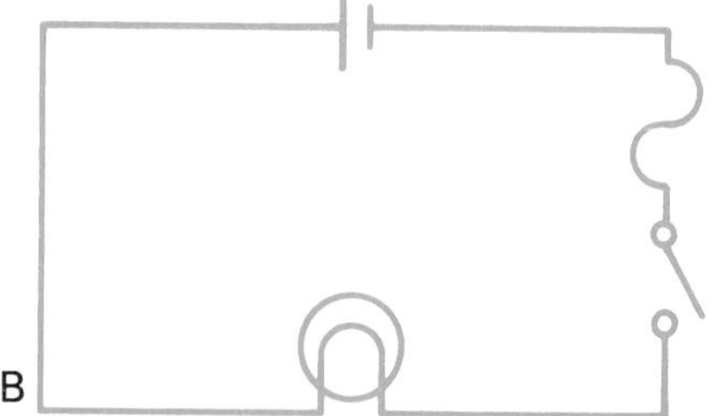

Figure 11.9. A fuse will stop flow of electricity if an overload occurs. A—A pictorial of a simple circuit with a fuse. B—A circuit diagram.

tacts so no current can pass.

Once a fuse "blows" it must be replaced. A circuit breaker's contacts can be reset.

Direction of current

Earlier, you learned that current is a "flow" of electricity in a conductor. The direction of this flow in a circuit can be shown either as "electron" flow or "conventional current," figure 11.10.

Electron flow is based on the electron theory. This theory states that current moves from negative to positive. Conventional current is based on an older theory of electricity. Early scientists assumed a current moved from positive to negative.

Both theories are acceptable. In this book, however, all explanations will be based on electron flow.

Series and parallel circuits

A circuit, as you have learned, is a pathway along which electricity may travel. A simple circuit shown in figure 11.11 contains only

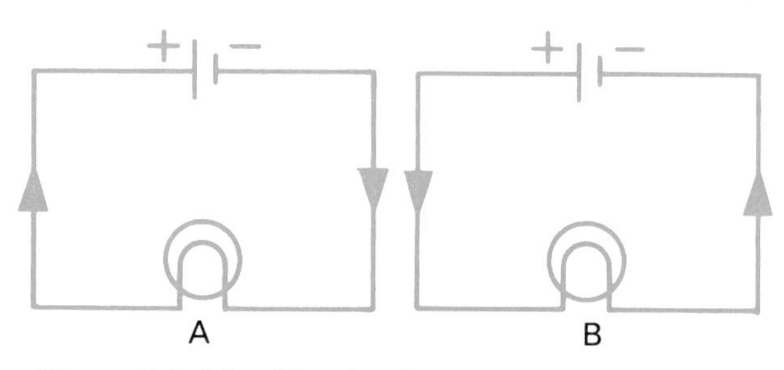

Figure 11.10. Circuit diagrams show the two theories of electric current. A—Electron flow is from negative to positive. B—Conventional current. Movement is from positive to negative.

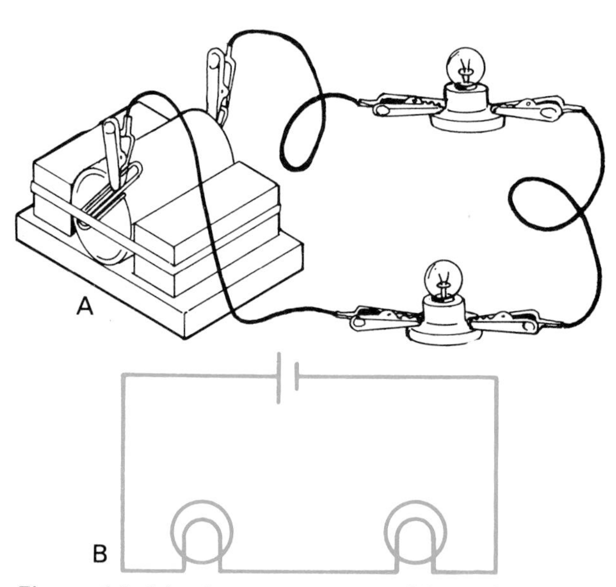

Figure 11.11. Lamps connected in series. Electricity flows through each bulb in turn.

one lamp. If two or more lamps are put into this circuit, they can be connected in one of two ways: in series or in parallel.

A **series circuit** is one in which lamps are connected so the same current enters each of them in turn. There is only one path for electron flow.

When lamps are connected in series, each gets part of the voltage. For example, three lamps connected to a 1.5 V cell each receive .5 V. Therefore, each bulb will be dimmer than if only one lamp is in the circuit. Current, however, remains the same across each resistance (lamp.)

Another thing about this circuit, if one lamp burns out, all of the lamps go out! This is because the circuit is open (broken). Electrons stop flowing.

Very few series circuits are used in our homes for this reason. Sometimes, however, Christmas tree lights are wired in series.

A **parallel circuit** is one having more than one path for electron flow, figure 11.12. You can see that when lamps are in parallel the current splits. It goes through each of the lamps without passing through any others first. In such a circuit, current varies. However, voltage will always be the same. If one bulb burns out, the circuit is not broken. Other bulbs will continue to burn. Most circuits in the home are of this type.

AND and OR gates

In the circuits just described, light bulbs were connected in series or in parallel. Switches may also be connected in series or in parallel. Look at figure 11.13. For the bulb to light, both switch A and switch B have to be closed. In figure 11.14, how many switches must be closed to close the circuit?

In **electronics,** switches connected in series are called an AND gate. Electronics is the use of electrically controlled parts to automatically control or change current in a circuit.

Can you think of a way in which we use AND gates? Look carefully at how an elevator works. Notice that there are two sets of doors. It is only when both the outside doors AND the inside doors are closed that the electrical circuits are complete. Only then will the elevator move.

Now, what happens when switches are connected in parallel, figure 11.15? Switch A OR switch B will complete the circuit and turn on the light. Switches in parallel are called OR gates.

Many homes contain an OR circuit. One example is a doorbell. Push-button switches are

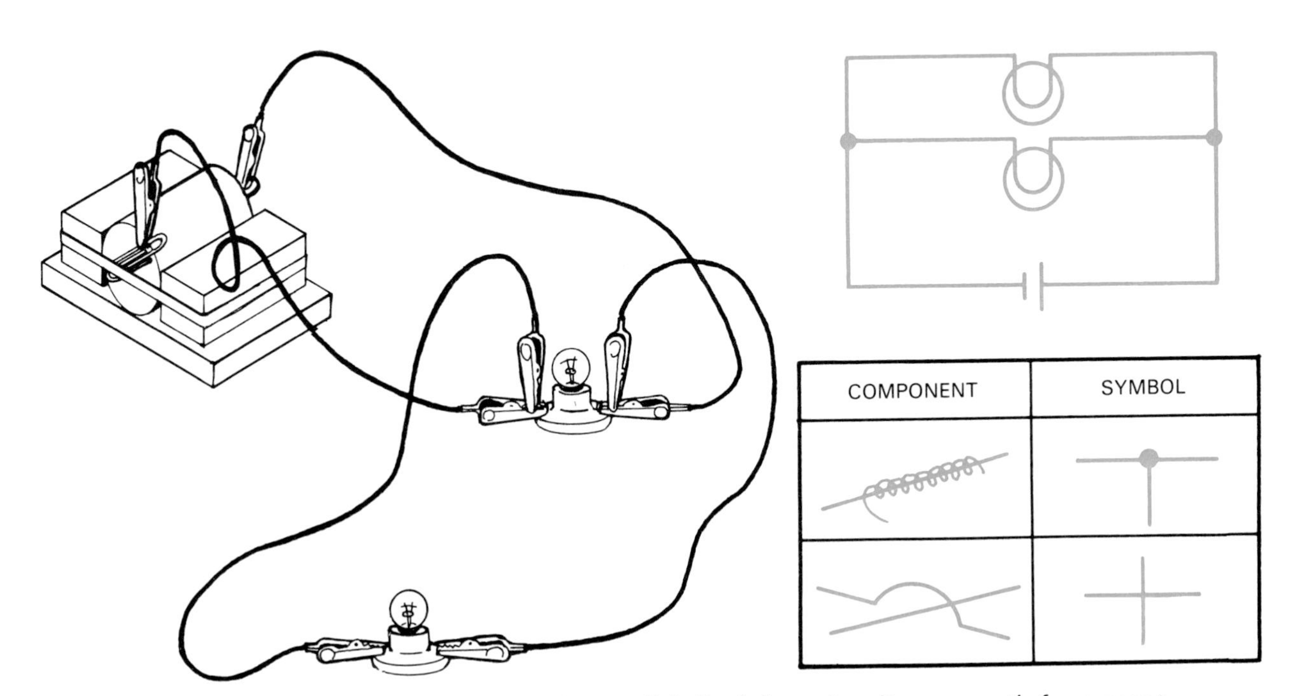

Figure 11.12. Lamps connected in parallel. Each lamp has its own path for current.

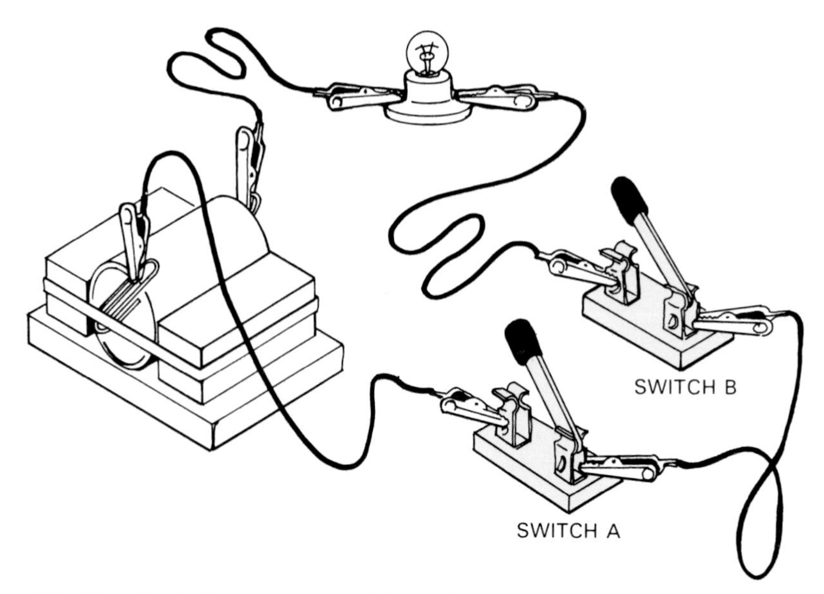

Figure 11.13. Switches connected in series. This circuit will conduct current only when both switches are closed.

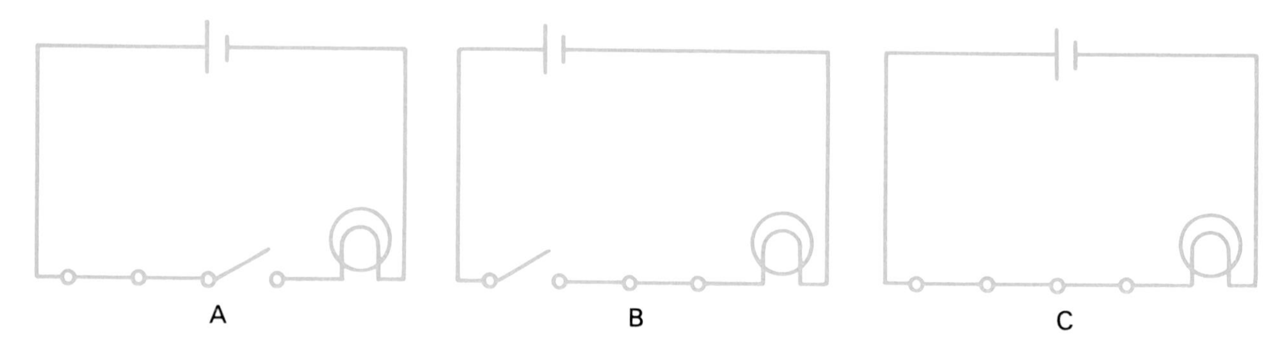

Figure 11.14. In which circuit will the bulb light?

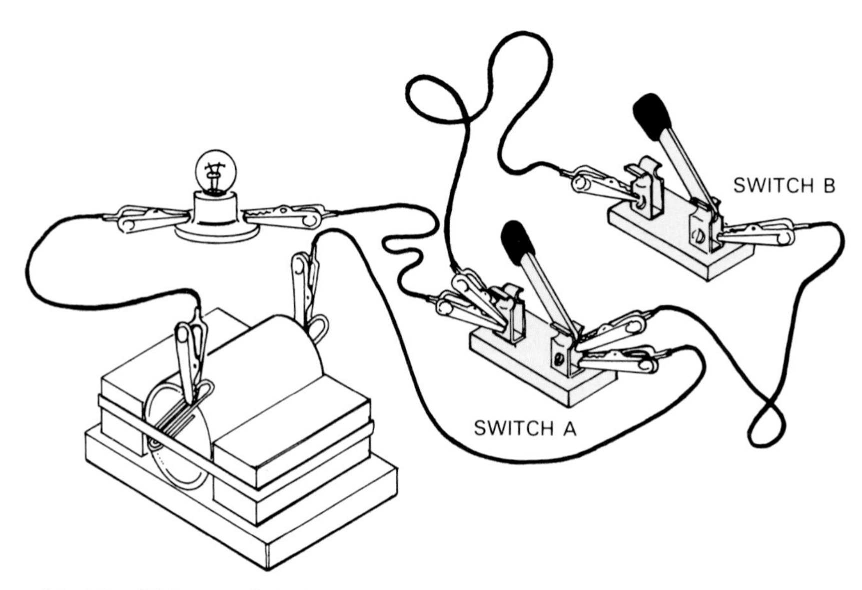

Figure 11.15. This parallel circuit will conduct current when either switch A or switch B is closed.

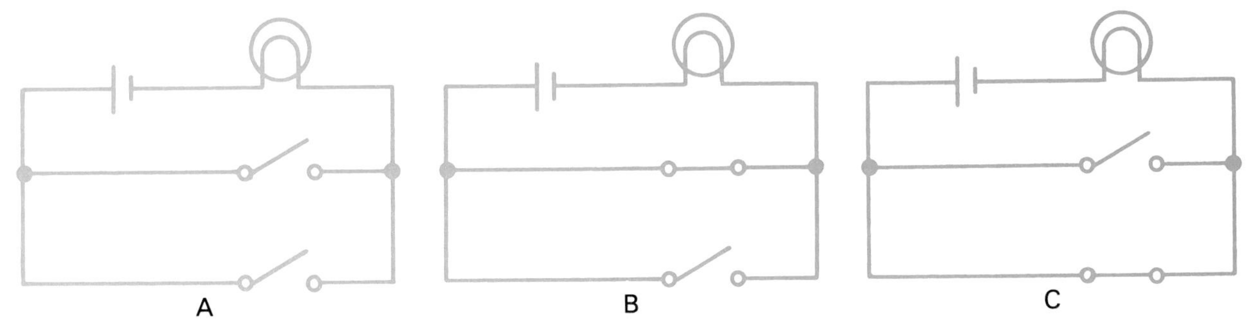

Figure 11.16. In which circuit or circuits is the bulb lit?

located at both the front and back doors. Pressing one OR the other will ring the bell. Look at the diagrams in figure 11.16.

■ CONDUCTORS AND INSULATORS

Materials that allow electric current to flow easily are called conductors. Copper, aluminum, silver, and most other metals are examples of good conductors. Copper is used most often for house wires. Low cost, strength, and low resistance to current make it a good choice, figure 11.17.

Materials that do not allow current to pass are called **insulators.** Glass, rubber, plastic, porcelain, and paper are good insulators. Insulators play an important part in controlling electricity. They are wrapped around a conductor to prevent it from passing current to another conductor. This keeps the current in the correct path.

One of the most important uses of insulators is to protect us from electric current, figure 11.18. Our bodies, especially when wet, will conduct electricity. If you touch a live wire by accident, you will get a dangerous shock. If your hands are wet, the shock may kill you.

Some materials are better conductors than others. Still, almost all materials have some resistance to the flow of electricity. Therefore, electricity produces heat as it forces its way through this resistance. The greater the resistance, the more heat produced. This can be used to advantage. A resistance wire can be used to produce heat. The most common type of resistance wire is an alloy of nickel and chromium. It is called nichrome wire. Sometimes the resistance wire becomes red hot as in an electric stove, toaster, and other heating applianes, figure 11.19. Sometimes the conductor becomes white hot. This is the case with an electric light bulb.

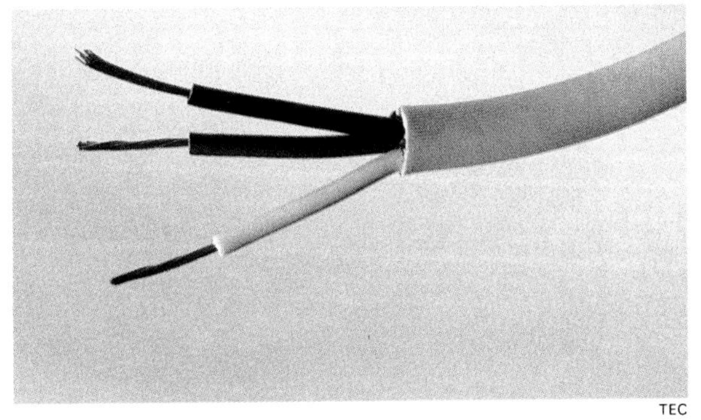

TEC

Figure 11.17. Copper is a good conductor of electricity.

ECRITEK

Figure 11.18. Insulators protect against electric shock. How is insulation being used in this photograph?

Semiconductors

Semiconductors are materials that do not conduct as well as copper or silver. Nor do

Figure 11.19. An electric heater works because an electric current makes a high-resistance wire red hot.

Class	Materials
Conductors	• Copper • Silver • Tungsten • Nichrome
Semiconductors	• Silicon • Germanium
Insulators	• Rubber • Nylon • Glass • PVC • Porcelain • Mica

Figure 11.20. Semiconductors are materials that have some properties of both conductors and insulators.

they insulate as well as rubber and glass, figure 11.20. They have some characteristics of each.

These traits are useful in electronic circuits. They allow electron flow only under certain conditions. The most common semiconductor materials are silicon and germanium. Transistors and other electronic components are made from them.

Superconductivity

For many years technologists have dreamed of producing a material that will conduct elec-

Figure 11.21. Comparing electricity with water. A—Water behind a faucet is always under pressure. B—Electricity behind an outlet is also under pressure. C—To use water, you must attach a hose and turn on the faucet. D—To use electricity, you must connect a wire to an outlet.

tricity without resistance. Such a material is called a superconductor. Until recently superconductivity was possible only at low temperatures close to absolute zero, −459 °F (−273 °C). Very recently, materials have been discovered that work at higher temperatures. Researchers believe that, within a short time, superconductor materials will be found to operate at room temperatures.

When these superconductors become widely available, they will revolutionize the electronics industry. The absence of electrical resistance reduces the amount of heat produced. This allows components to be packed more closely together. This reduces the size of components and products. Superconductors will also enable computers to operate at much greater speeds.

□ MEASURING ELECTRICAL ENERGY

When measuring electrical energy, three terms are important: **volts, ohms,** and **amperes.** One way to understand their meaning is to compare electricity to water. Water flows through a hose under pressure. Electricity inside a wire is also under pressure. With electricity the pressure is called voltage and is measured in volts. If pressure is high, more water, or electricity will flow, figure 11.21.

As the water flows through the pipe it meets resistance. The smaller and longer the pipe, the greater the resistance. So it is with electricity. Flow is affected by diameter and length of the wire. Also, electricity flows more easily through some materials than through others, figure 11.22. Electrical resistance is measured in ohms.

The amount of water that flows out of the end of the pipe in a given period depends on pressure and resistance, figure 11.23. The higher the pressure and the weaker the resistance, the greater the amount of water leaving the hose.

Ohm's Law

The amount of electricity that passes a point in a conductor in a given period also depends on pressure and resistance. This electrical current is measured in amperes (A). The higher the pressure (volts) and the weaker the resistance (ohms) the higher the current. This relationship between volts, ohms, and

Figure 11.22. Left. The greater the resistance, the smaller the flow of water. Right. Why does an air conditioner require a larger diameter wire than a hair dryer?

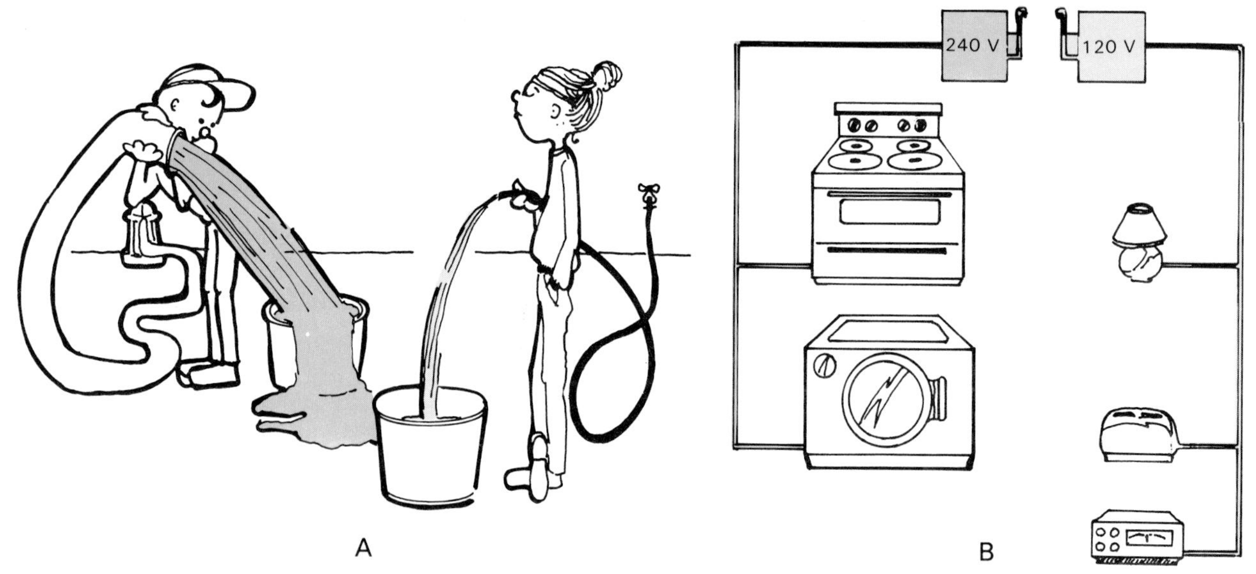

Figure 11.23. Comparing high pressure and low resistance of water and electricity. A—A large hose and high pressure equal heavy water flow. B—High voltage and a large conductor equal high electron flow.

amperes is described by a formula known as Ohm's Law. Ohm's Law is written as:

$$\text{current} = \frac{\text{voltage}}{\text{resistance}} \text{ or amperes} = \frac{\text{volts}}{\text{ohms}} \text{ or}$$

$$I = \frac{V}{R}$$

Watts

Another unit of electrical measurement is the **watt.** A watt is the unit used to measure the work performed by an electric current. To calculate the power (P) in watts, multiply the voltage by the current.

$$\text{Watts} = \text{voltage} \times \text{current or Watts} = \text{Volts} \times \text{amperes or } P = V \times I$$

The monthly electricity bill for your home is based upon the number of watts used, figure 11.24. The utility company provides a meter for each home. It measures how many watts are used. The watt is a small unit, so the basic unit used by power companies is a kilowatt. (This is equal to 1000 watts.)

Most appliances are being constantly switched on and off. The electricity used is measured over periods of one hour. The unit is therefore called one kilowatt hour (kWh). A kilowatt hour means 1000 watts used for a period of one hour.

TEC

Figure 11.24. This meter measures the amount of electricity used in kilowatt hours (kWh).

Figure 11.25 shows the dials of a typical electrical energy meter. When reading the dials, be careful because some of them revolve clockwise and some revolve counterclockwise. To read a dial, write down the number the pointer has passed. In figure 11.25 the correct reading is 23642.

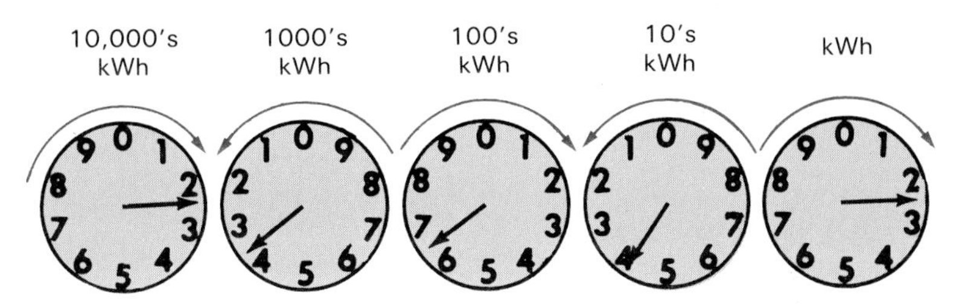

Figure 11.25. What is the reading of the meter at your home?

☐ ELECTRONICS

In this chapter, you have learned about the flow of electrons in a circuit. **Electronics** is the technology of controlling electron flow. Electrons can be used to control, detect, indicate, measure, and provide power. These functions are carried out by using a variety of electronic components.

Resistors

Resistors are electrical devices that control how much current flows through a circuit. Resistors make it more difficult for a current to flow. In figure 11.26 bulb A will be brighter than bulb B. The current in bulb B is smaller because there is a resistor in that loop of the circuit.

Resistors are made in many sizes and shapes, figure 11.27. All do the same thing; they limit (or resist) current. In a typical carbon composition resistor, powdered carbon is mixed with a gluelike binder. The resistance is changed by changing the ratio of carbon particles to binder. The greater the amount of carbon, the less the resistance, figure 11.28.

Resistors are made in a wide range of values. (This is the degree to which they limit the flow of electrons.) Resistors are often quite small. It would be difficult to write their values on them. To overcome this problem, resistors are usually marked with four colored bands, figure 11.29.

You can calculate the value of a resistor from the first three bands. To read the value, hold the resistor with the colored bands to the

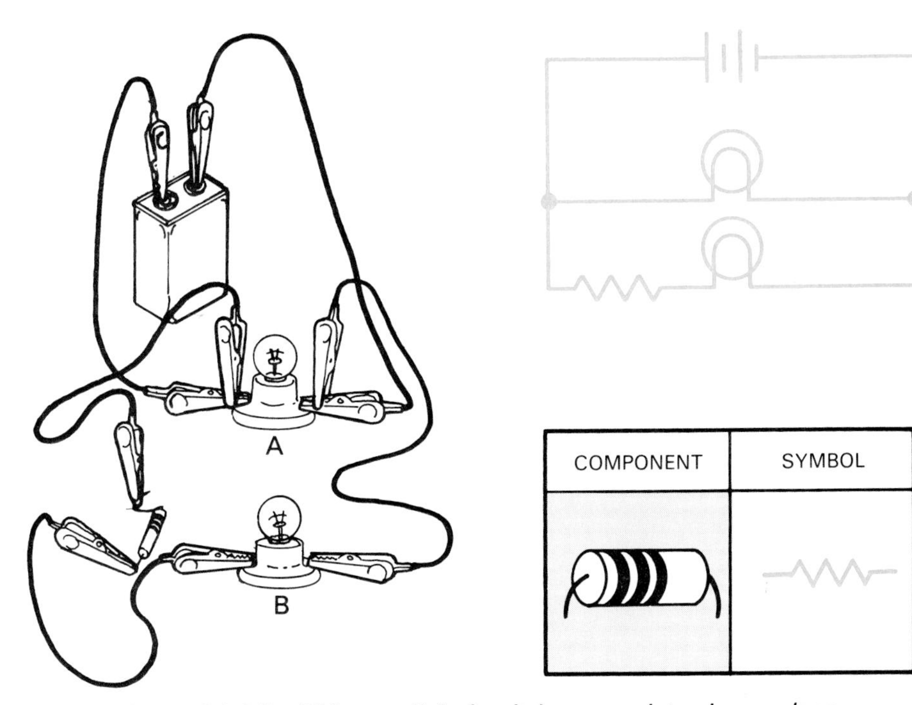

Figure 11-26. This parallel circuit has a resistor in one loop.

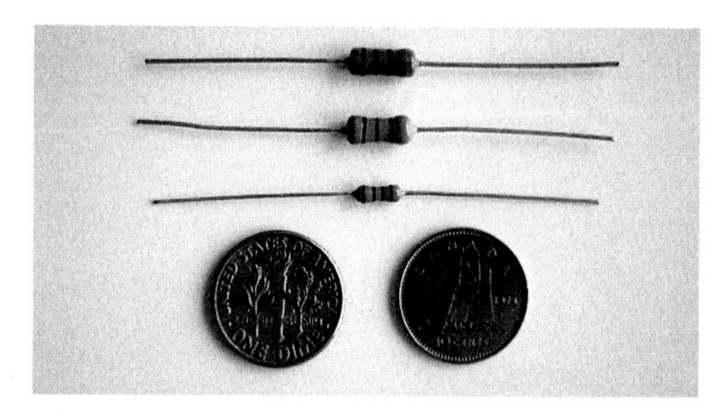

Figure 11.27. Resistors come in different shapes and sizes. They are often quite small.

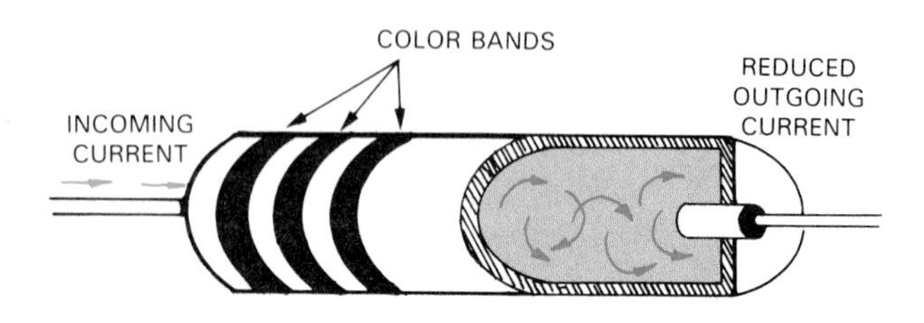

Figure 11.28. Inside a resistor. Note how current is having difficulty moving through it.

left. Then using the table in figure 11.29, calculate the value of the resistor.

In figure 11.30 the first band of the transistor is red. The table shows this means the first number in the resistance value is 2. The second band is yellow. The second number in the resistance value is 4. The third band is brown. This means that one zero follows the first two numbers. The value of this resistor is, therefore, 240 Ω (ohms). If the value were larger, for example, 47,000 Ω, its value would be written as 47 k (this is a simpler way of writing 47 k Ω. This k stands for ''kilo'' or thousand). A resistor whose value is 47,000,000 Ω is written as 47 M. (This stands for 47 M Ω. The M means ''mega'' or million.)

Resistors have a fourth band that is usually silver or gold. It indicates the accuracy or tolerance of the resistor. The fourth band of the resistor in figure 11.30 is silver, indicating that the resistor has a tolerance of ± 10 percent. Ten percent of 47,000 is 4,700. Therefore the actual value is between 42,300 and 51,700 Ω.

Color	1st Band	2nd Band	3rd Band	4th Band
Black	0	0	Ω	
Brown	1	1	1 zero	
Red	2	2	2 zeros	
Orange	3	3	3 zeros	
Yellow	4	4	4 zeros	
Green	5	5	5 zeros	
Blue	6	6	6 zeros	
Violet	7	7	7 zeros	
Gray	8	8	8 zeros	
White	9	9	9 zeros	
Gold				5
Silver				10
None				20
	BAND 1	BAND 2	BAND 3	BAND 4

Figure 11.29. Reading resistor color bands.

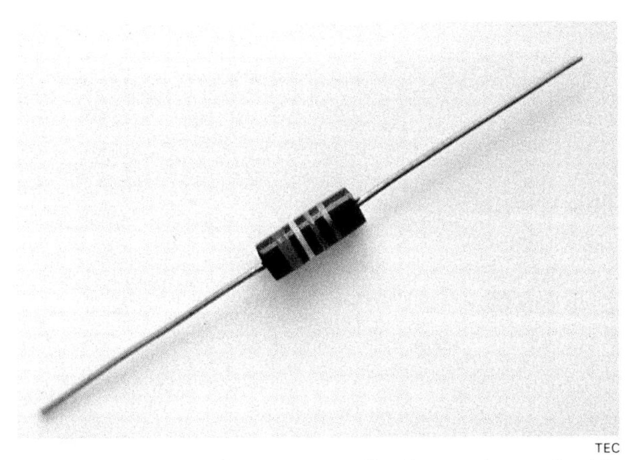

Figure 11.30. Can you tell the value of this resistor? Use the chart in figure 11.29.

Variable resistors are of two types: rheostats and potentiometers. Rheostats lower or raise the current in a circuit, figure 11.31. The dimmer switch used for car dashboard lights is one example. Potentiometers lower or raise the voltage, figure 11.32. They are used as volume controls in stereos.

Diodes

Diodes are devices that allow current to flow in one direction only. There are two ends to a diode: positive and negative. A dark band indicates the negative end, figure 11.33.

Figure 11.34 shows two bulbs in parallel. Each has a diode in its loop of the circuit. On-

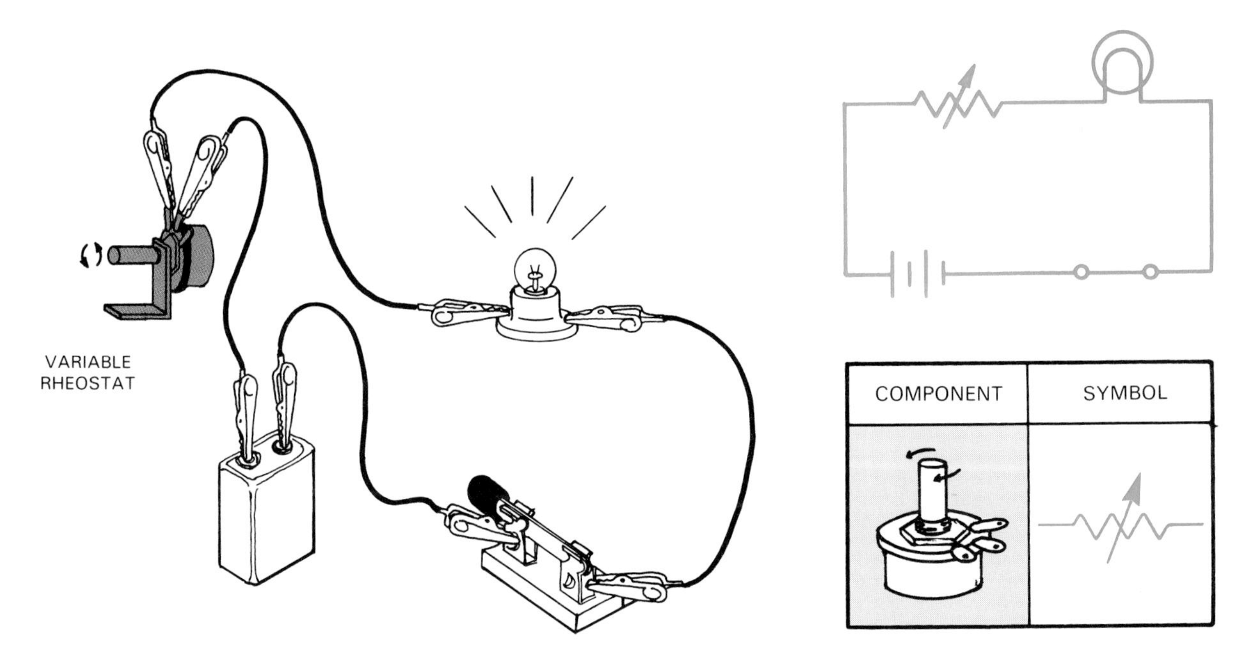

Figure 11.31. A rheostat is connected in series.

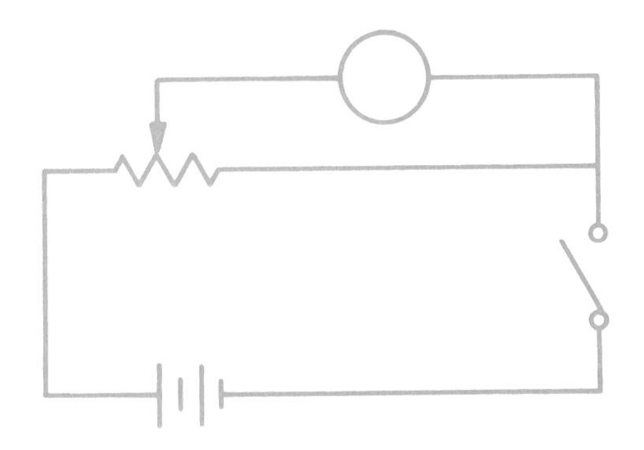

Figure 11.32. A potentiometer is connected in parallel.

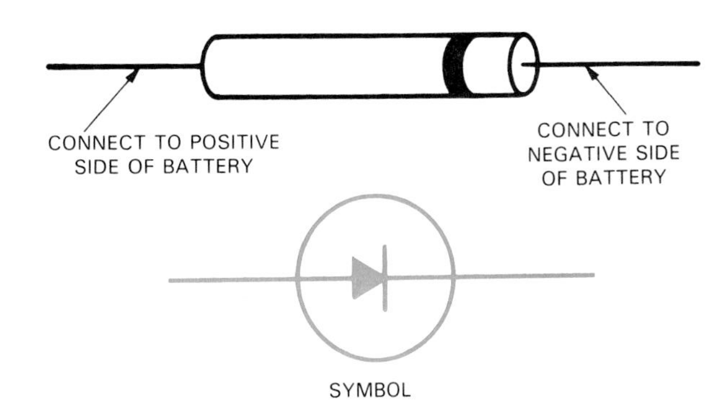

Figure 11.33. A diode. How would you connect it in a circuit?

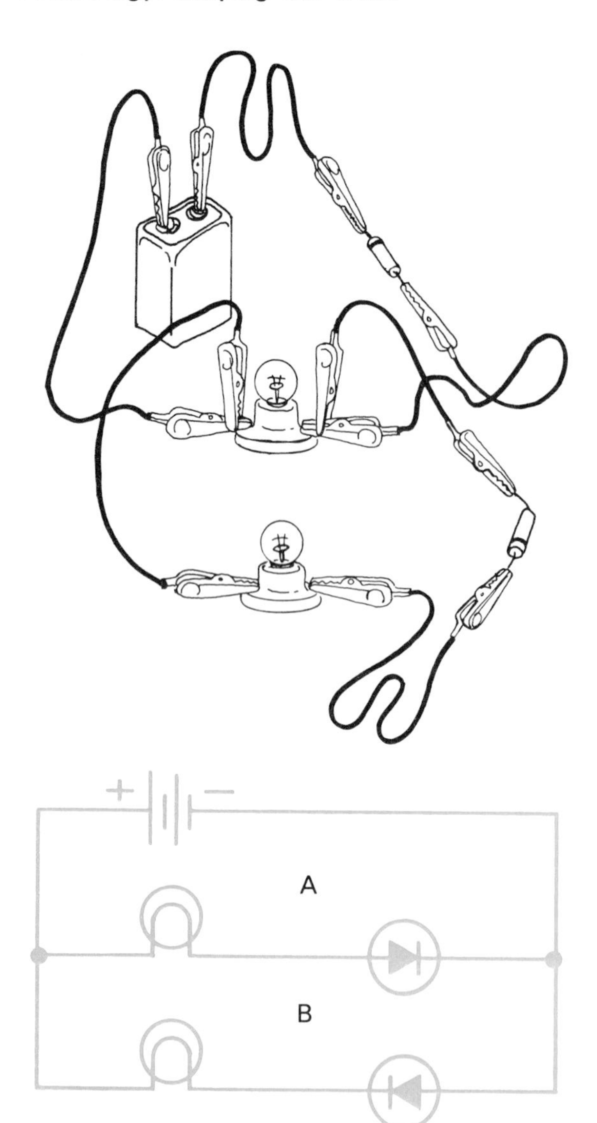

Figure 11.34. A circuit is set up with two lamps in parallel. Each has a diode, but only one lamp will light. Explain why.

ly bulb A will light because the diode next to it is positioned correctly to allow electrons to flow. Diodes are most commonly used as rectifiers. They are used to change alternating current to direct current.

Another type of diode is a light emitting diode (LED). LEDs also only conduct in one direction. They need less current to make them glow than do bulbs but they are not as bright. LEDs, therefore, are used where the brightness of the bulb is not important. They show that electrical equipment is turned on and working, figure 11.35.

ECRITEK

Figure 11.35. How many light emitting diodes do you see in this sound system?

Capacitors

Capacitors are designed to store an electrical charge. The simplest capacitor is made of two metal plates (conductors) separated by an insulator, figures 11.36 and 11.37. The insulator may be air, but is often a thin sheet of plastic.

A capacitor will smooth the pulsating current from a DC rectifier into a steady direct current. A capacitor connected in a DC circuit can store a charge for a considerable time after the voltage to the circuit has been switched off. NEVER touch the leads of a capacitor before it has been discharged.

Transistors

A **transistor,** figure 11.38, switches an electric current on and off. It is like an electric light switch; the switching, however, is done by voltage, instead of by hand. A transistor can

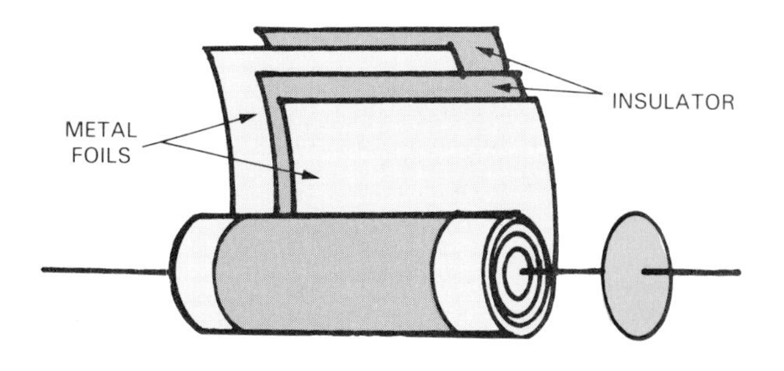

Figure 11.36. A capacitor is a "sandwich" made up of conductors and insulators.

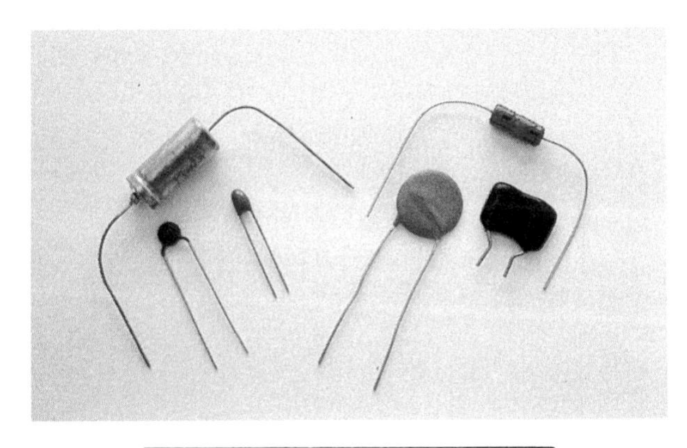

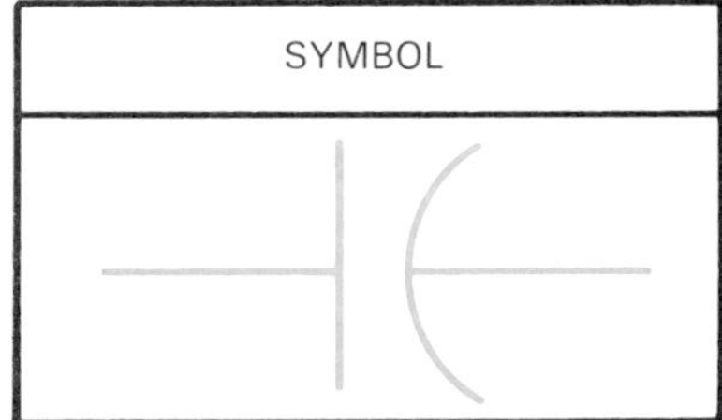

Figure 11.37. Different types of capacitors.

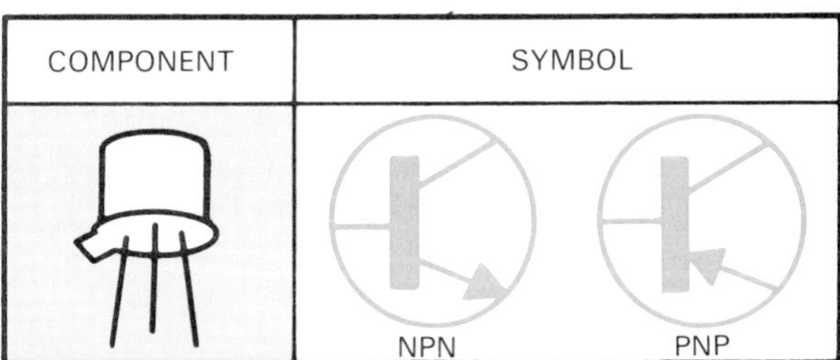

Figure 11.38. There are two types of transistor, a PNP and an NPN. Negative voltage controls the output of a PNP transistor. Positive voltage controls the NPN transistor.

be used to switch the current to an electric lamp on and off. The lamp will not light unless electrons can flow through the circuit by way of the transistor.

All transistors have three terminals: a collector (c), a base (b), and an emitter (e). An electric current will flow through the transistor only when an electrical voltage is applied to the base. For the base to have a voltage it must be connected to the positive side of the battery, figure 11.39. If connected to the negative side of the battery, the base will have a low voltage and the lamp will not light, figure 11.40. Therefore, the lamp can be switched on and off by changing the voltage on the base of the transistor.

Many transistors are fast switches. They are used in timing, counting, and computer circuits. In these circuits, the signal is either present (on) or is not present (off). Amplifying transistors are used in radios and stereos

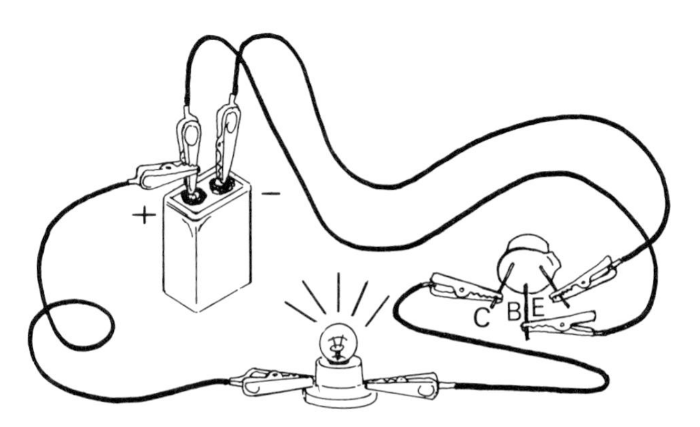

Figure 11.39. The base terminal of the transistor is connected to the positive terminal of the battery. The lamp will light.

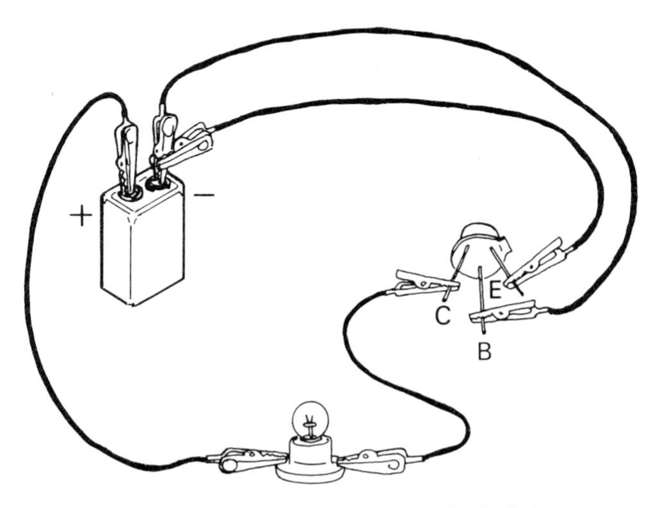

Figure 11.40. With the base terminal of the transistor connected to the negative terminal of the battery, the lamp will not light.

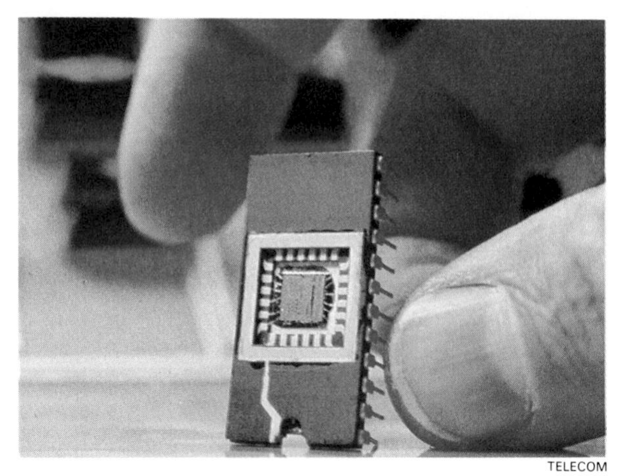
TELECOM

Figure 11.41. An integrated circuit. This term is usually shortened to "IC."

where a weak signal must be amplified (increased) in order to be heard over a speaker.

Integrated circuits

The electronic circuits described in this chapter have been built using separate components. These components have included resistors, capacitors, diodes, and transistors. Since the early 1960s it has been possible to build single electronic components, called **integrated circuits.** These replace a whole group of separate components. Today, one integrated circuit may contain the equivalent of about 1,000,000 components.

Most integrated circuits are built on a tiny piece of silicon called a silicon chip, figure 11.41. Resistors, capacitors, diodes, and transistors, with their connections, are formed in miniature on the surface of the chip. We owe our advances in computer technology to the development of integrated circuits.

SUMMARY

In order for any electrical device to operate, it must contain one or more circuits. An electric circuit is a continuous path from a source, through a load, and back to the source. Circuit diagrams use symbols to show components. There are three basic types of circuits: simple, series, and parallel. Direction of current flow in a circuit may be shown in terms of electron flow or conventional current flow.

Materials that allow electrons to flow are called conductors. Those that do not allow electron flow are called insulators. Between these two are materials known as semiconductors. They allow current under certain conditions.

Electrical energy is measured in terms of volts, ohms, and amperes. Work performed by an electrical current is measured in watts.

Electronics is the study of the way in which electron flow may be controlled. Control is carried out using electronic components. These include resistors, diodes, capacitors, and transistors. Integrated circuits replace a large number of these separate components.

KEY TERMS

Ampere
Capacitor
Diode
Electric circuit
Electronics
Insulator
Integrated circuit
Load
Motor
Ohm
Overload
Parallel circuit
Resistance
Resistor
Series circuit
Semiconductor
Short circuit
Transistors
Volt
Watt

TEST YOUR KNOWLEDGE

Write your answers to these review questions on a separate sheet of paper.

1. An electric circuit can be defined as ____ ________________
2. Draw the symbol for (a) a light bulb, (b) a dry cell, (c) a fuse, and (d) a switch.
3. Draw a circuit diagram of a simple circuit which includes a light bulb, dry cell, fuse, and switch.
4. If too many appliances are plugged into the same receptacle the circuit may become ________.
5. If two bare wires carrying a current touch each other a ________ ________ occurs.
6. Describe how a fuse protects a circuit from overload.
7. An alternative to a fuse for protecting the circuits in a home is a ________ ________.
8. Using diagrams and notes, explain the difference between electron flow and conventional current flow.

9. Draw a circuit diagram to show two light bulbs and a dry cell that are connected in series.
10. Draw a circuit diagram to show two light bulbs and a dry cell connected in parallel.
11. Is it better to connect your Christmas tree lights in series or parallel? Explain your answer.
12. A circuit in which two switches are connected in series is called a(n) ________.
    A circuit in which two switches are connected in parallel is called a(n) ________.
13. A conductor is a material that ______
    ________________________________.
    An insulator is a material that ______
    ________________________________.
    A semiconductor is a material that __
    ________________________________.
14. Name four materials which are conductors of electric current and four insulators.
15. Electric pressure is measured in ______.
    a. Joules.
    b. Watts.
    c. Ohms.
    d. Volts.
16. Electric resistance is measured in ______.
    a. Ohms.
    b. Amperes.
    c. Watts.
    d. Volts.
17. Electrical current is measured in ______.
    a. Joules.
    b. Volts.
    c. Amperes.
    d. Watts.
18. A portable electric heater with a resistance of 15 ohms is connected to a 120 volt AC outlet. The current flow in the circuit will be ________.
19. What is an integrated circuit?
20. Give the value of each of the following resistors.

	1st band	2nd band	3rd band	4th band	Value
(A)	red	green	yellow	silver	
(B)	orange	blue	brown	gold	
(C)	white	brown	red	none	
(D)	violet	green	orange	silver	

APPLY YOUR KNOWLEDGE

1. State the advantages of a parallel circuit over a series circuit.
2. Collect samples of different materials. Design and build a method to test each sample to determine if it is an insulator or a conductor.
3. Sketch three different electronic components. Use colors where appropriate. State the function of each component.
4. Design and build a model of an electromagnetic crane to pick up scrap iron.
5. Design and build a land vehicle using a 6-9 V DC motor. The objective is to minimize the time taken to travel a distance of 30 ft. (9 m).

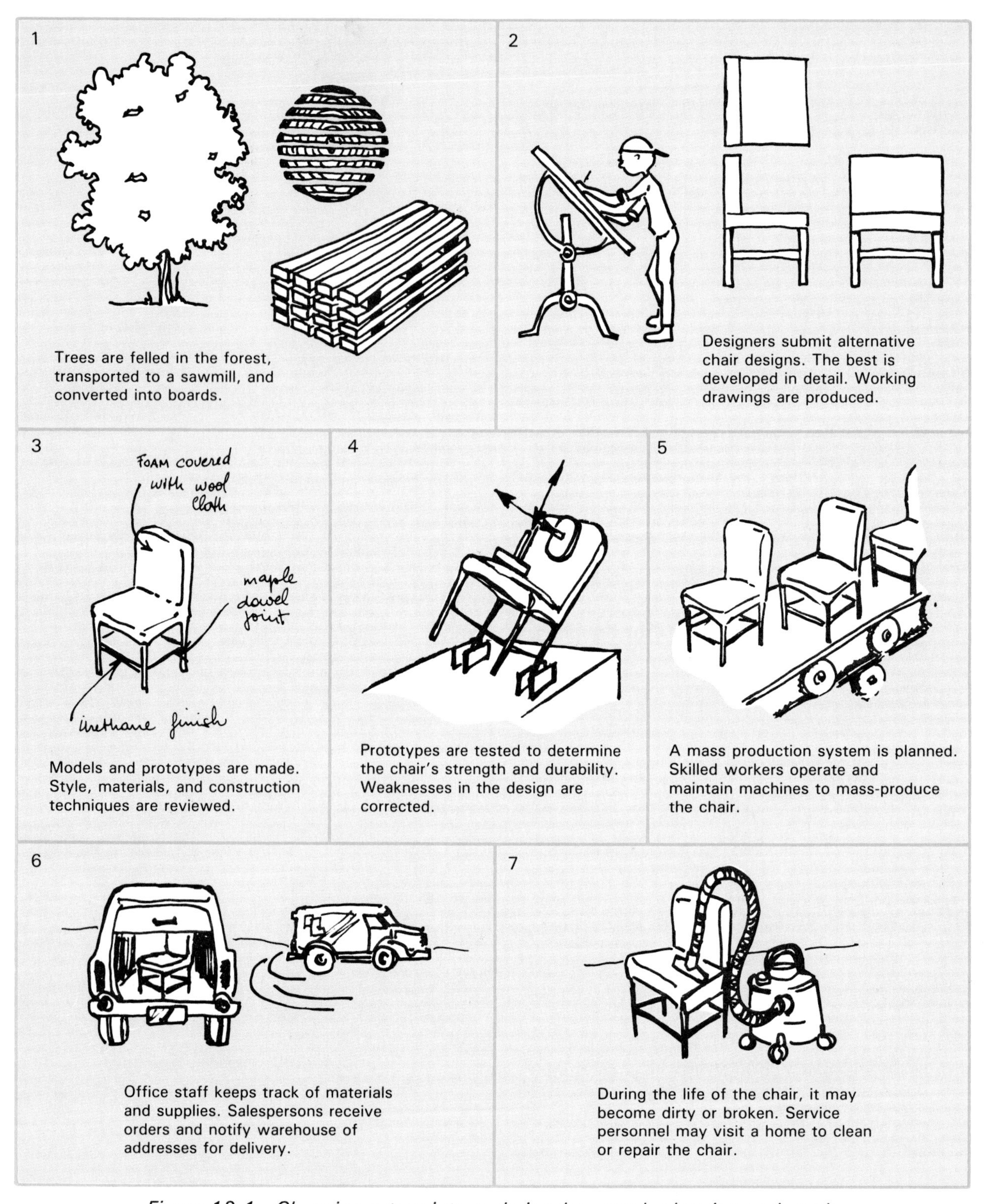

1
Trees are felled in the forest, transported to a sawmill, and converted into boards.
2
Designers submit alternative chair designs. The best is developed in detail. Working drawings are produced.
3
Foam covered with wool cloth
maple dowel joint
urethane finish
Models and prototypes are made. Style, materials, and construction techniques are reviewed.
4
Prototypes are tested to determine the chair's strength and durability. Weaknesses in the design are corrected.
5
A mass production system is planned. Skilled workers operate and maintain machines to mass-produce the chair.
6
Office staff keeps track of materials and supplies. Salespersons receive orders and notify warehouse of addresses for delivery.
7
During the life of the chair, it may become dirty or broken. Service personnel may visit a home to clean or repair the chair.

Figure 12.1. Changing a tree into a chair takes much planning and work.

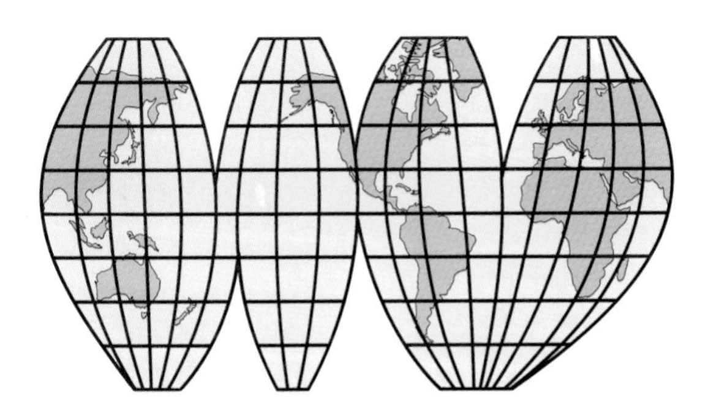

Chapter 12

The World of Work

OBJECTIVES

After reading this chapter you will be able to:

- Differentiate between primary and secondary materials processing.
- Compare traditional and modern manufacturing processes.
- State the purpose of mass production and identify its advantages and disadvantages.
- Describe the elements of a production system.
- List the major processes used to change standard stock into finished products.
- Describe how new technologies have replaced, outmoded, or created jobs.
- Explain the recent growth of the tertiary sector.
- Discuss the range of careers in the primary, secondary, and tertiary sectors.

The wood used to build the chairs you have in your home came from a tree growing in a forest. What is involved in changing a tree into a chair?

From figure 12.1 you can see that producing a chair from a tree involves many steps. The steps in producing any product are organized into three sectors:

- **Primary** (first) **sector:** obtaining and processing raw materials.
- **Secondary** (second) **sector:** changing raw and processed materials into a product each of us can use.
- **Tertiary** (third) **sector:** delivering and servicing the product.

■ THE PRIMARY SECTOR: PROCESSING RAW MATERIALS

The primary sector is concerned with obtaining and processing **raw materials.** These materials come from nature in one form or another. Some materials are renewable and can be reproduced continually. Others are nonrenewable; once used, they cannot be replaced.

Renewable materials

Renewable raw materials come from plants or animals, figures 12.2 through 12.7. Some are found in a wild state. Others are produced on farms. For instance, forests and tree farms provide wood for the lumber industry. The fishing industry harvests fish and other marine life from oceans, lakes, and waterways. Wild animals are hunted and trapped for their furs, hides, and meat.

Nonrenewable raw materials

Nonrenewable raw materials are of three types: fossil fuels, nonmetallic minerals, and metallic minerals. Fossil fuels include coal, peat, petroleum, and natural gas. These were once living organisms that decayed and were trapped in layers of sediment in the oceans and on land. Over millions of years the heat and pressure from the sediment created mineral fuels. In addition to providing fuel, petroleum is the raw material used in making many products. These include plastics, synthetic fibers, and drugs.

Nonmetallic minerals include such construction materials as sand, gravel, and building stone. They also include abrasive materials such as corundum and insulating materials such as asbestos. Among the metallic minerals are those from which iron, copper, and aluminum are extracted. Metallic minerals, also called ores, are removed from the earth

Figure 12.2. Lumberjacks harvest trees, a renewable raw material.

Figure 12.3. Logs are floated down rivers to be processed into usable form. Small logs may go to a mill to be processed into pulp for paper. Large logs will go to a sawmill to be sawed into lumber.

Figure 12.4. Log booms are towed by a tug boat.

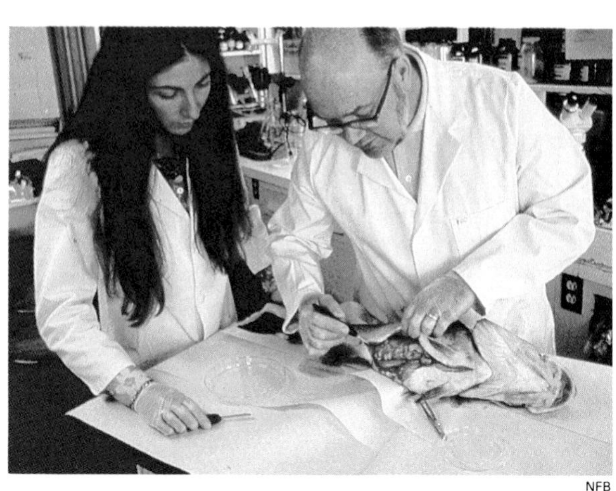

Figure 12.5. Biologists conduct research aimed at improving fish harvests.

Figure 12.6. Colombian fishermen hauling in their nets.

Figure 12.7. Trapping is another method of harvesting materials.

by surface mining or underground mining. Figures 12.8 through 12.11 show several methods of extracting nonrenewable materials.

Careers in the primary sector

People with careers in the primary sector frequently work outside, figure 12.12. (A **career** is an occupation, a way of making a living.) Farmers plant, fertilize, cultivate, and harvest crops. Ranchers breed, feed, and care for animals. Horticulturalists grow and maintain plants, shrubs, and trees. Forestry workers cut, transport, and process trees for papermaking, construction, and furniture manufacturing. Geologists locate minerals and fossil fuels. Miners and oil riggers operate the machinery to extract raw materials. Oceanographers study the plant and animal life of the oceans. Environmentalists try to find solutions to problems relating to land use, pollution, conservation of natural resources, and the preservation of wildlife.

■ THE SECONDARY SECTOR: MANUFACTURING PRODUCTS

The secondary sector changes raw and processed materials into useful products. It is concerned not only with the manufacture of products, but with the construction of structures. Today's manufactured products include

SEEDS

Figure 12.8. Coal lying near the earth's surface can be strip mined.

NFB

Figure 12.10. An open pit mine. This type of mining is cheaper and safer than underground mining.

SEEDS

Figure 12.9. Offshore drilling rigs drill wells in the ocean's floor.

TEC

Figure 12.11. Sometimes fossil fuels are locked in mineral deposits called tar sands. Strip mining removes these deposits from the ground.

computers, jet planes, glues, lasers, plastics, medications, photocopying machines, and bubble gum. Construction involves the building of structures that people use for living, working, traveling, and playing. Among these structures are houses, office towers, and sports stadiums. Also included is the construction of road tunnels, bridges, towers, and dams.

The secondary sector has changed in the last 200 years. It has evolved through three stages:

- The individual artisan.
- Mechanization and **mass production.**
- **Automation.**

The individual artisan

Before the 18th century, the entire production of articles was in the hands of individuals. One person, or a small group in a village, made products. They used hand methods. Each product evolved by trial and error. One generation passed on acquired skills to the next through an apprenticeship system. Each artisan was normally responsible for every step in the process. He or she did everything from obtaining the raw materials to completing the finished product. For instance, figure 12.13 illustrates the work of a chair bodger. The bodger, working alone, produced Windsor chairs.

Mechanization and mass production

The second stage in the evolution of the secondary sector divided the production process into specialized steps. Machinery replaced hand work. This was done to reduce the unit cost of the product. Through mass production, products could be made in large quantities. This change, occurring first in England, was known as the **Industrial Revolution.**

Early machines, powered by steam, replaced the muscle power of workers and animals. This steam was produced by burning coal. The English had known the uses of coal for several centuries. However, Watt's steam engine was the first machine to convert the chemical energy of coal into steam and then into mechanicl energy (energy of motion). Mechanical energy then powered machinery.

PRIMARY SECTOR CAREERS

OCCUPATIONAL CLUSTER	PROFESSIONALS	SKILLED WORKERS
Agriculture	agricultural economist agronomist animal physiologist botanist soil conservationist soil scientist veterinarian	animal breeder animal inspector beekeeper dairy farmer farm equipment operator field crop farmer fruit farmer livestock rancher poultry farmer ranch worker
Horticulture	landscape architect	greenhouse manager groundskeeper landscape contractor landscape gardener lawn service worker nursery worker tree surgeon
Forestry	botanist	chainsaw operator logger sawyer
Marine science	oceanographer marine biologist marine geologist	fisher fish farmer
Natural resource extraction	geologist petroleum engineer	miner oil rigger
Environmental control	conservationist ecologist meteorologist urban planner waste water treatment engineer	

Figure 12.12 Which of these primary sector careers would you choose?

Factories. As a result of the increased use of steam-powered machinery, production had to be located in larger buildings called factories. Towns and cities developed rapidly as people moved to live near these factories.

At first, factories had to locate near the coal fields. Moving the coal long distances was too costly. After the mid-nineteenth century, however, transportation became cheaper. Coal could be more readily transported. Factories could be built almost anywhere.

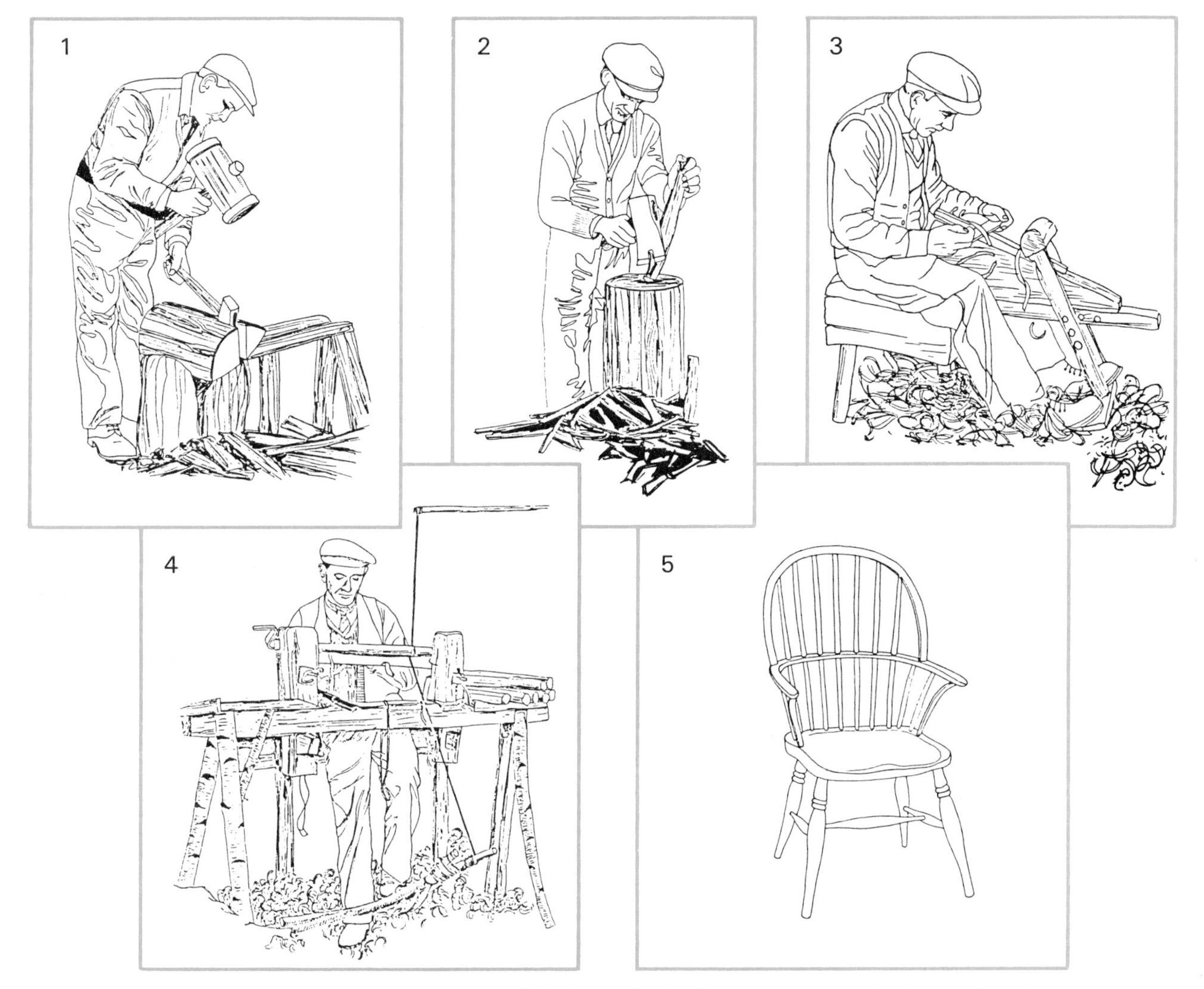

Figure 12.13. Steps in making a Windsor chair. One person made the entire chair.

The factory system expanded because it was thought to be more efficient. Efficiency means that good use is made of energy, time, and materials. This efficiency was achieved through the use of new methods of production. These included:

- The division of labor.
- The use of machines to build parts.
- The use of interchangeable parts.
- The introduction of mass production and assembly lines.

The division of labor means each person is assigned one specific task in the making of a product. Through constant repetition, the worker becomes skilled in that task. The task can be performed at a more rapid rate. With specialized machinery, the same part can be made again and again with no variation in size or shape.

Because more products could be produced in the same amount of time, the cost for producing each item dropped. This meant the item could be sold for less, too. That all parts were alike was important. It meant any new part could be substituted for one worn or broken. Such parts are said to be interchangeable. Each new part is identical to the old one. As we shall see later in this chapter, assembly lines further increased the efficiency of a factory.

Factory working conditions. When efficiency was the only concern of factory owners, working conditions were often poor. Machinery improved but little was done to improve life for the workers. The factories were built very close to one another. Their interiors were dark and dismal. No provisions were made for proper lighting. Children as young as four or five cleaned the floors. Older children worked 12 hours a day cleaning and maintaining machinery. They were poorly fed, ill-treated, and completely at the mercy of the factory owners.

Assembly lines. Use of assembly lines made factories more efficient. A continuous assembly line could quickly produce a large number of identical items. An assembly line allows assembly of parts in a planned sequence. This is possible only when parts are manufactured to uniform standards.

Assembly usually begins with one major part. Other parts are added as the product moves to other stations along the line.

Making a large number of products on an assembly line is called **mass production.** It was first used by Henry Ford in 1914. Ford's assemblers worked side by side in long lines. The parts were brought to them. Each worker had only one assigned task as the Model T cars moved slowly by on the assembly line. His line cut production time per car from 13 hours to one, figure 12.14.

Today, most automobile plants are huge assembly points. Each plant contains a number of small assembly lines. Some lines produce subassemblies. These are components having a number of parts. Engines, gear boxes, and suspensions are examples. Subassemblies are later assembled into complete automobiles.

While the assembly-line techniques pioneered by Ford increased the rate at which products could be produced, they were still limited. Humans could work only so fast. To produce products faster, one must use an automated assembly line.

Automation

In this stage, the worker builds, monitors, and maintains machines that make products. Computers or other machines control machine operations. Machines have been developed that:

- Perform many different shaping operations.
- Can receive a number of parts and assemble them at high speed in the correct sequence.

Typical is the machinery used to package macaroni, figures 12.15 through 12.18. The display packages have to be opened, filled, sealed, weighed, and boxed. Almost all of the operations are automated. Even on final inspection, machines check each package for the correct amount of macaroni.

FORD

Figure 12.14. Henry Ford perfected the first assembly line to produce his Model T car.

ECRITEK

Figure 12.15. An automated packaging line. Macaroni boxes arrive flat. They are opened and one end is sealed.

ECRITEK

Figure 12.16. The open ends of filled packages are glued and sealed.

Robots. The 1980s brought further development in automated manufacturing. Industrial **robots** revolutionized (completely changed) production lines. To understand the operation of a simple robot, imagine that you have been blindfolded and tied to a chair. You are able only to move one arm, and to rotate at the waist. Your own arm has joints at the shoulder, elbow, and wrist. Robots also have joints much like a waist, shoulder, elbow, and a wrist that can move in two or three directions. Each joint, or direction of movement, in a robot arm is called a **degree of freedom.** Most robots have five or six degrees of freedom, figure 12.19.

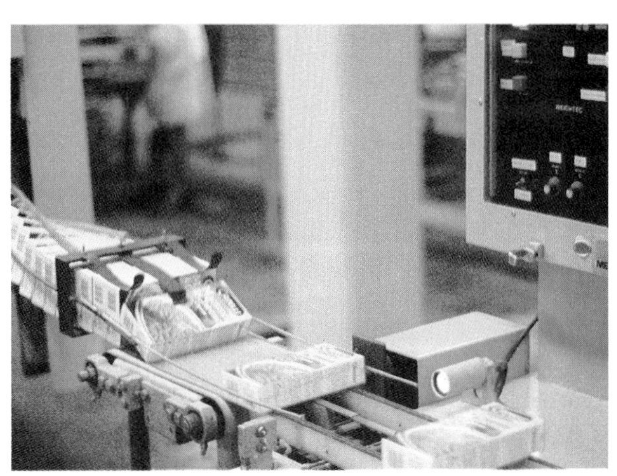

ECRITEK

Figure 12.17. Checking package weight. The machine removes from the conveyor packages that are either too light or too heavy.

ECRITEK

Figure 12.18. Packages are packed into cartons. Then they will move on to storage in preparation for shipment.

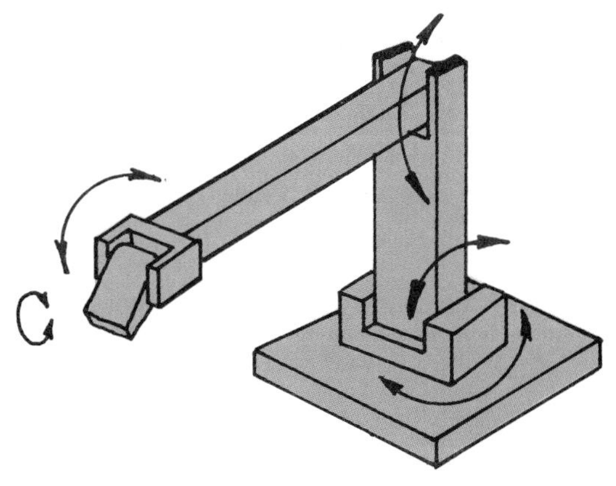

Figure 12.19. Arrows show the different movements of a robot arm. Each movement is known as a "degree of freedom."

Other robots can travel along a fixed path. They are used to transport materials to and from manufacturing operations.

Robots are computer controlled. They can be programmed to perform the following functions:

- Handling—loading and unloading components onto machines.
- Processing—machining, drilling, painting, and coating.
- Assembling—placing and locating a part in another compartment.
- Dismantling—breaking down an object into its component parts.
- Fixing—assembling objects permanently by welding or soldering.
- Performing dangerous operations.
- Transporting materials and parts or delivering mail.

A robot can be "taught" to perform an operation by leading it through the sequence of moves it has to follow. This is like taking someone by the hand to guide them through a strange place. As the robot arm is moved, it is possible to record positions into the computer memory by pushing a button or trigger on the robot arm.

Advantages of automation. Robots have several advantages over humans. They work better in hot, noisy, or dangerous situations. Robots do not take coffee breaks, go home ill, or sleep. They continue working 24 hours a day and will operate for thousands of hours before they require maintenance, figure 12.10.

Automation, including robotics, has certain advantages for industry:

- It has improved the quality of work. A machine set up to produce one product to a high standard will continue to produce parts to the same standard.
- It has increased production. Thus, the cost of each item is reduced. Each worker can produce more products in the same amount of time.
- It has decreased the amount of waste material and the number of parts that need to be scrapped. Each machine produces parts accurately.
- It has reduced the amount of time needed to teach a worker a new task when a new product is being produced. Each worker is responsible for only one small task.

Disadvantages of automation. Automation brings savings in time and money. Production increases. Still, it has some disadvantages.

- Now and then, a machine will malfunction (not operate as planned). It might have been incorrectly set up. Products will be inaccurate or substandard. Such mistakes, if not detected, can be very costly. Thousands of defective items may have been produced.
- The repetition of a small task for weeks or months can be boring. Workers often lose interest in such work. Dissatisfied with these conditions, they may negotiate (talk over and exchange ideas) either with their union or their employer. These negotiations, relating to working conditions, wage rates, or other benefits, may not be successful. If they do not lead to agreements or contracts, work stoppages (strikes) may occur.
- There is a loss of jobs as machines and robots take over tasks in the workplace.

Certain industries are more likely to be affected by automation. Those concerned with the assembly of finished goods, such as the car industry, rely more and more on automation. Without it, they cannot compete for world markets. They must modernize to keep costs down and improve quality.

How automation affects jobs. The prospect of being replaced by a machine frightens or angers many people. Yet, in the long run,

FORD

Figure 12.20. Robots are programmed to assemble automobiles on this automated assembly line.

machines are likely to create more jobs than they destroy. Anyone doubting this should think about times when similar revolutions have taken place. In the late 19th century, farm mechanization displaced more than two-thirds of the farm hands. Until 1880 more than half of all workers in advanced nations worked on farms. At that time, tractors and other machines were developed. These enabled one person to produce what 10 or more had done. In the last 100 years, about 90 percent of the farm jobs have disappeared.

Workers leaving the farms entered factories. They soon found that there, too, machines were replacing them.

Ford's mass production was another labor-saving technology. Again, it was feared that the assembly line would cause unemployment. This was not the case. Mass production created mass markets. Goods once affordable only to the rich, and thus made only by the thousands, could be bought by most people. Products began to sell in the millions.

Today, automation and industrial robots, figure 12.21, are as threatening to workers as earlier technology.

CNC and FMS. In many factories computers control machinery. The computer performs only a simple task. However, once its program of instructions has been written, it can do the same job again and again. Provided they do not break down, automatic machines can work day and night. Computer-controlled machines are often called CNC machine tools. The initials stand for **Computer Numerical Control.**

The most modern factories of today group their machine tools together into FMS cells. The letters FMS stand for **Flexible Manufacturing System.** An FMS cell consists of a number of CNC machine tools. Each machine is supplied with metal pieces, called "blanks." These are carried either on moving conveyor belts, or by special robots called Automated Guided Vehicles (AGVs).

As the metal blanks arrive at the FMS cell, robots lift them from the conveyor or AGV onto the machine tools. Each CNC machine tool completes its task. A robot transfers the machined blank to the next machine in the FMS cell. In this way, each metal blank is machined into a finished product. Then, a robot lifts the completed product onto another AGV or conveyor. It is either shipped or stored. A major advantage of an FMS cell is that it can be easily reprogrammed. It can then manufacture a totally different part or product.

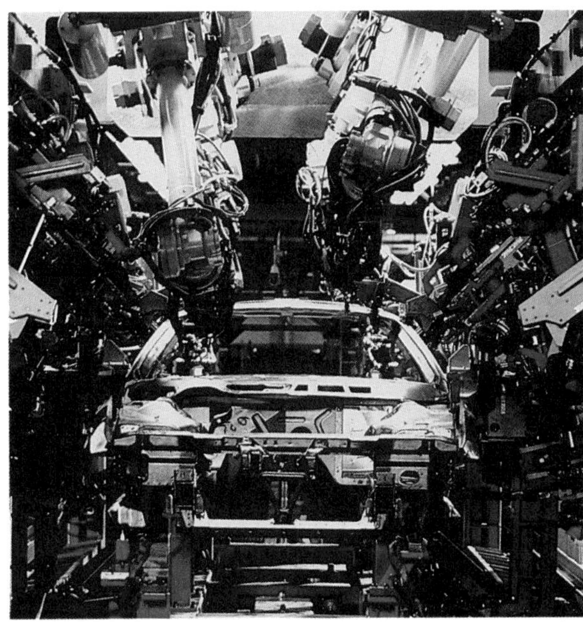

FORD

Figure 12.21. This robogate is equipped with 10 robots allowing the metal structure of the vehicle to be welded with high precision.

A factory with a number of FMS cells may have as few as three people operating it during the daytime. At night, only one or two people need be present. They check the computers and supervise the loading of the blanks. They also supervise removal of the finished parts.

Factories vary greatly in the extent to which they have been automated. There is also great variety in the type of product they make. However, all factories are similar in two respects. They must have a management system and a production system.

The management system

Operating a factory requires the combined effort of a team of people. The team may include:

- A president who has the responsibility for all decisions.
- A number of vice-presidents who are in charge of major divisions of the company.
- Managers who direct particular operations.

- Department heads who supervise one major activity.
- Supervisors who direct the workers.
- Workers who complete tasks assigned to them.

Everyone needs to know who is responsible for each job. A chart is drawn to show lines of authority, figure 12.22.

The production system

The production system involves five basic operations.

- Designing—making original plans and drawings of products that satisfy consumer demands.
- Planning—organizing a system in which personnel, materials, and equipment can work together.
- Tooling up—acquiring and setting up tools and machines for production.
- Controlling production—using machines to make the product.
- Packaging and distribution—packaging, storing, and transporting the product to wholesalers and retailers.

Designing. Customers want or need a variety of products. Manufacturers satisfy these needs. Before starting to design a product, a manufacturer will determine what buyers want. This will involve market research. The product will not be made if the number of potential sales is not large enough to make a profit after costs are paid.

Designing is the responsibility of engineers, drafters, and, sometimes, industrial designers. Designers first learn the needs of potential customers. Relying on research, they make preliminary sketches of the product. Next, they rework and refine these sketches, figure 12.23.

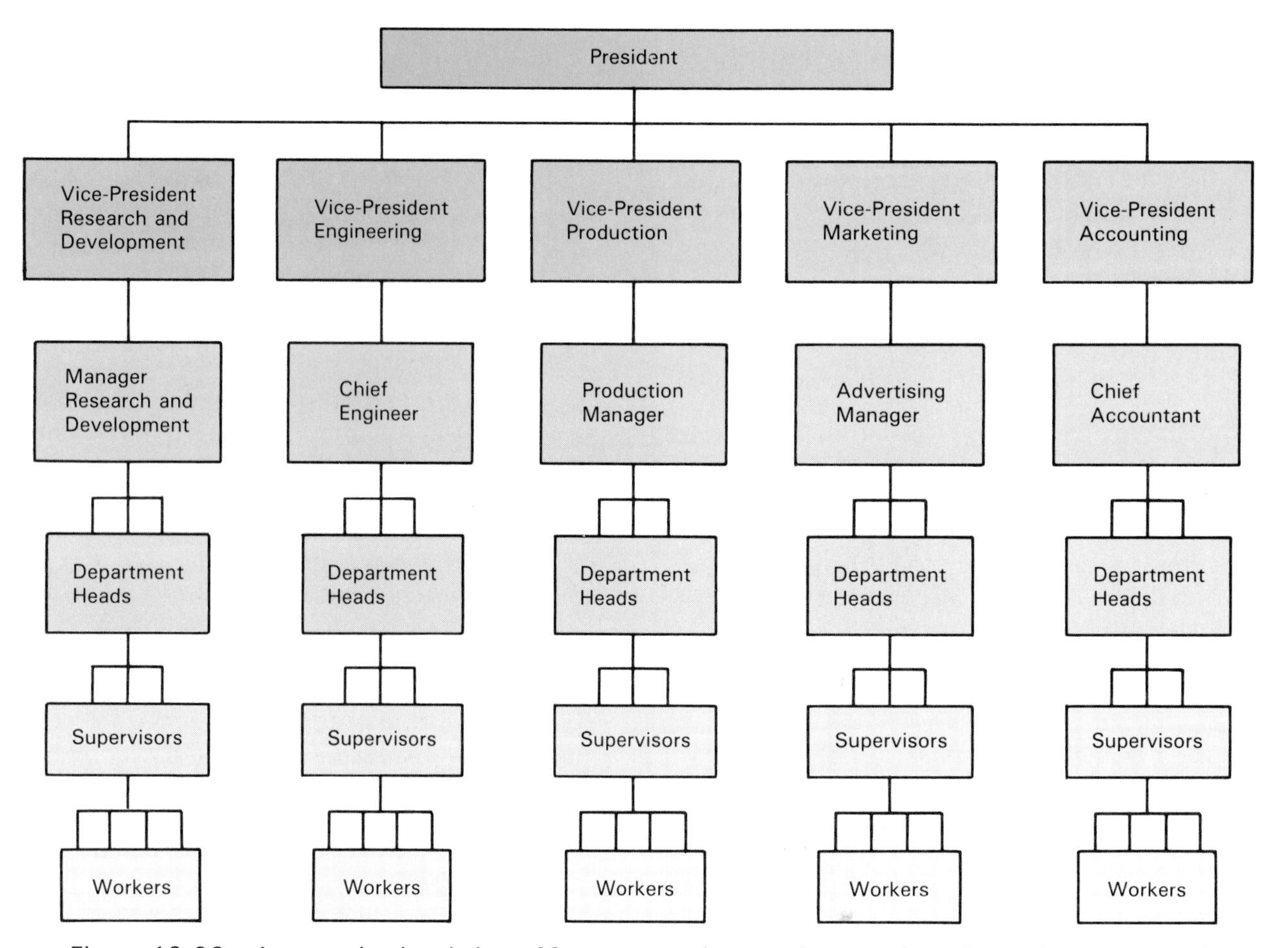

Figure 12.22. An organizational chart. Many companies use them to show lines of authority and responsibility.

CADKEY

Figure 12.23. Many designers do all of their sketches, reworking, and refining on a computer.

The designers must consult with the engineering staff. They will determine what production methods and processes to use. When a basic design has been approved, drafters produce scaled or full-size drawings. In many cases, model makers will build a three-dimensional model. For instance, a model of a new aircraft design may be tested. This helps determine its aerodynamic characteristics. See figure 12.24.

The information gained from the sketches, drawings, and models is then converted into working drawings and specifications. These drawings show the exact size and shape of each part. Specifications include information about:

- The materials to be used.
- The number of parts to be made.
- The operations needed to produce the part.
- The level of accuracy required.

Full-scale working models and prototypes are sometimes made, figure 12.25. These help identify weaknesses or errors in the design.

CANADAIR

Figure 12.24. A model aircraft is tested to determine the effect of crash landing on water.

CANADAIR

Figure 12.25. A full-scale working model of a plane is under construction.

Planning. Production must be planned. Personnel, materials, and equipment must be combined to ensure a smooth operation. This planning includes:

- Selecting and ordering equipment, machines, and processes.
- Finding the best way for people to work together.
- Determining how long each manufacturing operation will take.
- Gathering information on production costs.

An engineer will consider a variety of ways to complete each operation, figure 12.26. He or she will select the most efficient one. The decision will be based on the time involved, the cost, and the quality and quantity of the product.

Tooling up. Tools and machines may sometimes be purchased. At other times, tool and die makers must design and make them.

Tooling up involves four steps:

- Deciding which tools, equipment, and machines will be needed.
- Selecting and ordering machines, tools, and equipment from manufacturers.
- Designing and making special equipment.
- Supervising the installation of machines, tools, and equipment and the organization of a trial production run, figure 12.27.

Controlling production. Production must be organized. Then materials and parts move efficiently from one operation to the next. This sequence begins in the receiving and storage area. This is where raw materials are kept. From storage, materials move through various stages of processing and assembling. A variety of procedures may be used. These may include machining, bending, shearing, folding, forming, and casting, figures 12.28 and 12.29.

Each of the machines used in processing and assembly will need a power supply. Each will also need a system for controlling fumes, dust, and waste.

Production control finds ways to reduce waste. One type of waste is the space excess material takes up in storage. Another waste is time lost when workers or machines have nothing to do.

CANADAIR

Figure 12.26. Engineers calculate costs of production before a product is made.

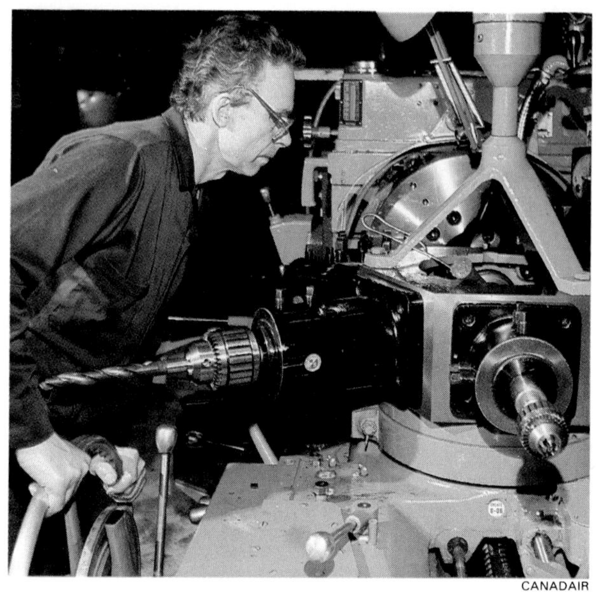

CANADAIR

Figure 12.27. A machinist checks out the setup of a production machine to make certain parts are machined accurately.

Production control is also in charge of:

- Ordering, routing, and scheduling materials.
- Dispatching job and work orders.
- Recording the performance of workers and machines.
- Taking corrective action when the production flow is interrupted.

Throughout the production process, a wide variety of inspection tools are used. Gauges check the sizes of parts. X rays check the internal structure of metal parts. The amount of inspection varies. In the manufacture of aircraft, it is important to check every part. However, for most consumer products, it is enough to check a small number of items from a large batch.

Packaging and distribution. Products being shipped must be protected. Various forms of packaging are used. Packaged products must also be labeled. Bubble packaging, boxes, cartons, and crates are methods used to protect the product. They provide insulation and protection against moisture, weather, and rough handling. Labeling is needed so that the contents can be recognized by the consumer. Labels and other kinds of markings show the product, the name of the manufacturer, quantity, directions for use and care, and provide other special information.

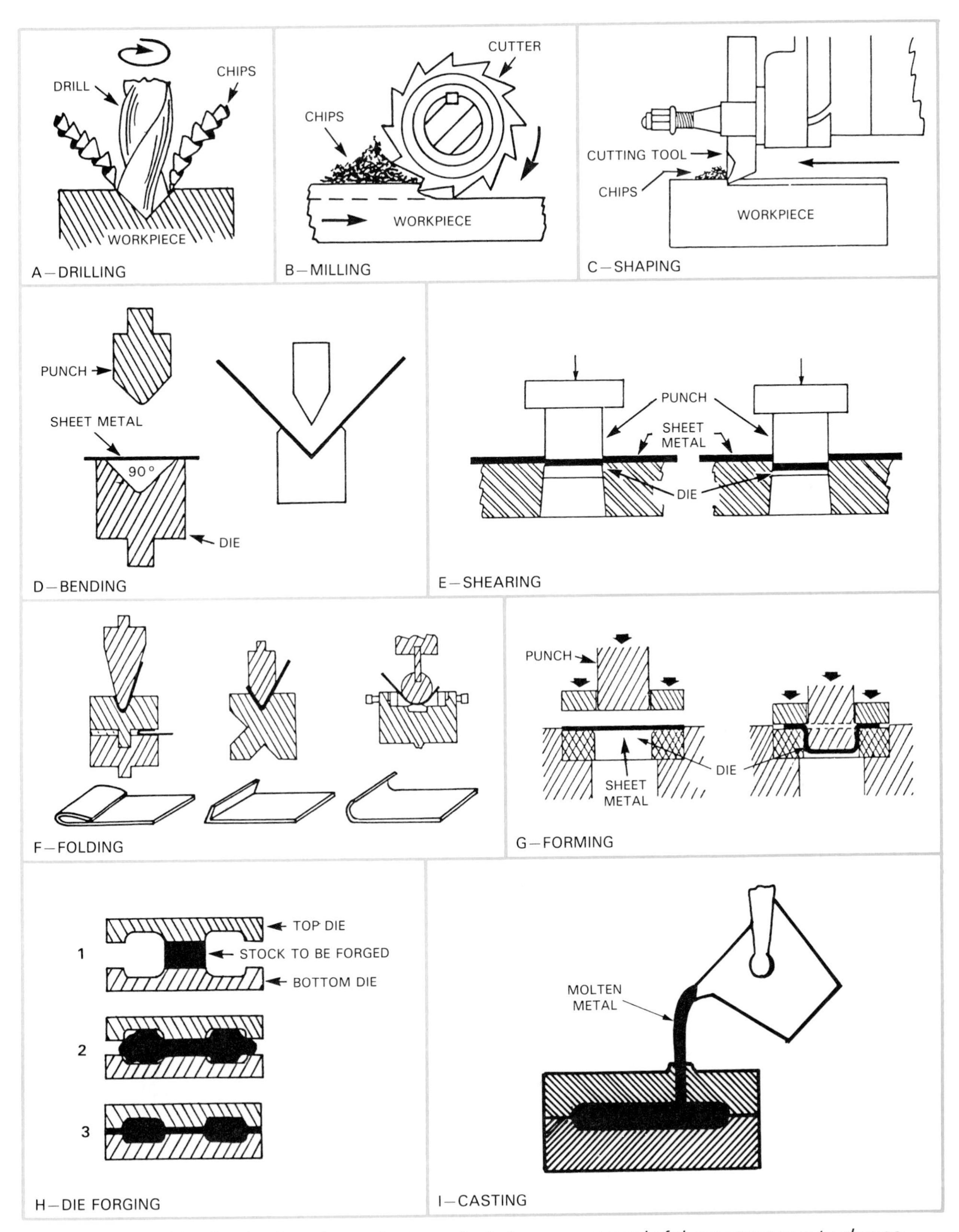

Figure 12.28. Production of a product may include one or several of these processes to change the shape or size.

Figure 12.29. Production control at an aircraft factory. A—Machinists shape parts. B—Quality control checkers inspect them. C—Other workers assemble parts. D—Aircraft nears completion.

Packaged and labeled products are usually warehoused (stored) to await shipment, figure 12.30. They are organized in quantities that are convenient for handling, sorting, and counting. The machinery for handling bulk shipments includes: conveyor belts, forklift trucks, and pallets. Pallets are wooden platforms on which the packaged products are placed. Then, a forklift can pick up the loaded pallet. Products may be loaded onto trucks, railroad cars, and ships. These vehicles transport products to wholesalers and retailers.

Careers in the secondary sector

Most people employed in the secondary sector work in factories or on construction sites, figure 12.31. Engineers carry out research to find new ways of changing raw materials into products. Market research analysts determine whether or not there is a market or need for the product. Factory workers install, operate, and maintain the machines that make the products. Management personnel oversee the production. They also ensure that the workplace is safe for workers.

CANADAIR

Figure 12.30. Parts and products must be warehoused until transported to wholesalers and retailers.

SECONDARY SECTOR CAREERS

OCCUPATIONAL CLUSTER	PROFESSIONALS	SKILLED WORKERS
Manufacturing	drafter industrial designer industrial engineer laboratory technician market research analyst production manager quality controller safety inspector supervisor	assembler cutter foundry worker instrument maker jeweler machinist model maker painter pattern maker press operator sheet metal worker tool and die maker upholsterer welder
Construction	architect civil engineer drafter electrical engineer soil technologist surveyor	bricklayer building inspector cabinetmaker carpenter cement mason contractor electrician floor covering installer glazier laborer painter paperhanger pipe fitter plasterer plumber roofer stone mason

Figure 12.31. How many people do you know who hold any of these jobs?

People who work in the construction industry are responsible for the planning and building of homes, bridges, industrial plants, dams, hospitals, highways, pipelines, and shopping centers. Civil engineers and surveyors perform many tasks. They design and lay out structures. They estimate costs and prepare materials specifications. They schedule and inspect work. They survey building sites. They often supervise the work at a job site to see that the structure follows their plans. Skilled workers specialize in a trade, for example, plumbing or carpentry. Laborers and hod carriers assist the skilled workers.

■ THE TERTIARY SECTOR: PROVIDING SERVICES

When was the last time you bought a hamburger? Visited a library? Needed to have a checkup at the doctor's or dentist's office? When did you last take clothes to be dry-cleaned or go to an amusement park? When you did any of these things, you made use of services in the **tertiary** (third) **sector.**

The tertiary sector is concerned with the servicing of products. It also provides services that add to the personal comfort, pleasure, and enjoyment of people.

Most people think of the service industry in a limited way. They see service as installing, maintaining, repairing, or altering (changing) products or structures. For instance, many homes are heated by an oil furnace. Once the distributor installs the furnace in the home, it will require maintenance. A truck delivers oil. Once a year, a service person will visit to perform routine maintenance. A 24-hour emergency service is available in case of a breakdown. Alterations, such as the addition of electric coils, may be made to the furnace to provide another method of heating. A heat pump may also be added to economize on fuel and provide air conditioning.

People who perform these services work for private companies. These companies aim to provide good service to their customers. They must compete with other companies who are providing similar services. They want to remain in business. To do so, they must also make a profit for their owners or shareholders.

Government agencies provide some services. These include education, health, and public works. This type of service is normally financed through local, provincial, state, or federal taxes.

Growth in jobs

During the last 20 years the number of service jobs has about doubled. There are reasons for this.

First, some workers cannot be more productive. A dentist is such a person. She or he can treat only a limited number of patients each hour. It is impossible to treat two patients at the same time. Thus, as population increases, the number of dentists needs to increase.

Second, many homes now have two wage earners. When people have more money, they consume more services but fewer goods. For instance, during his or her lifetime, a consumer may purchase three new television sets. The same person, however, may go to the bank once a week. She or he may eat at a restaurant twice weekly and use public transportation every day. Most products are purchased infrequently. However, many services are used daily.

Finally, the service sector faces little foreign competition. A television may have been manufactured in Japan; local people provide all services.

Looking to the future

The growth in the service sector is likely to continue for three reasons. First, the aging population will need more health-care.

Second, many businesses are now franchised. This means that one or more persons form a local company and enter an agreement with a parent company to use its name and provide its services. Many of the fast-food chains, like McDonalds, are franchised. There may be branches throughout the world. Franchised businesses generally grow quickly. Failure rates are also lower as a result of sound management practices.

Third, North American people like starting new small businesses providing services. This enthusiasm is likely to continue.

A person who organizes, manages, and assumes the risks of starting a business is called an **entrepreneur.** Entrepreneurs play a number of roles in the economy. They can:

- Create new products and services in response to consumer demand.
- Tailor the business to suit local needs and offer a quality of service which might not be available from a large corporation.
- Help to maintain or lower prices through competition.
- Provide employment opportunities.
- Help contribute to the economic growth of a country and improve the country's place in international competition.

Careers in the tertiary sector

People employed in the tertiary sector work in one of the following occupational clusters.

- Business and office.
- Communications.
- Health.
- Hospitality and recreation.
- Marketing and finance.
- Personal services.
- Public and social services.
- Transportation.

Business and office. Employees in this occupational cluster are involved in five areas. These include: administration, management, accounting, secretarial, and clerical tasks. See figure 12.32. Professional employees solve problems, analyze data, and make major administrative decisions, figure 12.33. They also

TERTIARY SECTOR CAREERS

OCCUPATIONAL CLUSTER	PROFESSIONAL	SKILLED WORKERS
Business and office	accountant actuary lawyer personnel manager programmer systems analyst underwriter	bank teller bookkeeper buyer business machine mechanic business machine operator cashier clerk file clerk receptionist secretary stenographer telephonist typist

Figure 12.32. Every company in the service field requires a business and office work force.

MOBIL CORP.

Figure 12.33. A systems analyst plans ways of using a computer to improve the efficiency of a company.

prepare reports, design computer systems, and oversee matters of finance.

Office workers keep businesses and organizations running smoothly, figure 12.34. Clerical workers maintain accurate records and files. They also operate office machines. Some ship and receive merchandise.

TEC

Figure 12.34. Skillful secretaries who pay attention to details are appreciated in business offices.

Communications. Some jobs in communications require creative skills. Among these are: writing, editing, and producing information. Other jobs demand technical skills, figure 12.35. People in these jobs operate, maintain, and repair equipment. Figures 12.36 and 12.37 show people at work in communication jobs.

Health. Employees in health occupations keep people healthy. They also help people recover from injuries or illness, figure 12.38. Physicians and other medical practitioners, such as optometrists, diagnose illnesses, and provide treatment. Nurses carry out doctors' orders. They also see to the day-to-day care of the ill and injured. Medical laboratory workers conduct tests to discover the cause of patients' illness. Many other people work behind the scenes in hospitals and clinics. They provide information and support to doctors, nurses, patients, and visitors. Figures 12.39 and 12.40 show people at work in health occupations.

Hospitality and recreation. People who work in hospitality and recreation occupations help others enjoy their leisure time. A travel agent helps with travel arrangements. People who work in hotels provide comfortable lodgings. Amusement and recreation employees provide fun activities for your enjoyment. Figure 12.41 shows a variety of occupations in this field.

TERTIARY SECTOR CAREERS

OCCUPATIONAL CLUSTER	PROFESSIONAL	SKILLED WORKERS
Communications	announcer cartographer commercial illustrator director drafter editor newspaper reporter photographer technical illustrator	bindery worker broadcast technician camera operator compositor disc jockey film editor lithographer press operator photoengraver radio dispatcher sign painter telephone operator telephone repairer television programmer

Figure 12.35. Which of these careers in communications would you choose?

Figure 12.36. Photographers are skilled at recording scenes on film. They must understand cameras and how to properly expose film.

Figure 12.37. When communication equipment fails to work, repairers must be able to fix it.

TERTIARY SECTOR CAREERS

OCCUPATIONAL CLUSTER	PROFESSIONALS	SKILLED WORKERS
Health	cardiologist chiropractor dentist gynecologist music therapist neurologist obstetrician optometrist orthopedic surgeon pharmacist physical therapist physician psychiatrist radiologist speech pathologist surgeon veterinarian	dental assistant dental hygienist dental lab technician dispensing optician licensed practical nurse nursing aide nursing assistant orderly parametic X-ray technician

Figure 12.38. Which of these occupations in the health field would you choose?

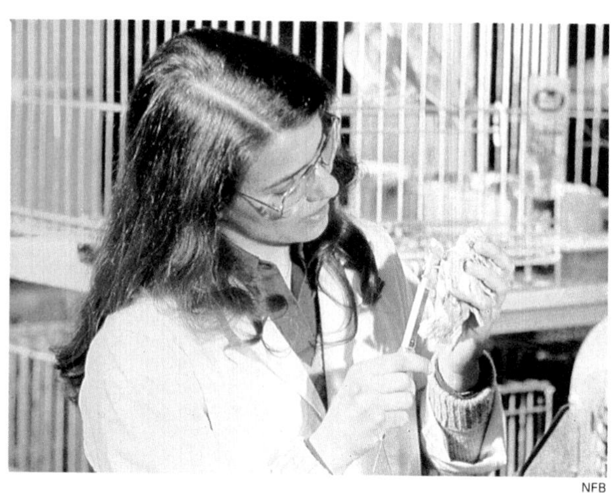

NFB

Figure 12.39. Veterinarians specialize in health care of animals.

Many jobs in this cluster are related to food, figure 12.42. Some people work directly with customers. Included are waiters and bartenders. Others, such as cooks and chefs, work behind the scenes.

In the performing arts, actors, musicians, and dancers dedicate themselves to their work, figure 12.43. They are assisted by stagehands, designers, electricians, and costume makers.

Marketing and distribution. Employees in this cluster buy, promote, sell, and deliver

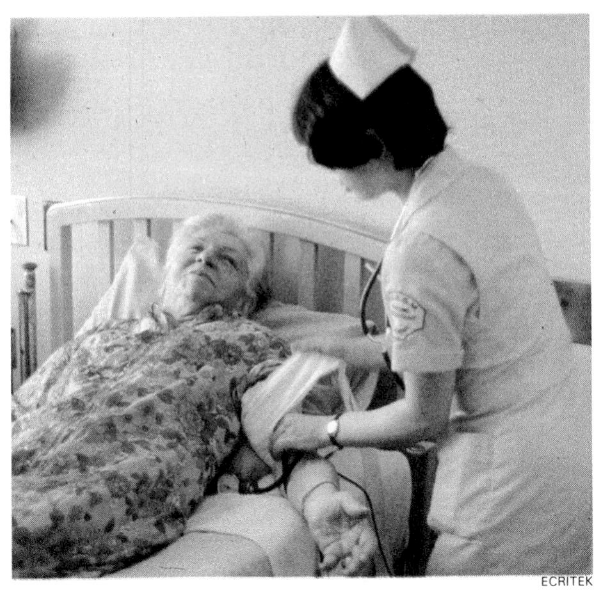

ECRITEK

Figure 12.40. A nurse provides care for those who are ill.

TERTIARY SECTORS CAREERS

OCCUPATIONAL CLUSTER	PROFESSIONALS	SKILLED WORKERS
Hospitality and recreation	actor/actress choreographer conductor dancer dietitian director hotel manager music director musician pop singer producer sports teaching pro	baker bartender bell captain camp manager chef choral singer cook cruise director executive housekeeper greenskeeper hotel desk clerk lifeguard park caretaker recreation leader ticket seller tour guide travel agent waiter/waitress

Figure 12.41. Hospitality and recreation careers. Which involve working directly with people? Which involve working behind the scenes?

goods and services, figure 12.44. Manufacturers employ buyers. They are responsible for purchasing materials and supplies required to produce products. The products are sold by the manufacturer to wholesalers. Wholesalers sell to retailers who, in turn, sell to consumers.

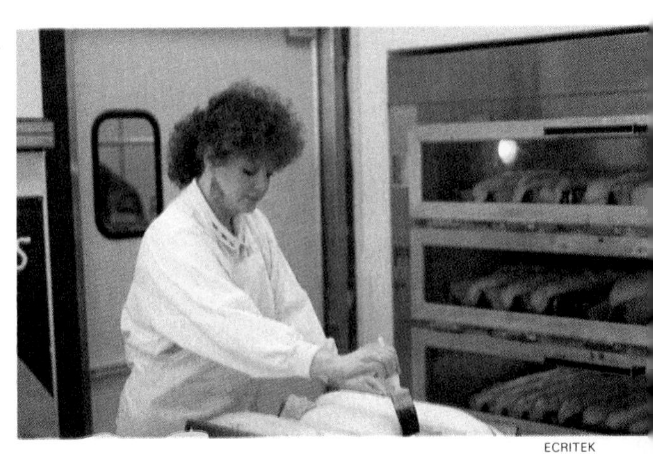

ECRITEK

Figure 12.42. A baker at work

T. DIAB

Figure 12.43. Actors and actresses spend many hours preparing for performances.

TERTIARY SECTOR CAREERS

OCCUPATIONAL CLUSTER	PROFESSIONALS	SKILLED WORKERS
Marketing and distribution	ad copy writer advertising manager bank manager insurance agent insurance investigator loan officer market research analyst model real estate agent	buyer customer service representative fork lift operator loader packer purchasing agent sales clerk sales representative shipping clerk stock clerk store manager truck driver warehouse person window dresser

Figure 12.44. Can you think of other careers in marketing and distribution?

Most jobs in marketing involve meeting people, figures 12.45 and 12.46. There are others, however, that involve the movement, storage and inventory of products. Some employees must be able to organize large amounts of data. Others who work for insurance companies and banks are involved in finance.

Personal services. Some people who work in this occupational cluster are concerned with the physical appearance of their customers. They help with personal grooming or physical conditioning. Others assist with tasks around the home. Still others help keep places where people live, work, or play safe and clean.

HUXLEY

Figure 12.45. Insurance agents sell insurance protection to clients.

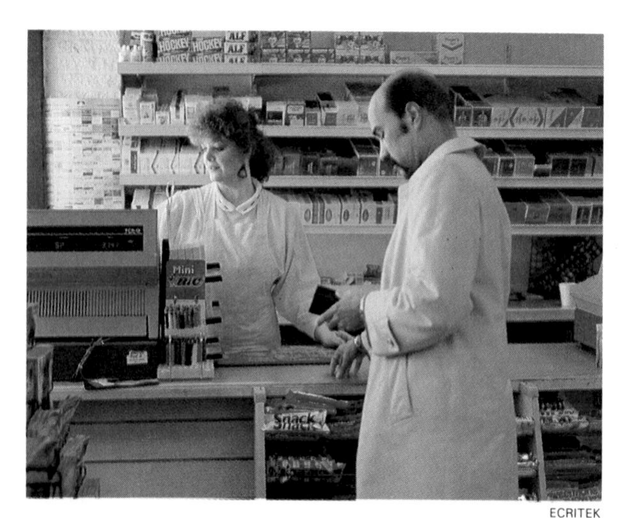

ECRITEK

Figure 12.46. A sales clerk. People in sales occupations must enjoy meeting people.

Figure 12.47 lists some jobs in this cluster. Figures 12.48 and 12.49 are examples of occupations in this field.

Public and social services. People working in the public and social services cluster provide services to everyone in a community. In

TERTIARY SECTOR CAREERS

OCCUPATIONAL CLUSTER	PROFESSIONALS	SKILLED WORKERS
Personal services	home nurse	animal trainer barber building superintendent butler chauffeur companion cosmetic demonstrator cosmetologist custodian exercise instructor exterminator funeral director hair stylist housekeeper kennel manager laundry worker manicurist nanny pedicurist seamstress shoe repairer tailor

Figure 12.47. Many people find satisfying careers providing personal services to others.

ECRITEK

Figure 12.48. Chauffeurs provide a personal service by being drivers for families, business people, and government officials.

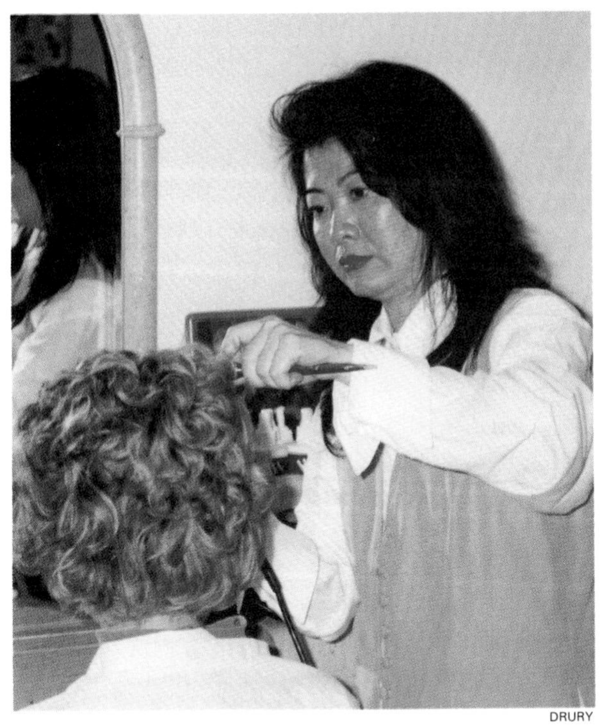

DRURY

Figure 12.49. Hair dressers must be skilled and enjoy working with clients.

Figure 12.51. The clergy may be called upon to teach, provide counseling, and look after people's spiritual welfare.

TERTIARY SECTOR CAREERS

OCCUPATIONAL CLUSTER	PROFESSIONALS	SKILLED WORKERS
Public and social services	coroner guidance counselor lawyer librarian minister notary police officer teacher	armed service personnel community youth worker firefighter health and safety inspector letter carrier mail clerk sanitation worker tax audit clerk

Figure 12.50. Every community has public and social programs.

CANADAIR

Figure 12.52. Firefighters must respond to emergency situations.

some careers, helping people stay safe and comfortable is the chief task. Other service workers help solve personal problems. Still others teach, figures 12.50 through 12.52.

Transportation. Transportation is meant to move people and materials. Employees in this cluster drive buses, taxis, trains, and trucks. They fly aircraft, or pilot ships. Other people provide services for customers. They make arrangements for their travel and for the movement of cargo. Other workers keep transportation equipment in good working order. See figures 12.53 through 12.55.

TERTIARY SECTOR CAREERS

OCCUPATIONAL CLUSTER	PROFESSIONALS	SKILLED WORKERS
Transportation	air traffic controller captain co-pilot pilot ship's officer	able seaman aircraft mechanic automobile mechanic boatswain bus driver chief mate diesel mechanic dispatcher flight attendant locomotive engineer merchant marine passenger car conductor reservation agent service station attendant taxi driver ticket agent travel agent truck driver

Fig. 12.53. Transportation workers. Which of the people working in these careers have provided you with a service?

NFB

Figure 12.54. Helicopter pilots move people and material short distances by air.

SUMMARY

The steps in producing any product are organized into three sectors: primary, secondary, and tertiary. The primary sector is concerned with obtaining and processing raw

ECRITEK

Figure 12.55. A travel agent handles airline reservations.

materials. Some of these materials are renewable. Other materials are nonrenewable. Careers in the primary sector include those in agriculture, horticulture, forestry, marine science, natural resource extraction, and environmental control.

The secondary sector is concerned with the manufacture of products and construction of structures. It has evolved through three stages: the individual artisan, mechanization and mass production, and automation.

Prior to the eighteenth century the individual artisan was responsible for every step in producing a finished product. Mechanization and mass production occurred during the Industrial Revolution. At this stage there were three important changes. Products were made with machinery. The machinery was powered by engines. The products were made in factories. The efficiency of factories was increased by the use of assembly lines and later by automation. In an automated factory the worker builds, monitors, and maintains machines that make the products. The latest development in automated production and assembly is the use of robots.

A production system involves five basic steps: designing, planning, tooling up, controlling production, and packaging and distribution. Careers in the secondary sector include those in manufacturing and construction.

The tertiary sector is concerned with the servicing of products and providing services. Installation, maintenance, repair, or the alteration of products are provided by private companies. Government agencies are responsible for education, health, and public works. Careers in the tertiary sector include those in: business and office, communications, health, hospitality and recreation, marketing and finance, personal services, public and social services, and transportation.

KEY TERMS

Artisan
Assembly line
Automation
Career
Computer Numerical Control (CNC)
Degree of freedom
Entrepreneur
Factory
Flexible Manufacturing System (FMS)
Industrial Revolution
Mass production
Primary sector
Production system
Raw materials
Robot
Secondary sector
Tertiary sector

TEST YOUR KNOWLEDGE

Write your answers to these review questions on a separate sheet of paper.

1. Look at figure 12.1 which describes the steps used to build a chair. Choose a simple object in your home. Describe the steps in its manufacture, from raw material to finished product.
2. When producing products, what is the main activity in each of the following sectors?
   Primary Sector.
   Secondary Sector.
   Tertiary Sector.
3. List three renewable and three nonrenewable raw materials.
4. Choose a career in the primary sector. Describe a typical day in the life of someone working in that career. Try to talk to someone working in your chosen career. Also use the library resources.
5. List the three stages through which the secondary sector has evolved.
6. Until the Industrial Revolution artisans worked alone or in small groups in villages. How did the Industrial Revolution change this?
7. How does the division of labor enable products to be made at a faster rate than by hand?
8. What is the advantage of products made with interchangeable parts?
9. What were some of the problems experienced by workers in early factories?
10. Mass production was first used by ________ ________in the year ______.
11. Describe the difference between a mass production assembly line and an automated assembly line.
12. List five jobs that robots can perform in the manufacture of a product.
13. State three advantges of robots over human workers.
14. Describe the five basic steps in the production system.
15. Choose a career in the secondary sector. Describe a typical day in the life of someone working in that career. Try to talk to someone working in your chosen career. Also use the library resources.
16. List five services provided by the tertiary sector that you have used.
17. State three reasons for the rapid growth in the number of tertiary sector jobs in the last 20 years.
18. Identify an entrepreneur in your neighborhood. State what new product or service is being offered. Describe how the product or service is helping you or your family. How many people are being employed?
19. Choose a career in the tertiary sector. Describe a typical day in the life of someone working in that career. Try to talk to someone working in your chosen career. Also use the library resources.
20. List the names of 10 adults whom you know. Name their jobs and state the occupational cluster and sector for each.

APPLY YOUR KNOWLEDGE

1. Select a raw material to research. Find out:
   a. Where it is found.
   b. In what form it is found in its natural state.
   c. How it is extracted, harvested, or farmed.
   d. How it is transported.
   e. How it is processed or refined.
   f. How new manufactured goods have

increased production or changed its processing.

2. List the advantages of robots compared to human workers. List five tasks that a robot can perform better than a human worker.
3. Imagine that in your working life you started as an artisan. You then moved to a factory and worked on a mass production line. Your last job was in a fully automated factory where the work was done by robots. Describe the advantages and disadvantages of each of your three jobs.
4. Choose a small, common mass-produced item. For each of the five steps in the production system, describe the activities that would occur to mass-produce the item you have chosen.
5. Refer to figure 12.28. What process would be used to:
   a. Cut a 1 in. (25 mm) diameter hole in a sheet of plastic?
   b. Bend a piece of sheet metal into a 90° angle?
   c. Reduce the diameter of a steel rod?
   d. Produce a flat surface on a block of aluminum?
   e. Make a large number of 1 in. (25 mm) diameter identity tag disks?
6. Research how new technologies have replaced, outmoded, or created new jobs in your hometown.
7. List the services in the tertiary sector that you have used in the past week.
8. Select one career from the primary, secondary, or tertiary sectors. Investigate the career to discover:
   a. The education or training required.
   b. Whether or not there will be a demand in the future for workers.
   c. The promotion possibilities.
   d. The salary, pension, vacation, and other benefits.
   e. The number of people currently employed in this type of work.

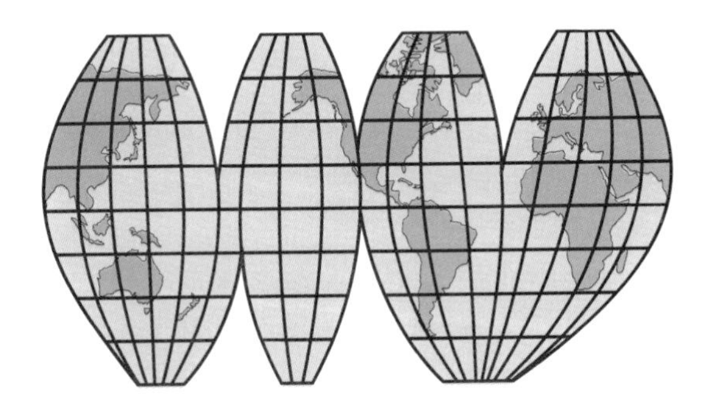

Chapter 13

Communication in an Information Society

OBJECTIVES

After reading this chapter you will be able to:

- Describe the characteristics of the information society.
- Compare "old" and "new" information technologies.
- Describe the role and importance of communications technology in society.
- List the input and output devices used with computers.
- Understand the importance of microelectronics to information processing.
- Identify the major components of a telecommunications system.

Since the 1970s we have seen major changes in the activities of the tertiary (service) sector. A larger percentage of service workers are now employed in communication. They create, process, and distribute information. **Information** is a collection of words or figures that have meaning or that can be combined to have meaning.

■ INFORMATION TECHNOLOGY

In our lives we deal with huge amounts of information. For instance, every day you use telephone numbers to call friends. You refer to a bus schedule, and class schedules. You can remember some of this information, but much of it must be stored, processed, and communicated by machines. The technologies used in storing, processing, and communicating information are referred to as **information technology.**

The "old" information technology relied on telephones, the postal service, printed materials, and film. Much of the equipment used was mechanical. Today, we rely more and more on electronics to handle information. Machines with moving parts have almost entirely disappeared. Replacing them is a flow of electrons.

The "new" information technology uses a variety of equipment. Included are computers, satellites, microprocessors, and television. These are frequently linked. For instance, the combined technologies of the computer, telephone, and television may be merged into a single communications system, figure 13.1. This system can transmit data. It also provides instantaneous interaction between people and computers.

HUXLEY

Figure 13.1. A telecommunication system. A personal computer processes information fed into it. Information can be sent or received by phone.

Such a communication system is used in modern offices, figure 13.2. Secretaries type and correct documents on a word processor. A disk stores input. A simple computer command causes a number of copies to be automatically printed. Meanwhile, the secretary may be speaking directly to the other side of the world on a push-button telephone. If the number is busy, the phone will automatically redial until it is free.

Phototelegraphy permits the sending of telex and facsimile pictures direct to clients. Incoming telephone calls received while the secretary is out of the office can be answered by a machine. Messages may be printed on a telex. Is overnight delivery by regular mail too slow? A letter can be fed into a document scanner for transmission over land lines. At the receiving end, a laser printer will print copies for immediate delivery.

Today's department store is another example of how machines are linked in a system. There a computer terminal is combined with a cash register. The system adds the consumer's bill. At the same time it sends information on the sale to the store's headquarters where another computer uses the information to monitor (keep track of) inventory. Payment for these purchases may be electronically processed. Another system that is known as an automatic teller offers round-the-clock banking services, figure 13.3.

JACK KLASEY

Figure 13.2. Modern office equipment includes communications equipment such as fax machines.

This rapid exchange of information between people and machines and also directly from one machine to another has a name. The term used is **information technology.**

Information technology relies on three complex technologies that have recently converged (come together). They are: computers, microelectronics, and telecommunications. These three are combined into systems that enable us to:

- Create, collect, select, and transform information.
- Send and receive information.
- Store information.
- Retrieve and display information.

ECRITEK

Figure 13.3. What impact does round-the-clock banking have on shopping habits?

The following are some examples of how this takes place:

- The head office of a chain of stores can receive, collect, and store sales information transmitted from the electronic cash registers in each store, figure 13.4.
- A night watchman responsible for the security of a large building can use a closed-circuit television system. It receives, displays, and stores information about the condition of different parts of the building.
- Hospitals use electronic machines for various purposes. Machines can measure the pulse, heartbeat, and other vital signs

ECRITEK

Figure 13.4. How does the computer "read" the cost of each item being purchased?

of a patient. This information can be displayed on a screen. A nurse can "read" the screen to keep track of a patient's condition, figure 13.5.

- Suppose a rock group makes a recording of its music. The recording equipment collects information. Sounds from individual performers and from each instrument or vocalist are placed on separate recording tracks. Afterwards, individual sounds are selected and mixed. This produces the final recording, figure 13.6.
- An engineer or drafter feeds data into a CAD system. (CAD stands for computer-aided design.) The machine turns the information into three dimensional drawings, figure 13.7. The engineer can then check the design of machine parts.

Let us now look at the three technologies that make all of these systems possible:

- **Computers.**
- **Microelectronics.**
- **Telecommunications.**

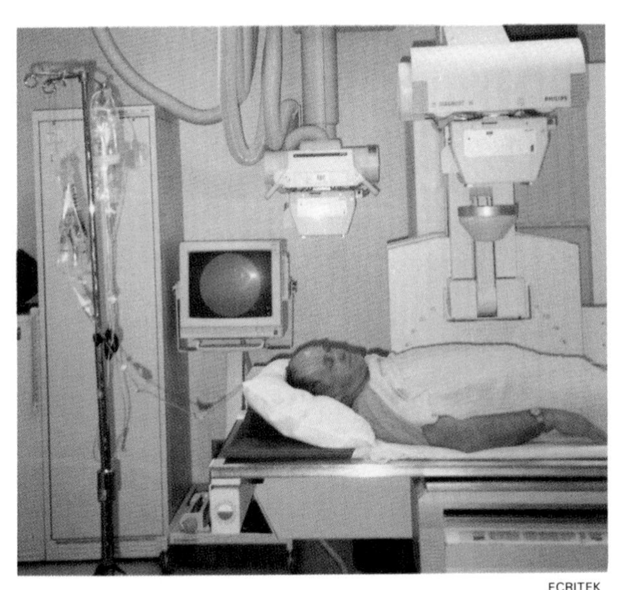

ECRITEK

Figure 13.5. X-ray machines are important diagnostic tools for doctors.

NFB

Figure 13.6. Recording and mixing unit in a recording studio. The equipment collects, stores, and retrieves the music for the final recording.

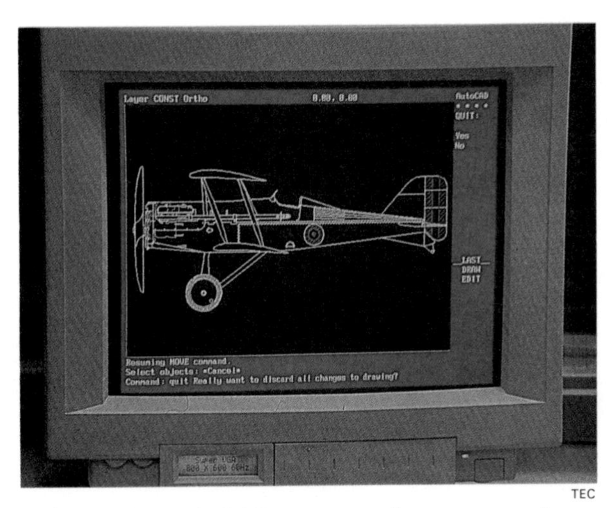

TEC

Figure 13.7. A CAD system allows an engineer or drafter to see designs in two and three dimensions.

■ COMPUTERS

What is a computer? Many people think of a computer as almost human. It seems to have a "brain" that allows it to think. Computers do not have brains. They cannot think for themselves. They are primarily machines for doing arithmetic at high speeds. However, the really important thinking is done by the humans who feed them data.

The modern computer is primarily a calculating machine. It can also store a vast amount of data, however. It can be programmed (instructed) to carry out logical operations. For instance, it can transfer data from one part of the machine to another. On command, it will sort this data and compare it with other data. From the comparison, it is able to provide new information.

The computing process involves three main stages: the input stage, the central processing stage, and the output stage. Figure 13.8 shows these three stages. It also shows a variety of devices used for each.

A computer may receive and process many types of information:

- Sales information from electronic cash registers.
- Pulse rate, heartbeat, and vital signs for a patient in a hospital.
- Sounds from musical instruments and vocalists.
- Three-dimensional engineering drawings.

INPUT	CENTRAL PROCESSING UNIT	OUTPUT
Digitizer Disk Graphics tablet Joy stick Keyboard Light pen Speech Touch sensitive screen	Uses instructions (programs), and data stored in memory, to carry out calculations.	Disk Machines Monitor Other computers Plotter Printer Robot Voice synthesizer

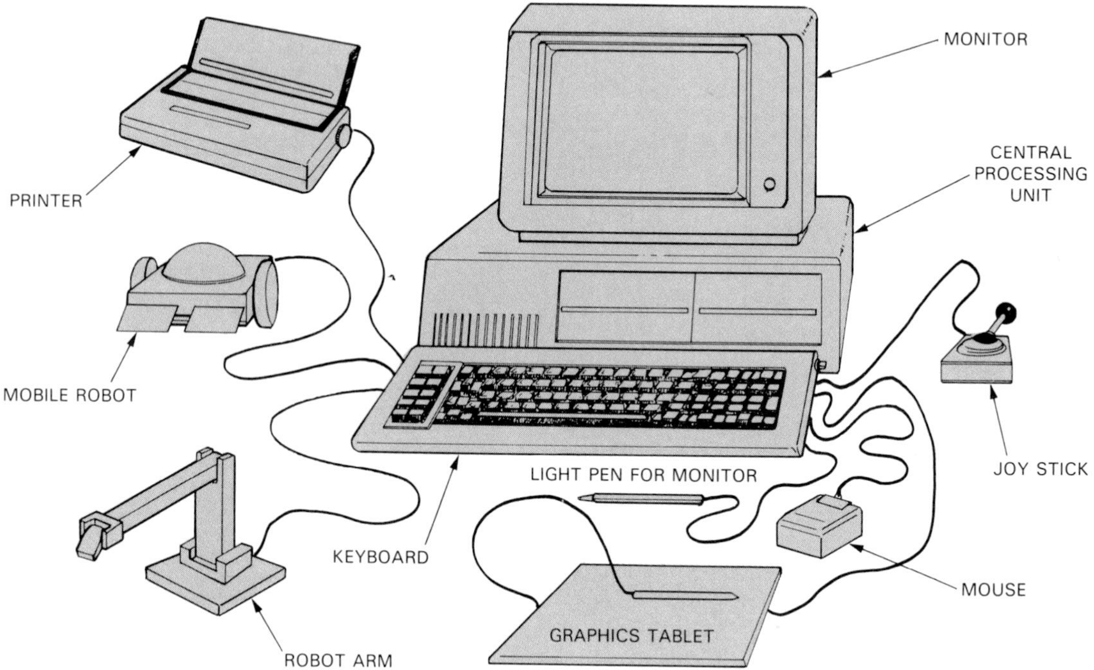

Figure 13.8. This is a computer with its three stages and various devices.

This information must be changed (coded) into a form that the computer can handle, however. Any set of symbols that represents information is correctly called data. The letters A, B, C, are data, for example.

Computers use a simple code made from electrical signals. There are only two signals in this code, on and off. These are written as 1s and 0s. This is called a **binary** digital code. Binary means two and digital means number. Inside the computer a high voltage is used to represent a 1. A low voltage represents a 0.

To provide for a code with more than two elements, signals are combined to produce patterns. Look at figure 13.9.

Imagine a set of four light bulbs. Each one can be turned on or off. Each of these bulbs can be assigned a value. Since, in the binary system, each digit has a value twice as large as the one on its right, the bulbs must be assigned values of 8, 4, 2 and 1. When a bulb is off it represents 0. When it is switched on a bulb represents the value assigned to that position. These values can be added together. In figure 13.9 the first bulb and the third bulb are on. This represents the number 5 (1 + 4). If the first bulb is off and the second, third, and fourth are on, the number represented is 14 (2 + 4 + 8). This group of four bulbs can represent any number from 0 (all off) to 15 (all on).

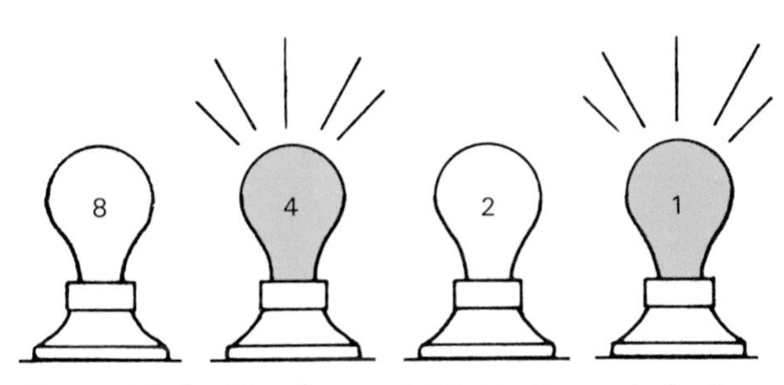

Figure 13.9. The first and third bulbs switched on would represent the number five.

Each on or off signal is known as a **bit,** short for binary digit. The example above uses a four-bit word. Since computers need to work with numbers larger than 15, an eight-bit system is used, figure 13.10.

■ MICROELECTRONICS

The second of the technologies that led to the growth of information technology is microelectronics. Microelectronics is the result of miniaturization (make tiny). This is the process by which the switches and circuits of processors and their accessories are made incredibly small. This miniaturization has resulted from the invention of new manufacturing processes and the use of new materials.

Of all the electronic elements resulting from these new processes and materials, the most important is the microprocessor or "computer on a chip." This is commonly referred to simply as a chip, figure 13.11. A chip is a tiny flake of a substance called silicon. It is covered

PLACE VALUE	128	64	32	16	8	4	2	1
BINARY NO.	1	1	0	1	1	0	1	0

128
64
16
8
2
218

Figure 13.10. The eight-bit word, 11011010, represents the number 218.

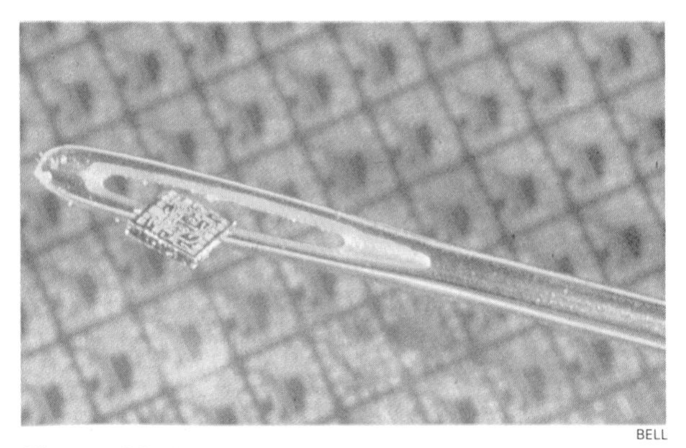
BELL

Figure 13.11. A microprocessor, also known as a "chip." See how it is "dwarfed" by the eye of a needle.

with microscopic electric circuits. These chips may be mass-produced in the tens of thousands. Thus, a single chip costs less than a textbook. Still, it contains most of the switches and circuits needed by a computer. Chips are becoming cheaper and more powerful each year and are reaching into every area of modern life.

Microprocessors provide a machine with decision-making ability, memory for instructions, and self-adjusting controls. In pacemakers, the miniature components on a chip time heart beats. Microprocessors also set thermostats, switch VCRs on and off, pump gas, and control car engines. Robots rely on them. So do the on-board computers in satellites.

■ TELECOMMUNICATIONS

The two technologies of computers and microelectronics are responsible for the processing of data. Moving data between points is the task of a telecommunications system. The sending and receiving of information makes use of a variety of devices and systems. The following are the three main types.

- Those depending on electrical waves or pulses transmitted through wires.
- Those that depend on the transmission of light pulses, produced by lasers, along optical fibers.
- Those that depend on electromagnetic waves or pulses broadcast through the atmosphere.

The telephone

The telephone is the most common telecommunications system using wire for transmission. This system has been adapted to permit connections between computers.

For the most part, telephone links are made by means of copper cables. They carry the messages in the form of electrical pulses. These cables are bulky, however. In many areas they already fill the cable ducts that carry them. Also, they are made of copper which is becoming expensive and scarce.

Glass cables

A more modern way of sending messages is to use light. It is transmitted along thin strands of glass called **optical fibers,** figure 13.12. These fibers can carry laser-coded messages over long distances. An optical fiber four-thousandths of an inch (0.1 mm) in diameter is capable of carrying 2000 two-way telephone conversations at once.

At the transmitter end of an optical-cable network, telephone signals are converted into pulses of **laser light.** These pulses travel through the glass fibers to the receiver end. There the pulses are converted back into electrical signals that are carried to a telephone.

Laser light

Laser light is a form of radiation. It has been boosted to a high level of energy. Laser light produces a strong, narrow beam of light. Nor-

BELL

Figure 13.12. Optical fibers are replacing copper cables in communications. A thin glass fiber cable can handle the same number of telephone messages as the thicker copper cable.

mally, this light travels only in a straight line. The glass fibers, however, bend the laser light so it will follow the cable wherever it is used.

Lasers have made possible such popular conveniences as automatic supermarket checkouts, fiber-optic communication, and a new generation of printing devices. Lasers are even behind the compact disc used in entertainment. Lasers have also changed many aspects of medicine. Laser surgery is commonplace.

Wave communication

Telecommunication messages are not only sent along copper and fiber-optic cables. Messages may be converted to electromagnetic waves or pulses. These are sent through the atmosphere. They are beamed into space and bounced off orbiting satellites. Communication satellites can handle more information than cables. They can also transmit it faster. See figure 13.13.

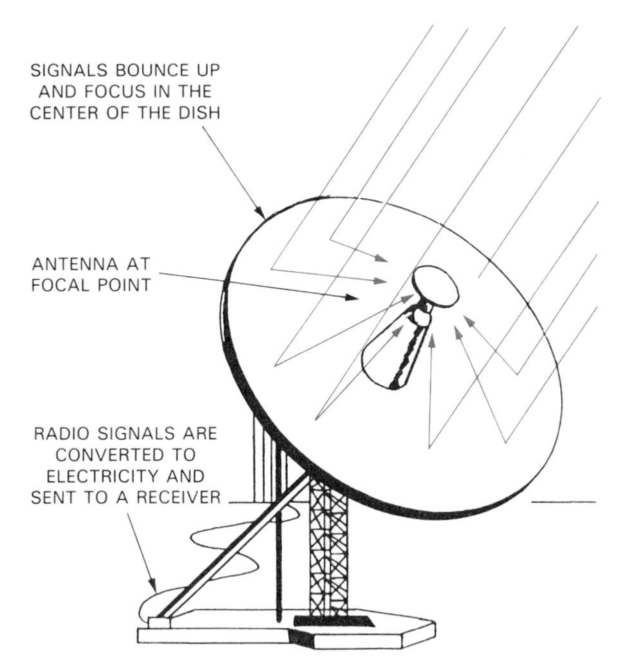

Figure 13.14. Satellite signals are picked up by a dish-shaped antenna.

Figure 13.13. A communication satellite is being launched from the cargo bay of the space shuttle.

Microwaves. Microwaves are electromagnetic waves with a short wavelength. Microwave signals can be beamed in a very small area. The signal can be very intense. Therefore, antennas to "catch" the signal can be made very small.

Information is beamed to and from satellites as microwave signals. These signals must be picked up by special dish-shaped antenna, figure 13.14. The antenna concentrates the weak signals. The microwaves bounce off the inside of the dish to a focal point. These signals are then converted to electricity. The current travels to a receiver such as a TV monitor.

SUMMARY

Most workers in the tertiary sector are now employed in the creation, processing, and distribution of information. Most of this information must be stored, processed, and communicated by machines. The "old" information technology relied on telephones, the postal service, telegraph, printed materials, and film. The "new" information technology uses computers, microprocessors, and telecommunications.

Computers are primarily machines for doing arithmetic at high speeds. They can also store a vast amount of data and carry out logical operations. Information to be handled by a computer must be changed into the binary digital code.

Microelectronics makes use of tiny switches and circuits. The most important component is the microprocessor (the "chip").

Telecommunications involves the moving of data between points. Three systems are used. These systems depend on electrical waves or pulses transmitted through wires, the transmission of light pulses along optical fibers, or electromagnetic waves broadcast through the atmosphere.

KEY TERMS

Binary
Bit
Chip
Computer
Information
Information technology
Laser light
Microelectronics
Microprocessor
Optical fibers
Telecommunications

TEST YOUR KNOWLEDGE

Write your answers to these review questions on a separate sheet of paper.

1. A collection of words or figures that have meaning is defined as ________.
2. The technologies used in storing, processing, and communicating information are, together, referred to as ________ ________.
3. What are the differences between the old and the new information technologies?
4. Copy and complete the chart below describing the equipment a secretary in a modern office would use to complete the tasks listed.

TASK	EQUIPMENT
Prepare a letter	
Store information	
Make a telephone connection to an engaged number	
Send a copy of a picture to Australia as rapidly as possible	

5. Give three examples of computerized systems that affect our daily lives.
6. What are the three major technologies that make up information technology?
7. Copy and complete the chart below. (Refer to Figure 13.10.)

	Binary number	Number
A		18
B		112
C	10001001	
D	1001101	

8. The common term for a microprocessor is ________.
9. What is the major task of a telecommunications system?
10. Telecommunications technology uses three different systems to transmit data. Describe the devices used by each system.

APPLY YOUR KNOWLEDGE

1. From your own community, give three examples of "old" information technology and three examples of "new" information technology.
2. Which of the communication systems described in this chapter do you or your parents use each week?
3. List the input and output devices used with computers. Give an example of the use of each.
4. Look through one issue of a newspaper. Cut out all the references to microelectronics. Try to find references from each section of the newspaper, including the advertisements.
5. Draw a block diagram to show the major components of a telecommunications system.

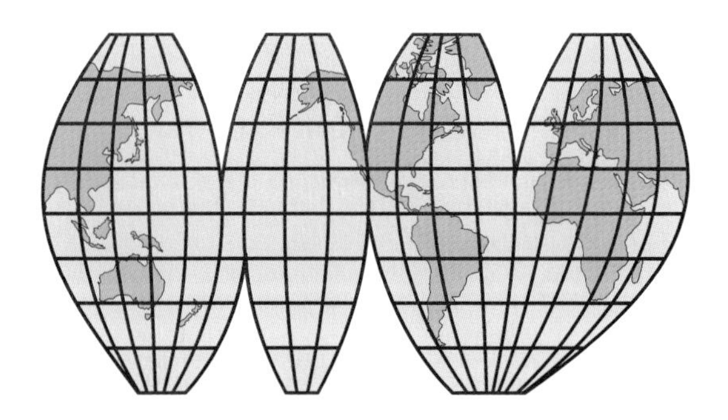

Chapter 14

Past, Present, and Future

OBJECTIVES

After reading this chapter you will be able to:
- O Describe examples of early technology.
- O List major inventions and innovations in different historical periods.
- O Cite examples of the technological change that is occurring at an ever-increasing rate.
- O Discuss the impact of technology on the individual and society.
- O Assess the environmental impact of various technologies.
- O Give examples of leading-edge technologies.

■ THE BEGINNING OF TECHNOLOGY

About two million years ago, prehistoric humans made the first tools. They discovered that when a large pebble is struck with great force against another stone, pieces flake off. Sharp cutting edges are formed, figure 14.1.

The list of stone tools about 300 to 400 thousand years ago included hand axes, points, clubs, and scrapers. The points were made of flint, roughly shaped, figure 14.2.

The bow and arrow was invented about 10,000 years ago. At the same time, small pieces of flint were attached to bone and wood to make knives and spears.

Until 8000 years ago, humans had obtained their food by hunting and gathering. They were nomadic, moving around in search of food. In 6000 B.C. they discovered agriculture and domesticated (tamed) animals. Stone tools become highly refined. About 2000 B.C., stone tools gradually gave way to tools cast in copper and bronze, figure 14.3.

ECRITEK

Figure 14.1. For what purposes did prehistoric people use these stone tools?

Figure 14.2. A spear point made from flint. Edges were shaped by chipping away the stone from both sides.

Figure 14.3. About 4000 years ago, humans learned how to cast tools from metal. These are examples of their craft.

Later still, about 1000 B.C., iron began to replace copper and bronze. Iron could be heated and shaped as desired.

People of Egypt and Mesopotamia were the first to use simple pulleys and levers. Later, Greek and Roman engineers developed these machines further.

The progress of technology was generally slow until about two centuries ago. In the 1700s inventions occurred more rapidly. Many of these inventions were machines to make products. They were run by a new power source, the steam engine. This mechanization led to the Industrial Revolution. Applied to railways and ships, steam power completely changed transportation. The invention of the telegraph and the telephone made long distance communication fast and easy.

These developments during the Industrial Revolution laid the foundations for modern technology. This technology has led to offshore oil rigs and nuclear reactors, combine harvesters and automated milking parlors, synthetic fibers and industrial robots, suspension bridges and hydroelectric dams, telephones with storage memory, compact discs, microwave ovens, digital sound systems, portable computers, remote control television sets, and space shuttles.

From the Stone Age to the present day, invention has been one of the most important human activities. Some of the most important inventions are summarized in figure 14.4. What conclusion can you draw about the rate at which inventions have occurred?

■ THE PRESENT: HOW TECHNOLOGY AFFECTS US

For the most part, technology has improved the quality of our lives. We live longer, healthier lives. We have more goods and services. Machines have done away with drudgery and hard physical labor. We have more leisure time. However, we have not always taken the time to study fully the impact of each technology on our lives or our environment. No matter what the benefits of any new technology, history shows that there is a cost for each advance.

Usually we see the benefits immediately. We like new technology; it makes our lives easier or more fun. Once we have a new technology it is hard to imagine life without it.

Benefits vs costs

However, the costs of using a new technology may only become known years after its introduction. For example, people have used chemicals such as DDT to increase food production. The chemicals killed unwanted insects that destroy crops. However, they also killed many birds, squirrels, cats, muskrats, rabbits, and pheasants. More recently, some of these chemicals have been found to cause cancer in humans.

Is there also a cost associated with television? The benefits deriving from this technology are easy to see. These include access to world news, live coverage of sports events, and famous celebrities to entertain us at home. What are the costs? What effect has television had on family life? Do people read in groups or as individuals? Are viewers affected by watching so much violence?

Unforeseen accidents. Another cost of technology, often a very high one, is the unforeseen accident. Bhopal, capital of the Indian state of Madhya Pradesh, used to be a peaceful city. This peace ended the night of December 2-3, 1984. A mysterious and deadly fog was discharged from a nearby pesticide

Period	Food	Shelter	Clothing	Defense	Transportation	Communication	Health
500,000–10,001 BC	Technological characteristics: tools of ivory, bone, wood, antler, and a variety of stones						THE SIX SIMPLE MACHINES WERE DEVELOPED DURING THESE PERIODS.
	-hoe -harpoon -nets for fishing	-cave painting -hides over frame -oil lamp	-animal skins -needle	-fire -spear -bow and arrow	-dugout	hieroglyphics	
10,000–1 BC	Technological characteristic: a basic understanding of metallurgy						
	-Archimedes' water screw -waterwheel -quern to grind corn -trained animals to pull plows -pottery -spoon -fishhook -sickle	-sun-dried mud hut -lock and key -rope and pulley -gear -brick -arch -nail -glass -bath	-vertical loom -cosmetics	-monumental stone buildings -bronze and iron weapons -knives -swords -sling	-sled -wheel -sail -boat -skis -harness (oxen)	-ink and paper -cuneiform writing -Phoenician alphabet -astrolabe -Etruscan alphabet -first coinage -papyrus -maps	-false teeth
AD 1–1399	Technological characteristics: small factories, foundries, forges, and mills by waterwheels						
	-aquaduct -windmill -horse collar -horseshoe -porcelain drinking vessels	-stained glass -Roman central heating -clock -dome -chimney	-spinning wheel -trousers -felt hat -button -lace	-crossbow -bronze-cast cannon -gunpowder -gun	-magnetic compass -Roman roads -Viking longboat -horse stirrup and saddle -rudder -skates	-movable type -paper from bamboo (Chinese) -stencil -pen -printing	-hospital -spectacles
AD 1400–1699	Technological characteristics: blast furnaces to melt and cast iron; the development of many scientific instruments						
	-ice cream -pressure cooker -bottle cork	-wallpaper -watch -theodolite -water closet (toilet) -thermostat -barometer -surveying instrument	-knitting machine -umbrella	-artillery shell -naval mine -hand grenade -rifle -submarine	-diving bell -dredger -telescope -wheelchair	-Gutenberg's press -arithmetic signs (+ − = ×) -newspaper -envelope -calculating machine	-toothbrush -artificial limbs -microscope -thermometer -innoculation
AD 1700–1849	Technological characteristics: first mechanized factories and the start of mass production; the use of all machine tools						
	-canned food -threshing machine -carbonated water -steam tractor -reaper -seed drill -sandwich -fertilizer	-electricity -street lighting (gas) -iron frame building -fire extinguishers -cement -matches -central heating	-spinning jenny -power loom -cotton gin -sewing machine -waterproof coat -dry cleaning	-Winchester rifle -standard parts for guns -machine gun -shrapnel	-pneumatic tire -bicycle -lifeboat -locomotive -hot-air balloon -sextant -roller skates	-metric system -photography -lithography -typewriter -Morse's telegraph -Braille -steel pen -eraser -postage stamp	-bifocals -anesthetics -sedatives -porcelain false teeth -ambulance -vaccination -plastic surgery -stethoscope -blood transfusion

Figure 14.4 Over 500,000 years of technology are represented in this chart. What will be invented in the next 100 years? *(Chart continued, next page.)*

Period	Food	Shelter	Clothing	Defense	Transportation	Communication	Health
1850–1899	Technological characteristic: use of magnetism to produce electricity						
	-barbed wire -refrigeration -condensed milk -milking machine -margarine -cola -breakfast cereal	-high-rise building -plastics -linoleum -electric lighting	-jeans -man-made fibers -zipper -aniline dyes	-dynamite -submarine -automatic machine gun -torpedo	-hang glider -airship -glider -clipper ship -diesel engine -automobile -helicopter -modern bicycle -motorcycle	-telephone -typewriter -cinematography -wireless telegraph -radio -postcard -fountain pen	-hyperdermic syringe -pasteurization -dental drill -antiseptics -incubator -X ray -aspirin
1900–1945	Technological characteristic: conquest of the skies						
	-tea bag -frozen food -insecticide (DDT) -combine harvester -supermarket	-prestressed concrete -air conditioning -fluorescent lighting -vacuum cleaner	-electric washing machine -nylon -artificial silk	-poison gas -tank -radar -gas mask -aerial bomb	-aircraft -tracked vehicles -safety glass -seaplane -traffic lights -helicopter -jet aircraft -subway system	-motion pictures -television -xerox -ballpoint pen	-electro-cardiograph -hearing aid -blood transfusion -chemotherapy -insulin -iron lung -kidney machine
1946–present	Technological characteristics: use of atomic energy; of computers; and of microprocessors to control apparatus						
	-synthetic fertilizers -microwave oven -nonstick pan -domestic deep freezer -foods for use in space -new ''miracle'' strains of rice & wheat -ultrasonics to detect fish -cloning	-solar panel -geodesic dome -synthetic turf -fiberglass insulation -lightweight modular dwellings -space station	-synthetic fibers -automatic clothes dryer -permanent creases in clothes -metallized fabric	-atomic bomb -ejection seat in aircraft -portable atomic weapon -rocket -hydrogen bomb -ICBM	-aqualung -hovercraft -lunar vehicle -nuclear submarine -monorail train -space shuttle -ultra high-speed train -VTOL aircraft -space station -fuel injected engines -magnetic levitation trains -space orbiters	-photo typesetting -radar -transistor -satellite -instant camera -long-playing record -laser beam -fiber optics -computer network -hologram camera -silicon chip -videotape -microprocessor -pocket calculator -integrated circuits -digitized typesetting -desktop publishing -compact disc player -laptop computer -space telescope	-artificial voice box -heart-lung machine -artificial heart and other organs -equipment for organ transplants -high speed dental drill

Figure 14.4. (Continued.)

factory. This fog began to settle into a section of Bhopal called Khazi Camp. Hundreds of families lived along a road bordering the factory. The fog was a gas, methyl isocyanate, used in pesticide production. This cloud of gas changed the lives of more than 250,000 people; 2000 of them died. Nobody knows exactly what effect the exposure will have on future generations.

Sometimes workers see the introduction of a new technology as a threat. Groups or individuals then resist change. Ned Lud, an English millworker, in 1799 destroyed two textile machines belonging to his employer. By smashing these labor-saving devices, he had hoped to avoid unemployment. Was Ned Lud's fear of technology justified? In this case, probably not. History shows that advances in technology usually produced more jobs than were lost. For example, when Henry Ford used an assembly line to mass-produce cars, the cost of each car was greatly reduced. More people could then afford to buy cars. This, in turn, led to an increase in the number of people required for service and repair.

Serious problems

On the other hand, it is sometimes wise to be cautious about the introduction of a new technology. For example, society today is increasingly concerned about the generation of electricity from nuclear fission. The experience of Chernobyl and Three-Mile Island, where radioactivity escaped into the atmosphere, have taught us sobering lessons.

The damage resulting from the accidents at Bhopal, Chernobyl, and Three-Mile Island was immediately obvious. Equally serious but less obvious damage is occurring constantly all around us.

Pollution in the home. Inside the home, air is **polluted** by materials we use and the things we do, figure 14.5. Plastics, adhesives, and insulation materials used in the construction

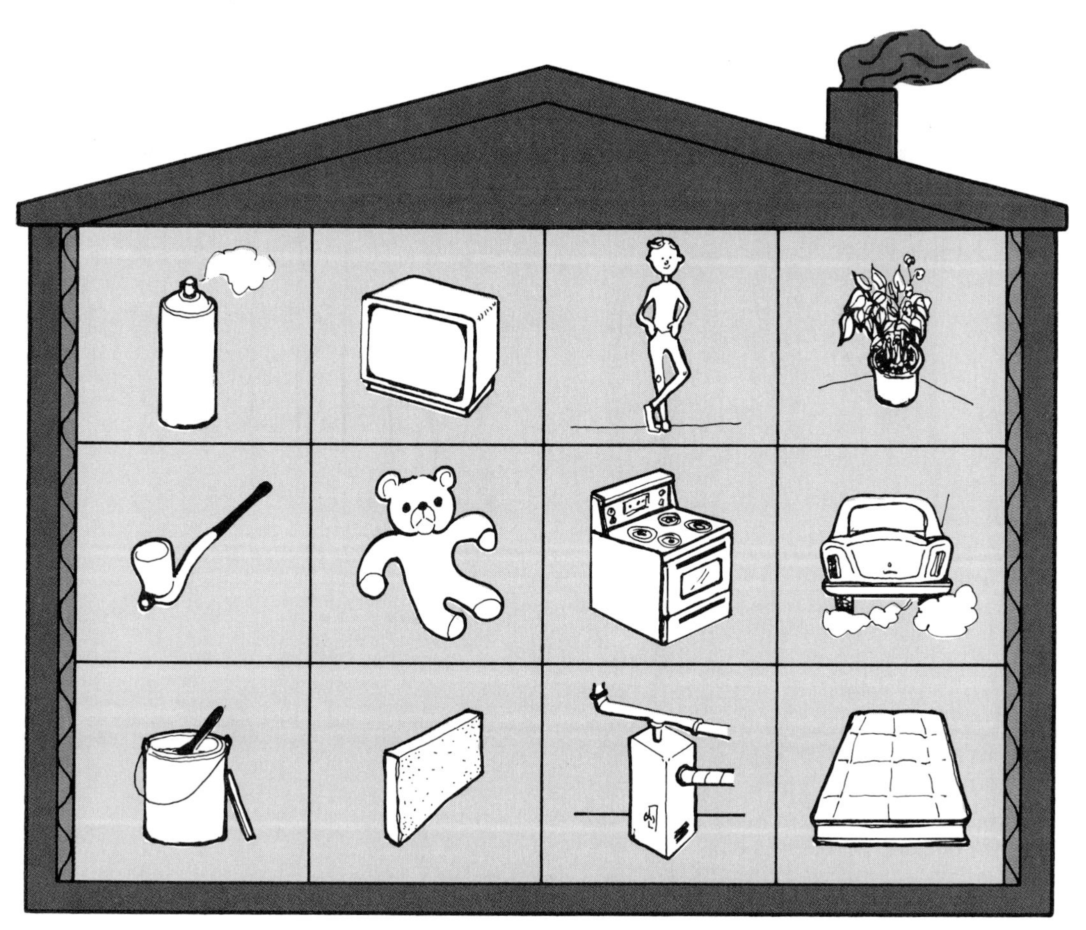

Figure 14.5. How many of the items pictured here are polluting the air in your home?

and furnishing of a house slowly disintegrate (break down). This causes them to give off harmful gases which pollute the air. To this list add tobacco smoke, carbon monoxide (from cook stoves and furnaces) and molds caused by condensation and poor ventilation.

Noise is sound we do not want to hear. It is a pollutant that affects people in the home, at work, and during leisure time. Experts estimate that the average noise level in North America is doubling every 10 years. For many people, the level is already high enough to cause hearing loss, nervousness, hypertension, ulcers, and suicidal tendencies.

Pollution outside the home. Outside the home, various forms of pollution affect our air, water, and soil. Ozone is a form of oxygen that exists throughout the atmophere. Nothing on earth can live without it. The **ozone layer** forms a protective shield that absorbs much of the sun's ultraviolet radiation. This radiation, if allowed to reach the earth, would cause skin cancer and threaten many life forms. We have evidence that the chlorofluorocarbons used in refrigeration, synthetic foam, and aerosol spray are destroying ozone in the atmopshere. Experts generally agree that every 1 percent reduction in the ozone layer could increase skin cancer in humans from 3 to 4 percent. Greater reductions could:

- Affect fish production and plankton, a vital element in the aquatic food chain.
- Damage food crops and plants.

Air pollution occurs when unwanted airborne matter, such as smoke, is added to the earth's atmosphere. In the United States alone, industry pumps out at least 250 million tons (227 million tonnes) of noxious waste each year. This form of pollution may affect the health of humans or animals and may damage materials. The first noticeable effect of air pollution is reduced visibility. If pollution is severe, it may even cause death. The great smog of 1952 in London, England, was responsible for thousands of deaths. Air pollution also corrodes metal structures, such as bridges, and blackens the stone facings of buildings.

Burning fossil fuels (oil and coal) is the major source of carbon dioxide pollution in the air. A buildup of carbon dioxide prevents solar radiation from escaping back into space. The effect is like a greenhouse; the atmosphere warms up. It is estimated that doubling the carbon dioxide would cause global temperatures to rise between 4 ° and 7 °F (2 ° and 5 °C). What effect would this change have on the polar ice caps, forests, and agriculture?

Acid rain. When rain mixes in the air with pollutants from burning fossil fuels the pollutants return to earth as **acid rain.** Severe damage occurs to forests, agricultural lands, lakes, machinery, and buildings. The air pollution may take place great distances from the problem area. In North America certain areas are particularly sensitive to the impact of acid rain. Sources of sulfur dioxide emissions are shown by black dots in figure 14.6. Those areas likely to be affected are shown in red and include important forest and fishing areas.

Lake and river pollution. Lakes and rivers are also polluted when undesirable foreign matter is introduced into them. There are four sources of water pollution — natural, sewage, thermal, and industrial, figure 14.7.

Natural pollution comes from the runoff from fertilized land. Sewage pollution is caused by the dumping of untreated human waste. Thermal pollution is the adding of heated water to a river or lake. The heat is usually the result of steam or electrical power

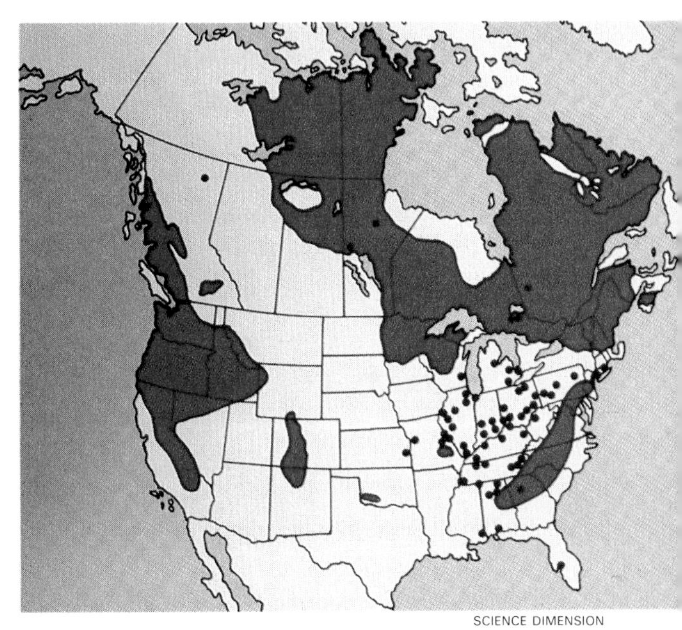

SCIENCE DIMENSION

Figure 14.6. Is acid rain affecting the area where you live?

generation. Industrial wastes that pollute include:
- Waste blood, food, and animal matter from processing plants.
- The chemical effluents (wastes) from petroleum, pulp and paper, and manufacturing industries.

Solid waste. The problem of solid waste disposal is becoming acute. What do we do with our garbage? About half the cities in the United States have filled up their local dump sites. As new sites are located farther away from a town or city, the costs of trash and garbage disposal increases.

■ ATTACKING POLLUTION PROBLEMS

One approach to the problem of solid waste disposal is **recycling.** This involves separating the different types of wastes and reclaiming those that can be reused. **Recycling** also helps to conserve our limited natural resources. The aluminum from empty soft drink cans can be remelted and made into new products. Old newspapers and magazines can be reprocessed into pulp to make new paper. Returnable glass bottles can be washed, sterilized and reused. Recycled plastic conserves petroleum.

Some products may be recycled by replacing the broken parts and keeping the product in use. A car engine can be rebuilt. The major parts can be remanufactured or replaced. Using cloth diapers is preferable to using disposables. In fact, the cost of using a diaper service is less than the cost of disposables.

Using the same material a number of times, less new material needs to be used. In some cases, not only does recycling conserve material but it also saves energy. Recycling aluminum requires only 5 percent of the energy needed to obtain the same amount from ore.

Many wastes are hazardous to our health. Before disposal, they must be treated:
- Solvents, pesticides, and paints can be almost completely destroyed in incinerators.
- Plant and animal wastes can be broken down by chemical treatment.
- Toxic metals may be treated by sealing them in ceramic.

CHRISTOPHARO

Figure 14.7. When lakes are polluted, fish die and the water becomes unfit to drink.

- PCBs, mercury, and lead can be digested by bacteria.
- Waste water containing small amounts of hazardous material can be injected into pockets of rock deep underground.

Reducing technology's negative effects

What can we do to reduce the negative effects of technology? Nearly everyone can become involved in three ways:
- Become aware that problems exist. One reason for this book and your technology course is to make you aware of the issues.
- Realize that technology is neither good nor bad. It depends on how it is used. For example, lasers are being used to perform very delicate surgery. They can also be made into highly destructive weapons.
- Each of us can become active in the solutions. We must be wiser users of technology. If we know that a certain technology is causing damage to the environment, or harms individuals, we must avoid using that technology. For example, we know that chlorofluorocarbons (CFCs), used in some aerosol sprays, are destroying the ozone layer. Ozone screens out harmful radiation from the sun. Without this protection, skin cancer and eye cataracts will increase. We should, therefore, avoid aerosol sprays that use this substance.

☐ THE FUTURE: CLUES FROM THE LEADING EDGE OF TECHNOLOGY

Have you heard of smart cards, **genetic engineering,** fiber optics, **expert systems,** and **artificial intelligence?** These are strange new words that are all part of the leading edge of technology. Let us look at a few of the technologies that will likely affect our lives in the near future.

Biotechnology

What is **biotechnology?** Biology is the study of living organisms. Technology is about solving problems to provide the goods and services we need. Biotechnology, then means the use of living organisms to make goods and servide services. These goods include chemicals, foods, fuels, and medicines. Services that depend on biotechnology include pollution control and waste treatment.

Farmers provide one of the most important goods—food. However, farming is a risky business. Weeds sometimes seem to grow faster than the crops. As if that weren't bad enough, the farmer also has to fight insects and plant diseases. An early frost in autumn or a late frost in spring can wipe out an entire crop. Having too much rain can cause as much damage as too little.

For as long as farmers have farmed, they have sought to overcome these problems. They have done so through selectively breeding animals and plants. For example, suppose a farmer wanted to increase the yield of soybeans. Plants with the highest yield would be crossed with plants that have a resistance to a particular problem such as drought. This method, however, takes generations. There is no guarantee of success.

Today, hardier plants and animals are possible in a single generation through genetic engineering. Genetic engineering is a technology used to alter the genetic material of living cells. Cells are the basic units of all living organisms. Inside a cell is a nucleus. It contains the hereditary information or genetic material of a cell. This genetic material determines, for example, whether a person is tall or short, whether a cow can produce more or less milk, and whether a plant can withstand drought.

Genetic material can be transferred from one type of living cell to another completely different type. When this is done, an entirely different organism is created. In this way, a high-yield crop that withstands drought could be quickly developed. There would be no need to wait generations, figure 14.8.

Through genetic engineering the following farm products may be developed:

- Frost- and drought-resistant crops.
- Disease-resistant plants that do not need pesticides.
- High-yield plant crops.
- Tastier and more nutritious foods.
- Plants that respond to biodegradable herbicides and pesticides.

Bioprocessing

A second branch of biotechnology is called **bioprocessing.** Bioprocessing uses microorganisms. Microorganisms are living cells that must eat to survive. Fortunately, some microorganisms will eat the plant and animal matter (biomass) in garbage. When this occurs, a gas called methane is produced. This gas can be used for cooking and heating.

In a similar way, sewage, containing human and industrial wastes, can be purified. The microorganisms eat some of the solid matter and, thereby, help to decompose (break down) the wastes. The capacity and efficiency of waste treatment plants can be increased by using genetically engineered microorganisms.

NFB

Figure 14.8. Carefully controlled experiments produce high-yield wheat.

Antibody production

A third branch of biotechnology is **antibody production.** When a foreign substance, such as a bacteria or virus, enters our body, the immune system recognizes it as an invader. This system responds by fighting the invader. It produces vast amounts of defense agents called antibodies. These attack and kill bacteria. Researchers have long sought ways to produce larger quantities of antibodies. They want antibodies selective in the virus or bacteria they destroy.

The goal of a new technology, called monoclonal antibody technology, is to isolate a certain antibody and to produce enough of it to be effective against a disease. Monoclonal antibodies will be used to attack cancer cells or virus organisms. Healthy cells will be unharmed. The technician in figure 14.9 is testing cells and isolating those that contain antibodies. The antibodies are cloned (identical reproductions made) in an incubator, as shown in figure 14.10.

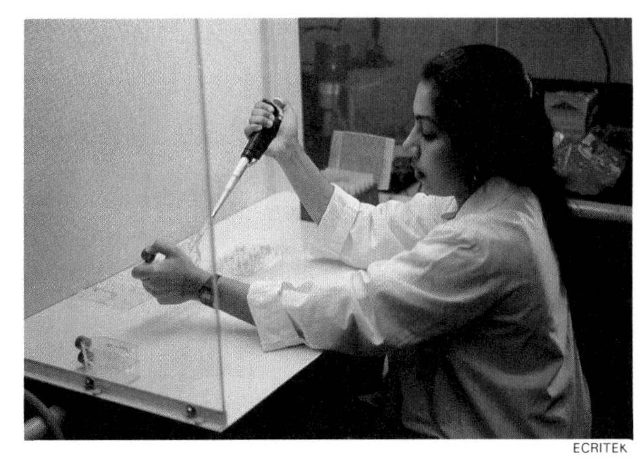

ECRITEK

Figure 14-9. A technician isolates a positive clone and tests a culture for positive antibodies.

ECRITEK

Figure 14.10. Incubators are used to grow antibodies.

Communication

The communications industry has had a period of rapid development and change from that shown in figure 14.11. Analysis of information is increasingly carried out by technical equipment, figure 14.12.

New telecommunication networks provide many new services to the consumer. Among them are interactive television, electronic mail, personalized computer communication service, pay-by-phone service,and synthetized voice messages. Interactive TV provides instantaneous two-way communications between viewers and a central computer. This permits the convenience of ordering merchandise at home with the touch of a button. The same technology allows pay-per-view TV. Without leaving his or her chair a viewer simply orders a movie or TV show from a video catalog.

BELL

Figure 14.11. How many operators are needed to handle a million long distance telephone calls? Thirty years ago it would have required thousands; today less than a hundred!

New materials

Important new materials are made possible through new technology. They offer advantages such as being easier to shape or to fasten. They may be corrosion-free, stronger, or lighter. New materials include: plastics, adhesives, and composites.

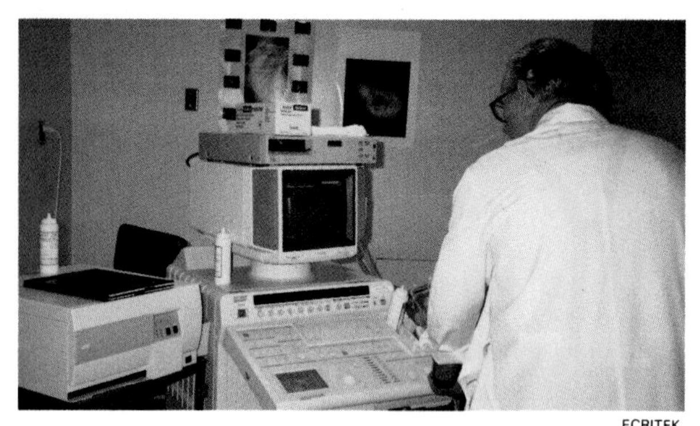

ECRITEK

Figure 14.12. Ultrasound machines can show images of internal organs immediately.

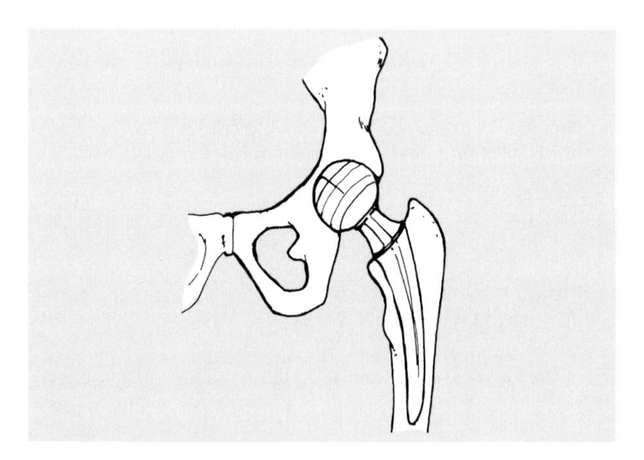

Figure 14.13. It is now common to replace damaged hip joints with a synthetic ball and socket. They are bonded to bone with an acrylic cement.

Figure 14.14. The windshield and many other parts of this motorcycle are made of tough plastics and composites.

Derived from petroleum, plastics have already replaced metals as the most common industrial material after wood and concrete. This is particularly noticeable in the automobile industry. Plastic is replacing metal at the rate of 10 percent a year. Body panels, seats, underbodies, fuel tanks, and even springs are expected to be made of plastic. Among the advantages of plastic parts is their ability to be molded into shapes that can be snapped together.

Plastic parts can be fused together using radio frequency waves or by bonding with adhesives. Metal parts are frequently welded, a more difficult and costly process.

Finishing processes can be eliminated with plastics; color and shine can be molded into the material. Plastics are by nature good electrical insulators. New plastics are being created, however, that will conduct electricity. Plastics are even being used in replacement parats for the human body, figure 14.13.

Composites. Composites are new materials made by combining two or more unlike materials. The composite gives better performance and reduces costs.

Recently, composites have been made by combining plastic resins with ultra-strong synthetic fibers such as Kevlar or graphite. These materials are now used throughout the mass transportation industry in the production of light rail cars, aircraft wings, boat hulls, and urban (city) transport vehicles, figure 14.14. Much of the sports equipment now in common use relies upon composites.

Adhesives. Robots' welding guns are being replaced by adhesive guns as more products are built from plastics, aluminum, and composites. Consumers favor them for their light weight and corrosion resistance.

Ceramics. The microelectronics industry depends upon new ceramic materials. These ceramics are related to those used to make earthenware dishes, tiles, and bathroom fixtures.

Engineers are now testing gas turbines and diesel engines made of ceramic. Such engines can operate at higher temperatures than those made of cast iron or aluminum. Hotter engines burn fuels more completely and convert more heat to power. The result is an increase of 30 percent in fuel efficiency. Ceramic engines perform better than metal engines at high temperatures.

Research into materials is important. Many of the recent Nobel prizes in physics have been awarded to materials scientists.

Structures and their systems

Some countries have been developing the technology and design skills to produce high quality housing assembly-line techniques. One leading manufacturer of factory-built homes can produce a house every 44 minutes.

More than one-third of the energy we use, including over 60 percent of the electricity, goes toward heating, cooling, and lighting buildings. New products to increase the energy efficiency of building are presently being developed. They include:

- Window panes separated by an air gap. The inner surface of one pane is coated with a thin film. The film has multiple layers of metal or metal oxide. While transparent to the sun's light and heat, the coating does not transmit long-wavelength infrared radiation. This is the type of radiation given off by warm indoor objects, figure 14.15.
- Plastic pipes that bring sunlight deep into a building's interior, reducing the need for artificial light. Buildings can be lit with "light pipes." Acrylic tubes distribute sunlight that has been collected and focused by rooftop mirrors.
- Superinsulated houses so snug that they can be heated by the warmth of bodies, light bulbs, and appliances.

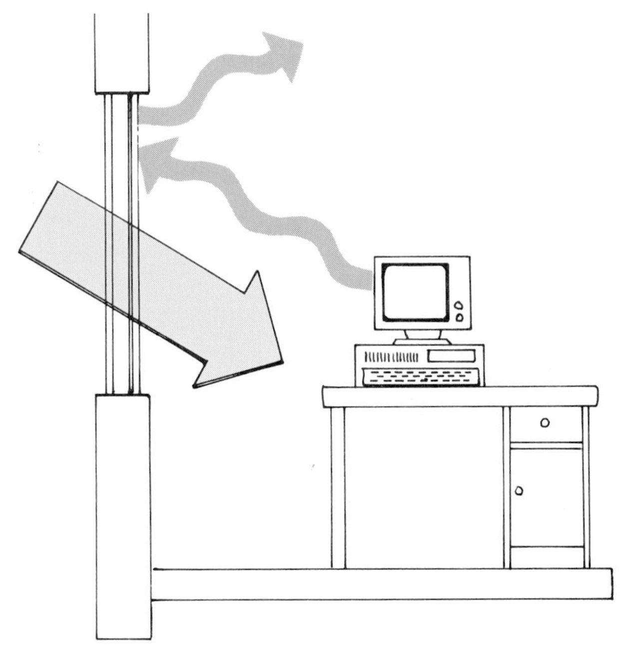

Figure 14.15. Windows of the future may work as heat "mirrors."

In the "smart house" of the future, all of the appliances and systems will be linked and controlled by microprocessors.

- Control of doors will be based upon voice recognition.
- Moisture sensors will determine when the lawn should be watered.
- Appliances will communicate with each other. For example, a message on the television screen will tell when the wash cycle is completed and clothes should be moved to the dryer.
- Electrical appliances, telephones, stereo speakers, and televisions will plug into the same outlet.
- Smart heating, cooling, and ventilation systems will respond to whether a room is empty or occupied. This will improve comfort and use less energy.

Space flight. The **space shuttle** will be the space transportation vehicle of the 1990s. The novel features of the shuttle, developed by NASA, are:

- It is a reusable system.
- It is capable of operating in space as well as flying in the Earth's atmosphere.

The shuttle can transport a maximum load of 65,036 lb. (29 500 kg) to an orbit 186.4 miles (300 km) above the earth. The propulsion unit consists of three engines burning liquid hydrogen and oxygen. There are also two gigantic solid fuel engines. These engines, called boosters, are used at lift-off.

On-board computers control all the functions of the shuttle. Insulating materials, such as the carbon-carbon tiles, keep temperatures to prescribed limits. The shuttle is also equipped with the "Canadarm" or Remote Manipulator System. This mechanical arm, figure 14.16, can be maneuvered at a distance by the astronauts. It is used for operations such as handling and release of payloads in the cargo bay. Another use is the recovery of satellites needing repairs.

NASA

Figure 14.16. This mechanical arm, fitted to the space shuttle, handles payloads or retrieves satellites in need of repair.

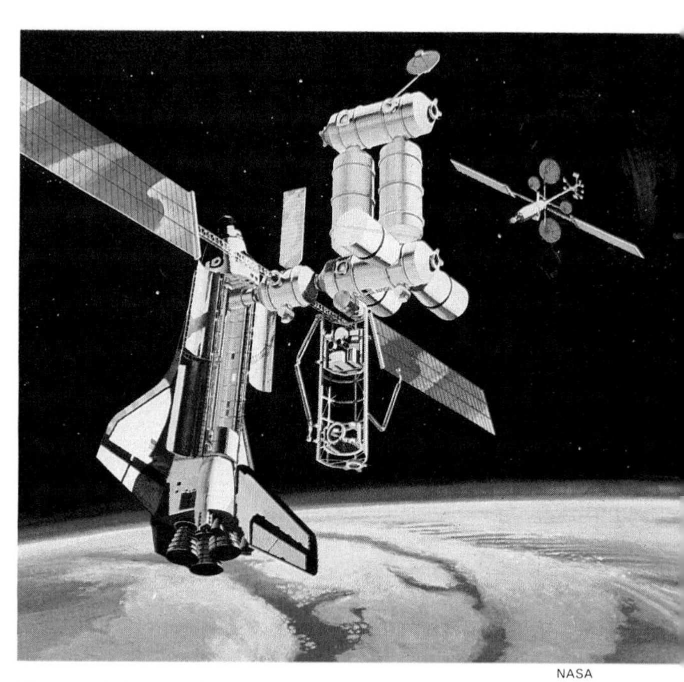

NASA

Figure 14.17. Space stations are the next step in the space program.

Living in space. The space station, orbiting at an altitude of 300 miles (480 km), is the next logical step in space. It is being developed for initial occupancy in the mid 1990s, figure 14.17. Pressurized modules are attached and connected by a series of external air locks and tunnels. The modules, each about the size of a large bus, serve as laboratories for basic research. There will be a "look up" observatory for astronomy, and a "look down" observatory for earth studies. There will also be a plant for manufacturing products not producible on earth, and a facility for earth communications, weather observation, and remote-sensing.

In the future, the space station can become a departure point, like the base camp of a mountain climb. It will be a base for such activities as building a permanent moon station and manned missions to Mars.

The space shuttle will help in the conquest of space. It will help enlarge our store of knowledge, and will have many positive applications for day-to-day life on Earth.

The development of tools, clothing, and vehicles for use in space has already resulted in a number of products we use daily. The Apollo lunar suit worn by moonwalking astronauts had special boots built into it. The material in the boots is now used in athletic shoes designed for improved shock absorption, energy return, and reduced foot fatigue.

Superconducting materials. The large-scale transmission of electrical power may, in the future, make use of new superconducting materials. Ordinary conductors, such as copper or aluminum wire, can be used to carry an electrical current. However, they offer some resistance to its flow. Because of this resistance, a part of the electricity is lost to heat. Conventional transmission lines lose power at a rate of one percent per 100 miles (160 km).

Superconductors are materials having no electrical resistance. The problem with them up until now is that they worked only at very low temperatures—several hundred degrees below zero. Researchers aim to develop superconductors that will operate at room temperatures. A major use of these superconductors will be in the large-scale transmission of electrical power. Superconducting transmission lines would largely eliminate energy loss.

Superconductors might also be used in energy storage. Large superconducting coils could stockpile electricity. Generators could, as a result, operate full-time while adjusting to variations in daily demand.

Computers

In just 25 years, computers have changed our lives, figure 14.18. They help make possible:

- Instantaneous, worldwide telecommunications.
- Accurate weather prediction.
- Numerical modeling of the world economy.

Even today's most powerful computers are not fast enough to handle some extremely complicated problems. To increase their power, single-processor computers must constantly add faster and more expensive circuitry. The fastest of these machines, supercomputers, run into physical barriers. Circuit switching speeds reach their top limits. Then, components must be placed closer and closer together to reduce the distance electrical signals must travel. At some point this closeness exceeds the ability of the circuits to scatter the generated heat.

Parallel computers. The future belongs to parallel computers. They will operate like a human brain; however, their speed will be much greater. A parallel computer will also handle many functions at the same time. It will do this by using a large number of processors.

Figure 14.18. Modern businesses use computers for many tasks.

The human brain is basically similar to this. It can monitor and guide many activities. It controls breathing, thinking, walking, and talking all at the same time. Without this ability we could not survive. In the year 2000 computers will parallel human thinking.

Expert systems. Hospitals rely upon a computer program called Mycin. This is an **expert system** used to diagnose blood infections. The program asks the user a series of directed questions about the patient. The questions cover laboratory test results, medical history, and particular symptoms. Answers lead the computer to a diagnosis (finding the problem). The computer is usually about as accurate as a human specialist. After diagnosis, Mycin considers patient weight, allergy history, and other factors. Finally, it recommends a proper medication.

Expert systems can handle a wide variety of problem-solving. They are used by:

- Salespeople for designing computer systems.
- Geologists for finding oil and minerals.
- Engineers for trouble-shooting problems with complicated equipment.
- Biologists and chemists for interpreting experiments.
- Manufacturers for staying on top of production schedules.
- Doctors for diagnosing heart defects.

Expert systems organize and deliver knowledge effectively. They cannot be said to ''think'' or ''know.'' They simply work.

Artificial intelligence. Computer technology is advancing rapidly. The next logical step is artificial intelligence (AI). Using AI, computers can learn, solve problems, and reason. A part of the AI research being done concerns designing machines that can ''see.'' This is no simple matter. Over half of the human brain is occupied with vision. Experts say that the equivalent of 10 billion calculations are made of an image before it reaches the human optic nerve. Already, computer vision systems can distinguish between parts. Vision will make industrial robots more flexible.

Advanced manufacturing technologies

Computer-integrated manufacturing **(CIM)** is the use of computers to carry on a whole range of manufacturing processes. These include design, evaluation, manufacturing, accounting, inventory control, and other business tasks. These processes are integrated (made one) and controlled by a central computer. Little human help is needed. Production equipment work stations are linked by a materials-handling system. This system moves parts from one work station to another. The new factory could be paperless. Information is shared among the work force through computer terminals. It is also more flexible.

Manufacturing cells. A manufacturing cell is a group of machines set up to produce a product, figure 14.19. Controlled by computer, the typical cell might include a lathe, a milling machine, and a drill. Robots transport material to and from the machines.

Cells can be changed quickly to meet productions needs. They readily accept new designs from a CAD system.

''Robot'' is a word for a machine designed to move objects. The word was first used by a Czech writer, Karel Capek. He coined it for his 1920 play, ''R.U.R. (Rossum's Universal Robots).'' Those used by today's industry combine mechanical, electrical, and hydraulic systems. The robot is controlled by a small computer.

CINCINNATI MILACRON

Figure 14.19. The modern factory may use machining centers or work cells. Robots can be seen in the right background. The control room is at the top left.

Robots bring materials to the machines during manufacture. They move objects from one manufacturing machine to another. They can shift, lift, and turn the parts just as a human operator would. Robots can also be fitted with tools to perform work on products being manufactured. Walking robots work in environments too hazardous or inaccessible for humans. They may handle ammunition, fight fires, or do routine chores in a nuclear power station, figure 14.20.

■ A LAST WORD

From earliest times to the present, men and women have tried to meet their needs using technology. As you have seen, technology is a process in which you:

- Identify a problem.
- Propose solutions.

SPAR

Figure 14.20. Remote manipulation and control system. It is designed to remove, inspect, and replace large components in the radioactive core of CANDU nuclear reactors.

- Choose the solution that responds best to the stated problem.
- Build a model or prototype.
- Test and compare the model or prototype with the originally stated purpose.

While technology, as a process, has not and will not change, the problems that we solve using this process change continually. For instance, a need of early humans was for a better club for hunting; a need today is to produce a fifth generation of computers that can "think" by themselves.

The technology of early humans was simple. Today's technology is often complex. A new technology may cause change throughout the world. We must learn to evaluate the effect that any new product or service has on our lives and our environment. While many technologies have beneficial results, we are increasingly aware that humans are capable of making mistakes. These mistakes may be costly for us and for future generations.

In the future, you will be a citizen with the responsibility for making decisions about the use of new technologies. Having read this book, you should be able to make more informed decisions.

SUMMARY

Although technology has existed since prehistoric humans roamed the world, it changed slowly until 200 years ago. For the most part, technology has improved the quality of our lives. However, the costs of using a new technology may only be apparent years after its introduction. To minimize the negative effects of technology, people must:

- Be made aware of the problems that exist.
- Realize that technology, by itself, is neither good nor bad.
- Help make decisions on its use.

Certain technologies show particular promise for the future. Biotechnology uses living organisms to make goods or provide services. The three branches of biotechnology are genetic engineering, bioprocessing, and antibody production.

The communications industry has combined voice, data, print, and image technologies to form complex systems.

The materials industry has undergone rapid change. Plastics have replaced metals as the most common industrial material after wood and concrete. The microelectronics industry relies heavily on ceramics. New materials are also being used to build structures and to increase their energy efficiency.

New developments in transportation on land, in air, and in space include electric cars, electronic maps, supersonic transport, space shuttles, and space stations.

Energy needs of the future may be met by greater use of electricity. The electricity will be transmitted through superconducting materials.

In just 25 years, computers have changed our lives. They are increasingly essential in bank transactions, expert systems, artificial intelligence, and all aspects of the production process.

KEY TERMS

Acid rain
Antibody production
Artificial intelligence
Bioprocessing
Biotechnology
CIM
Expert system
Genetic engineering
Ozone layer
Pollution
Recycling
Space shuttle

TEST YOUR KNOWLEDGE

Write your answers to these review questions on a separate sheet of paper.

1. What do you consider to be the most important invention in each of the following time periods? Tell why each invention is important.
   a. 500,000 to 10,000 BC.
   b. 10,000 BC to 1 BC.
   c. AD 1 to 1400.
   d. 1401 to 1700.
   e. 1701 to 1850.
   f. 1851 to 1900.
   g. 1901 to 1945.
   h. 1946 to present.
2. What statement can be made about the number of objects invented during different periods throughout history?
3. Name 10 technical objects that have been invented in your lifetime.
4. Technical objects are designed to help

humans, but they sometimes have negative side effects. One example is shown in the chart below. Copy and complete this chart, adding three more examples.

Technical Object	Intended Effect	Negative Side Effect
DDT	Destruction of unwanted insects	Kills birds, pets and wild-life and is poisonous to humans

5. List items that you throw away that could be recycled.
6. Biotechnology is defined as ________.
7. Genetic engineering is defined as __.
8. A superconductor is a material ______
9. Describe how your life would change if computers could completely understand and reproduce the human voice.
10. Name five technical objects you believe will be invented 20 years from now. Describe (a) how each will work, (b) who will benefit from their use, and (c) what potential problems may occur as a result of using them.

APPLY YOUR KNOWLEDGE

1. Make a model of a tool used prior to A.D.
2. Select what you consider to be the most important invention in each of the historical periods in figure 14.4. State the reasons for your choice.
3. Make a list of five products that have been invented since you were born. Is the number of products invented each year increasing or decreasing? Explain your answer.
4. Choose one product and discuss its impact, both positive and negative, on (a) society and (b) the environment.
5. Give one example of "leading edge" technology in each of the following:
   a. Biotechnology.
   b. Communication.
   c. Materials.
   d. Structures.
   e. Mechanisms.
   f. Energy.
   g. Computers.

   If possible find examples that are NOT described in the textbook.
6. From a science-fiction book, comic, TV show, or movie in which events occur in the future, identify three technical objects or systems that do not exist today. Describe how each one operates.

Dictionary of Terms

A

Abutments: the supports where a bridge arch meets the ground. They resist the outward thrust (push) and keep the bridge up.

Acid rain: an environmental problem that occurs when rain mixes in the air with pollutants from burning fossil fuels.

Acoustical properties: properties of a material that control how it reacts to sound waves.

Adobe: a mixture of clay and straw used as a structural material in the southwestern United States.

Alloy: term describing a material that is a mixture of metals.

Alternating current: electron flow that reverses direction on a regular basis.

Alternative solutions: different ways of solving a problem.

Ampere: the unit that is used to measure the amount of current.

Antibody production: a branch of biotechnology in which researchers attempt to isolate and produce antibodies (substances the body's immune system produces to attack and kill bacteria) effective against specific diseases.

Arch bridge: a type of bridge in which the compressive stress created by the load is spread over the arch as a whole.

Artificial intelligence: a technology that makes it possible for a computer to learn, solve problems, and reason.

Artisan: an individual responsible for every step in producing a finished product; also, one skilled at a trade or craft (woodworking, for example).

Assembly line: a continuous process that is used to quickly produce a large number of identical items by assembling parts in a planned sequence.

Atom: the smallest possible particle of matter.

Automation: the use of computers or automatic machines to control machine operations and make a product.

B

Balance: in design, the arrangement of mass equally (or appearing to be equal) over the space used. There are three types of balance: symmetrical, asymmetrical, and radial.

Batter boards: boards that support the lines set up to locate the building so excavating can begin for the foundation.

Beam: a horizontal structural member, usually used to support floor or roof joists.

Bending: a method of shaping sheet material by folding it like a sheet of paper.

Binary: a simple electrical signal code used by computers. There are only two signals in this code, on and off (written as 1s and 0s).

Biomass energy: energy from plants that can be burned or processed as fuel.

Bioprocessing: the use of microorganisms to break down and purify organic wastes.

Biotechnology: the use of living organisms to make goods or provide services.

Bit: an on or off signal, known as a binary digit, in the binary code used by computers.

C

CAD: new method of making drawings using a computer. CAD stands for "computer-aided design."

Cantilever: a beam capable of supporting a load at one end when the opposite end is anchored or fixed.

Capacitor: a device designed to store an electrical charge, consisting of two metal plates (conductors) separated by an insulator.

Career: an occupation or way of making a living.

Casting: a method of shaping parts or products by pouring liquid material into a mold.

Ceramic: a material used for making pottery, bricks, and other products. The name comes from the Greek *Kermos*, meaning "burnt stuff."

Chemical energy: energy locked away in different kinds of substances that is often released by burning.

Chemical joining: a method of fastening joints by using chemicals such as glues, adhesives, solvents, and cements.

Chemical properties: properties of a material that affect how it reacts to its surroundings.

Chip: a tiny flake of a substance called silicon, covered with microscopic electric circuits, that is used in computers. Also known as a *microprocessor*.

Chiseling: a technique used to shape material by cutting away the excess with a chisel and mallet.

CIM: abbreviation for computer-integrated manufacturing; the use of computers to carry on a whole range of manufacturing functions and processes.

Closed-loop system: a system that includes a feedback device to provide control.

Communication: the process of telling other people about your ideas.

Communication system: a means of transmitting and/or receiving information between two or more points.

Composite: combinations of different materials.

Compression: a squeezing force.

Computer: a machine that can be programmed (given a set of instructions) and that can store a vast amount of data. Computers can perform many mathematical operations very rapidly.

Computer numerical control (CNC): a system in which machine movements and operations are controlled by a computer program.

Conduction: movement of heat energy by passing from molecule to molecule in a solid.

Conductors: materials that will carry an electric current.

Construction line: thin, faint lines used to start a drawing.

Convection: movement that occurs when expanded warm liquid or gas rises above a cooler liquid or gas.

Corrosion: the result of a chemical reaction in which a material is changed by its environment.

D

Degree of freedom: the term used to describe each joint, or direction of movement, in a robot arm.

Design brief: a statement that describes clearly what problem a design must solve.

Design process: a careful and thoughtful means of solving a problem by working through a number of steps.

Designer: a person who creates and carries out plans for new products and structures.

Diesel engine: an engine in which air is squeezed to very high pressure and temperature inside the cylinder. Diesel fuel is injected and spontaneous ignition occurs.

Diodes: devices that allow current to flow in one direction only. They are most commonly used as rectifiers to change alternating current to direct current.

Direct current: current that does not change direction in an external circuit.

Drilling: a process used to make holes in wood, plastic, metal, and other materials.

Ductility: a material's ability to be pulled out under tension.

Dynamic load: load on a structure that is always changing.

E

Elasticity: the ability to stretch or flex but return to an original size or shape.

Electric circuit: a closed path around which electrons can move.

Electric current: the flow of electrons in a circuit.
Electric motor: a machine that changes electrical energy into mechanical energy.
Electrical energy: the movement of electrons from one atom to another.
Electrical properties: properties of a material that determine whether it will or will not conduct electricity.
Electrical system: the circuits that carry electricity in a product or for light, heat, and appliances throughout a home.
Electromagnet: a magnet that is energized by an electric current.
Electron: one of the particles making up an atom. It has a negative electrical charge.
Electronics: the use of electrically controlled parts to automatically control or change current in a circuit.
Elements of design: the things you see when you look at an object, including line, shape and form, texture, and color.
Engine: a machine composed of many mechanisms that converts a form of energy into useful work.
Entrepreneur: a person who organizes, manages, and assumes the risks of starting a business.
Ergonomics: the study of how a person, the products used, and the environment (our surroundings) can be best fitted together.
Expert system: a computer program that uses a large base of data to solve complex problems in a specific area of knowledge.
External combustion engine: an engine that burns fuel outside itself.

F

Factory: a building in which products are manufactured.
Fatigue: the ability to resist constant flexing or bending.
Feedback: information provided a manufacturer by consumers after they try samples of a product.
Ferrous: any metal or alloy that contains iron.
Filing: a process used to smooth a material (usually metal) by removing small amounts from the surface with a toothed tool called a file.
Finishing: a process that changes the surface of a product by treating it or placing a coating on it.
Flexible Manufacturing System: a grouping of machine tools, controlled by a computer program, that can perform a series of operations on a single manufactured part. By reprogramming the computer, the tools can make a different part.
Floor plan: a drawing that shows the arrangement of rooms in a building.
Footing: the lowest portion of a building's foundation. The foundation wall rests on top of the footing.
Form: a three-dimensional representation of an object.
Forming: changing the shape of sheet material (often through the use of a mold).
Foundation: the footing and foundation wall that support a building and spread its load over a large ground area.
Friction: a force that acts like a brake on moving objects.
Function: what an object does or how it works. A functional object or product solves the problem described in a design brief.

G

Gasoline engine: engine in which a mixture of air and gas is ignited, pushing down pistons to drive a crankshaft in a rotary motion.
Gear: a rotating wheel-like object with teeth around its rim. It is used to transmit force to a gear with matching teeth.
Generating station: a facility that uses an energy source to operate turbines and produce electricity.
Generators: machines that produce electricity when turned by an outside force, such as a turbine.
Genetic engineering: a technology used to alter the genetic material of living cells.
Geothermal energy: energy produced by hot rocks changing underground water to steam.

H

Hardness: the ability of a material to resist cuts, scratches, and dents.
Hardwood: trees that have broad leaves which they usually lose in the fall. They are also known as deciduous trees.

Harmony: a condition in which chosen colors or designs naturally go together.

Heat energy: energy that occurs as the atoms of a material become more active.

Heat joining: a process that melts the material itself or a bonding agent (such as solder) to secure a joint.

Heating system: the furnace and associated ducts or pipes used to distribute heated air or water to the different rooms of a home.

Hydraulics: the study of pressure in liquids.

Hydroelectricity: electricity produced by using the energy of moving water.

Hydrogen: one of the two elements in ordinary water. When separated from water, it is a very combustible gas that can be used as a fuel.

I

Inclined plane: a simple machine in the form of a sloping surface or ramp, used to move a load from one level to another.

Industrial Revolution: the change from a single artisan performing all manufacturing operations to dividing the production process into specialized steps in which machinery replaced hand work.

Information: a collection of words or figures that have meaning or that can be combined to have meaning.

Information technology: the equipment and systems used in storing, processing, and extending human ability to communicate information.

Insulation: material used in the walls and ceiling of a building to help control heat loss (in winter) and heat gain (in summer).

Insulators: materials that will not carry an electric current.

Integrated circuit: a single electronic component that replaces a whole group of separate components (one integrated circuit may contain the equivalent of about 1,000,000 separate components).

Internal combustion engine: an engine that burns fuel inside itself.

Investigation: the part of the design process during which information is gathered.

Isometric paper: paper that is used for sketching. It has a grid of vertical lines and other lines at 30° to the horizontal.

Isometric sketching: the simplest type of picturelike drawing; it is used to share an idea or record ideas for further discussion.

J

Jet engine: an engine that sucks in air at the front, squeezes it, mixes it with fuel, and ignites it. This creates a strong blast of hot gases that rush out of the back of the engine at great speed.

Joist: horizontal member of a house framework that supports the floor.

K

Kinetic energy: the energy an object has because it is moving.

L

Laminating: a process that involves gluing together several veneers (thin sheets of wood) to form a strong part.

Landscaping: designing the exterior space that surrounds a home.

Laser light: a form of radiation that has been boosted to a high level of energy, producing a strong, narrow beam of light.

Lever: a simple machine, consisting of a bar and fulcrum (pivot point), that can be used to increase force or decrease effort needed to move a load.

Light energy: energy from the sun that travels as a wave motion. Also called *radiant* energy.

Lines: design elements that describe the edges or contours (outlines) of shapes. They show how an object will look when it has been made.

Linkage: a system of levers used to transmit motion.

Load: the weight, mass, or force placed on a structure; in an electrical circuit, any current-using device.

M

Machines: devices that do some kind of work by changing or transmitting energy.

Magnetic: having magnetic properties; able to be attracted by a magnet.

Magnetic properties: properties of a material that determine whether it will or will not be attracted to a magnet.

Magnetism: the ability of a material to attract pieces of iron or steel.

Malleability: a material's ability to be forced (compressed) into shape.

Marking out: measuring and marking material to the dimensions shown on a drawing.

Mass production: the process of making products in large quantities to reduce the unit cost.

Mechanical advantage: in a simple machine, the ability to move a large resistance by applying a small effort.

Mechanical joining: the use of physical means, such as a bolt and nut, to assemble parts.

Mechanical properties: the ability of a material to withstand mechanical forces.

Mechanism: a way of changing one kind of effort into another kind of effort.

Microelectronics: the process by which the switches and circuits of processors and their accessories are made incredibly small.

Microprocessor: a tiny silicon wafer, covered with microscopic electric circuits, sometimes called a "computer on a chip."

Model: a full-size or small-scale simulation (likeness) of an object, used for testing and evaluation.

Modular construction: a building system that involves basic room units of different sizes and shapes that can be combined on site.

Molding: a method of making shapes by forcing liquid material into a shaped cavity.

Molecules: groups of atoms.

Moment: the turning force acting upon a lever, determined by the effort times the distance of the effort from the fulcrum.

Motor: an electrically powered device with a rotating shaft that provides the kinetic energy to operate machines and other devices.

N

Nonferrous: metals or alloys that do not have iron as their basic component.

Nonrenewable energy: energy from sources that will eventually be used up and cannot be replaced. Examples: coal, oil, and natural gas.

Nuclear energy: energy produced from the nucleus of atoms.

Nuclear fission: energy produced by the splitting of atomic nuclei. The process gives off heat, which is used to produce steam to run turbines and generate electricity.

Nuclear fusion: the same energy source that powers our sun and the stars. It requires high temperatures. So far, technologists have not been able to produce it on earth.

O

Object line: line that is darker and thicker than a construction line. Object lines are used to show the outline of an object.

Open-loop system: a system that is not controlled through use of a feedback device.

Optical fibers: thin strands of glass that can carry laser-light-coded messages over long distances.

Optical properties: a material's reaction to light.

Orthographic projection: kind of drawing that shows each surface of the object "square on" (at right angles to the surface).

Overload: a condition that occurs when devices on an electrical circuit demand more current than the circuit can safely carry. It will cause a fuse or circuit breaker to interrupt current flow.

Ozone layer: a protective shield in the earth's atmosphere that absorbs much of the sun's ultraviolet radiation.

P

Parallel circuit: a circuit that provides more than one path for electron flow.

Pattern: a design used to reproduce a shape many times.

Perspective sketching: sketching method that provides the most realistic picture of objects.

Physical properties: properties that give a material its size, density, porosity, and surface texture.

Pier: a structure used to support the center of a bridge.

Planing: a process for smoothing wood that uses a sharp blade to remove very thin shavings.

Plasticity: the ability to flow into a new shape under pressure and to remain in that shape when the force is removed.

Plumbing system: the means used to bring in clean water and dispose of waste water in a home.

Pneumatics: the study of pressure in gases.
Pollution: the presence of dangerous or unwanted materials in the air, water supply, or other areas of the environment.
Polymer: a chainlike molecule made up of smaller molecular units; the scientific name for plastic.
Post and lintel: the simplest form of a framed structure, with horizontal framing members (beams) supported by vertical members (posts).
Potential energy: energy that is stored until it is released (used).
Prefabrication: a system of building in which components (such as wall framing or roof trusses) are built in a factory, rather than on the job site.
Pre-production series: a small number of samples made up by a manufacturer to obtain feedback through testing with consumers.
Pressure: the effort applied to a given area.
Primary cell: a device that chemically stores electricity. Its electrode is gradually consumed (used up) during normal use. It cannot be recharged.
Primary color: the three most important colors of the spectrum — red, yellow, and blue.
Primary material: a natural material; a material that exists in nature.
Primary sector: the portion of the economy concerned with gathering and processing raw materials.
Problem: a situation or condition that can be solved or improved through the application of technology.
Production system: a system involving the five basic operations of making a product: designing, planning, tooling up, controlling production, and packaging and distributing the finished product.
Proportion: the relationship between the sizes of two things.
Prototype: the first working version of the designer's solution to a problem.
Pulley: a simple machine in the form of a wheel with a groove around its rim to accept a rope, chain, or belt. It is used to lift heavy objects.

R

Radiation: the particles or rays thrown off by unstable atomic nuclei; also, one of the methods by which heat travels.
Raw materials: materials that come from nature in one form or another.
Recycling: the process of separating different types of wastes, and reclaiming those that can be reused.
Reinforced concrete: concrete in which steel rods have been embedded to increase the concrete's resistance to tension.
Renewable energy: energy from sources that will always be available, such as the sun, wind, and water.
Resistance: opposition to the flow of electricity.
Resistor: an electrical device that controls the amount of current flow through a circuit by making that flow more difficult.
Rhythm: a quality or feeling of movement, provided by repeating patterns.
Robot: a computer-controlled device that can be ''taught'' to perform various production or material handling operations.

S

Safety: the practice of working in a way that will avoid injury or damage.
Sawing: the process of cutting material with a tool that has a row of teeth on its edge.
Scale drawing: drawing that is larger or smaller than the object by a fixed ratio.
Science: the field of study that is concerned with the laws of nature.
Scientists: persons whose field of study is the laws of nature.
Screw: a simple machine that is an inclined plane wrapped in the form of a cylinder.
Secondary cell: a device for storing electrical energy as a chemical; it can be charged, discharged, and recharged.
Secondary color: a color obtained by mixing equal parts of two primary colors.
Secondary sector: the sector of the economy that changes raw and processed materials into useful products.
Semiconductor: materials that allow electron flow only under certain conditions. They have some characteristics of conductors and some characteristics of insulators. Semiconductor devices, such as transistors, are widely used in electronic equipment.
Series circuit: a circuit that provides only one path for electron flow.

Shape: a two-dimensional representation of an object.

Shear: a sliding and separating force.

Shearing: a process used to cut thin material. A scissors-like tool, called shears or snips, is used.

Short circuit: a condition that occurs when bare wires in an electrical circuit accidentally touch, causing more current to flow than the circuit can safely carry. It will cause a fuse or circuit breaker to interrupt current flow.

Site: the land on which a building is to be built.

Softwoods: coniferous trees that retain their needlelike leaves and are commonly called evergreen trees.

Solar energy: energy from the sun, the most important of the alternative sources of energy.

Sound energy: a form of kinetic energy that moves at about 1100 ft. (331 m) per second.

Space shuttle: a reusable vehicle, developed by NASA, that is capable of operating in space as well as in the Earth's atmosphere.

Static load: a load that is unchanging or changes slowly.

Stays: cables that support a bridge deck from above.

Strain energy: the energy of deformation, possessed by materials that tend to return to their original shape after being stretched or compressed.

Structure: something that encloses and defines a space; also, an assembly of separate parts that is capable of supporting a load.

Strut: a rigid structural member that is in compression.

Style: an individual, unique way of designing an object or solving a problem; also, an identifiable set of common elements that give a common appearance to related objects.

Subfloor: a covering over joists that supports other floor coverings.

Subsystem: a smaller system that operates as a part of a larger system.

Superconductivity: a quality of a material that will allow it to conduct electricity without resistance.

Supersonic: faster than the speed of sound. Used to describe an aircraft that flies at such speeds.

Suspension bridge: a bridge in which the deck is suspended (hung) from hangers attached to a continuous cable. The cable passes over towers and is anchored to the ground at each end.

Symbol: simple picture or shape used as a means of communicating without using words.

System: a series of parts or objects connected together for a particular purpose.

T

Technologist: a person who solves problems by designing and making products or structures.

Technology: the knowledge and process used to solve problems by designing and making products or structures.

Telecommunications: the sending and receiving of information over a distance, making use of a variety of technological devices and systems.

Tension: a pulling force.

Tertiary sector: the economic sector that is concerned with the servicing of products and with providing services that add to the personal comfort, pleasure, and enjoyment of people.

Texture: a design element that determines the way a surface feels or looks.

Thermal expansion: expansion of matter caused by heat.

Thermal properties: properties that control how a material reacts to heat or cold.

Thermoplastics: materials that can be repeatedly softened by heating and hardened by cooling.

Thermosets: a material that assumes a permanent shape once heated.

Thrust: pushing power, based on the principle that for every action there is an equal and opposite reaction.

Tidal energy: energy produced by the rise or fall of tidal water in the oceans.

Tie: a rigid structural member that is in tension.

Torsion: a twisting force.

Toughness: the ability to resist breaking.

Transformer: an electrical device, consisting of a pair of coils, that changes voltage and amperage.

Transistors: electronic devices that switch electric currents on and off.

Transmission lines: lines that transport electricity to wherever it is needed.

Transportation: the process of moving people or material from a point of origin to a point of destination.

Truss: a structural element made up of a series of triangular frames.

Turbine: an energy converter that includes a large wheel with blades turned by water, a jet of steam, or hot gases.

Turbofan: a type of jet engine in which the gas stream drives a large fan located at the front of the engine. Thrust is as great as a simple jet but the engine is quieter.

Turboshaft: an engine that uses a stream of gases to drive turbine blades connected to a shaft which, in turn, is connected to rotors or propellers.

V

View: in orthographic projection, a drawing of the front, top, or right side of an object as seen looking at right angles to the surface.

Voltage: a measure of electrical pressure.

Voltaic cell: a simple cell consisting of copper and zinc rods immersed in a weak sulphuric acid solution.

W

Wall stud: a vertical framing member to which gypsum board, paneling, or other wall coverings are attached.

Watt: the unit used to measure the work performed by an electric current.

Wave energy: energy produced by the movement of ocean waves.

Wedge: a simple machine that consists of two inclined planes placed back-to-back.

Wheel-and-axle: a simple machine that is a special kind of lever. Effort applied to the outer edge of the wheel is transmitted through the axle.

Wind energy: one of the oldest sources of kinetic energy, used for centuries to grind grain and pump water. Today, wind is being used to spin wind turbines and generate electricity.

Work: measurement of the amount of effort needed to change one kind of energy into another.

Index

A

Acid rain, 258
Acoustical properties, 54
Alloy, 62
Alphabet of lines, 44, 45
Alternating current, 194-196
Alternative solutions, 15
AND and OR gates, 207-209
Antibody production, 261
Arch, 105
Artificial intelligence, 265
Atoms, 54
Automation, 226-229

B

Balance, 27
Bending,
 and forming plastics, 78, 79
 and forming sheet materials, 77
 sheet metal, 78
 wood, 77
Bioprocessing, 260
Biotechnology, 260
Blackboard, 58

C

Cantilever, 105
Capacitors, 216
Careers,
 primary sector, 223
 secondary sector, 234, 235
 tertiary sector, 236-242
Casting and molding, 79-81
Cells and batteries, 198-200
Ceramics, 64, 65
Chemical,
 energy, 166
 joining, 88-90
 properties, 54
Circuit, 190, 195
Circuits, protecting, 205
Closed-loop system, 120
Color, 25, 26
Communicating ideas, 35-49
Communication, 261
 forms of, 36-43
 information, 245-252
 system, 35, 36, 123
Composite materials, 51, 65-67
Computer-aided design, 47-49
Computers, 248, 249, 265
Conductors, 55, 190
 and insulators, 209-211
Construction, 109-125
Converted surface finishes, 91

D

Design brief, 14,16,17
Design process, 16-21
 choosing a solution, 19
 developing alternative solutions, 18, 19
 investigation, 17
 manufacturing, 20, 21
 testing and evaluation, 20
 the problem, 16
Dimensioning, 45
Diodes, 215, 216

Direct current, 196, 197
Direction of current, 206
Drawing techniques, 43-47
  alphabet of lines, 44, 45
  dimensioning, 45
  scale drawing, 46, 47
Drawings and their types, 37, 38
Drilling, 76

E

Electric circuits, 203-209
Electric,
  current, 190
  motors, 161, 197, 198
Electrical energy, 166
  measuring, 211, 212
Electrical properties, 55
Electrical system, 122, 123
Electricity, 189-191
  sources, 183-201
Electricity and electronics, use, 203-219
Electromagnets, 191
Electronics, 213-218
  capacitors, 216
  diodes, 215, 216
  integrated circuits, 218
  resistors, 213-215
  transistors, 216-218
Electrons, 189
Elements of design, 21-26
  color, 25, 26
  line, 21, 22
  shape and form, 22, 23
  texture, 23-25
Energy, 165-181
  conversion, 167, 168
  nonrenewable sources, 169, 170
  renewable sources, 171-180
  where it comes from, 169
Engines and motors, 153-162
Entrepreneur, 236
Ergonomics, 17
Expert systems, 265
External combustion engine, 154

F

Feedback, 16
Ferrous, 59
Finishing materials, 90-92
Form, 23
Forms of communication, 36-43
  drawings and their types, 37, 38
  hand signals and sounds, 36
  symbols and signs, 36, 37
  ways of communication, 37-43
Forms of energy, 166-168
  conversion, 167, 168
  losses during conversion, 168
Foundation, 115
Four-stroke diesel engine, 157
Four-stroke gasoline engine, 155, 156
Friction, 147, 148
Function, 17
Fuses and circuit breakers, 205, 206
Future, looking to, 236

G

Gasoline and diesel engines, 154
Gears, 140-143
Generating electricity, 183-185
Generation of electricity using magnetism, 194-197
Generators, 184
Geothermal energy, 177, 178
Glass cables, 250

H

Hand signals and sounds, 36
Hardboard, 58, 59
Hardwoods, 55
Harmony, 25
  and contrast, 29, 30
Heat energy, 166
Heat joining, 90
Heating system, 121, 122
House frame, 110, 111
Hydraulics and pneumatics, 144-147
Hydroelectricity, 174, 175, 184

I

Inclined plane, 136-138
Individual artisan, 224
Information, 245
Insulators, 55, 209
Integrated circuits, 218
Internal combustion engine, 153, 154
Introduction, technology, 7-12
Isometric paper, 38
Isometric sketching, 38-40

J

Jet engine, 160
Job growth, 236
Joining,
  chemical, 88-90

heat, 90
materials, 81-90
Joists, beams, 110

K

Kinetic energy, 165

L

Laminating, 77
Laser light, 250, 251
Levers, 129-131
Line, 21, 22

M

Machines, 127-151
Magnetic properties, 55
Magnetism, 191-194
and electric current, 193, 194
Magnets, natural and artificial, 191
Management system, 229, 230
Manufacturing cells, 266
Manufacturing products, 223-235
Manufacturing technologies, advanced, 265, 266
Marking out, 74, 75
Materials, 51-69
finishing, 90-92
joining, 81-90
processing, 71-94
shaping, 74-81
types of, 55-68
Measuring electrical energy, 211, 212
Mechanical,
joining, 81-88
properties, 52, 53
Mechanism and mass production, 224-226
Metals, 59-62
Microelectronics, 249, 250
Model making, 15, 16
Models and prototypes, 19, 20
Modes of transportation, 153
Modular construction, 110
Molecules, 54, 62
Moments and levers, 131, 132
Movement and rhythm, 31

New materials, 261, 262
Nonferrous, 59
Nonrenewable,
raw materials, 221-223
sources of energy, 169, 170
Nuclear,
energy, 176
fission, 176
fusion, 176

O

Ohm's Law, 211, 212
On-site transportation, 153
Open-loop system, 120
Optical properties, 54
Orthographic projection, 42, 43

P

Parallel,
circuit, 207
computers, 265
Particleboard, 58, 59
Pattern, 30, 31
Perspective sketching, 40, 41
Physical properties, 53
Pier, 102
Planning for a home, 111-118
finding and preparing a site, 113-115
inside space, 111-113
main parts, 115-118
Plastics, 62-64
Plumbing system, 123
Plywood, 58
Pollution,
in home, 257, 258
lake and river, 258, 259
outside home, 258
Pollution problems, attacking, 259
Polymer, 62
Post and lintel, 110
Potential energy, 165
Prefabrication, 110
Pressure, 143-147
hydraulics and pneumatics, 144-147
Primary,
cell, 198, 199
colors, 25
Primary sector,
careers, 223
processing raw materials, 221-223
Principles of design, 27-32
balance, 27
harmony and contrast, 29, 30

movement and rhythm, 31
pattern, 30, 31
proportion, 28, 29
unity and style, 32
Problem, 14
Problem solving,
asking questions, 14
defining need, 14
finding other solutions, 15
in technology, 13-34
Processing materials, 71-94
safety, 71-73
Production system, 230-234
Properties of materials, 53-55
acoustical, 54
chemical, 54
electrical, 55
magnetic, 55
mechanical, 52, 53
optical, 54
physical, 53
thermal, 53
Proportion, 28, 29
Pulley, 127, 132-134

R

Radiation, 176
Raw materials, processing, 221-223
Reinforced concrete, 105, 106
Renewable,
materials, 221, 222
sources of energy, 171-180
Resistors, 213-215
Robots, 227, 228
Rocket engine, 160

S

Safety, processing materials, 71-73
Sawing, 75
Scale drawing, 46, 47
Science and technology,
relationship, 10,11
Screw, 139, 140
Secondary cells, 199, 200
Secondary sector,
careers, 234, 235
manufacturing products, 223-235
Semiconductors, 209
Series and parallel circuits, 206, 207
Series circuit, 207
Services, providing, 235-242
Shape and form, 22, 23
Shaping materials, 74-81
marking out, 74, 75
Shearing and chiseling, 75, 76
Short circuit, 205
Simple machines, 128-140
inclined plane, 136-138
levers, 129-131
pulley, 132-134
screw, 139, 140
wedge, 138, 139
wheel and axle, 134-136
Softwoods, 55
Solar energy, 171, 172
Space flight, 263
Static and dynamic loads, 97-99
Stays, 103
Steam turbine, 158
Structure of a house, 109-111
Structures, 95-107
and their systems, 263-265
designing to withstand loads, 100-105
forces acting on, 100
Strut, 102
Superconducting materials, 264
Superconductivity, 210, 211
Surface coatings, 91, 92
Suspension bridges, 103, 104
Symbols and signs, 36, 37
System,
closed-loop, 120
definition, 120, 121
in structures, 118-123
management, 229, 230
open-loop, 120
production, 230-234

T

Technology, 8
beginning, 253, 254
both old and new, 10
effect on health, 9, 10
future, 260-266
how it affects us, 8
introduction, 7-12
old and new, 245-247
past, present, and future, 253-268
present, 254-259
understanding, 8-10
Telecommunications, 250, 251
Telephone, 250

Tertiary sector, careers, 236-242
Tertiary sector: providing services, 235-242
Texture, 23-25
Thermal properties, 53
Thermal-electric generating stations, 185, 186
Thermoplastics, 62
Thermoset, 62, 63
Tidal energy, 175, 176
Tie, 101, 102
Transformer, 189
Transistors, 216-218
Transmission,
 and distribution of electricity, 185-189
 lines, 185
Transportation, 153-164
 impact, 161, 162
Truss, 103
Turbofan and turboshaft engines, 154
Turbofan,
 engine, 159
 operation, 159, 160
Turboshaft engine, 160
Types of materials, 55-68
 ceramics, 64, 65
 composite materials, 65-67
 metals, 59-62
 plastics, 62-64
 woods, 55-59
Types of structures, 97

U

Understanding technology, 8-10
Unity and style, 32

V

Voltaic cell, 198

W

Watts, 212
Wave,
 communication, 251
 energy, 176
Ways of communication, 37-43
Wedge, 138, 139
Wheel and axle, 127, 134-136
Wind energy, 172-174
Woods, 55-59
 blackboard, 58
 hardboard, 58, 59
 particleboard, 58, 59
 plywood, 58
World of work, 221-244

ACKNOWLEDGMENTS

Illustrations by Nadia graphics